Direct Nuclear Reactions

Direct Nuclear Reactions

NORMAN K. GLENDENNING

Nuclear Science Division
Lawrence Berkeley Laboratory
University of California
Berkeley, California

1983

ACADEMIC PRESS
A Subsidiary of Harcourt Brace Jovanovich, Publishers
New York London
Paris San Diego San Francisco São Paulo Sydney Tokyo Toronto

ACADEMIC PRESS, INC.
111 Fifth Avenue, New York, New York 10003

United Kingdom Edition published by
ACADEMIC PRESS, INC. (LONDON) LTD.
24/28 Oval Road, London NW1 7DX

Library of Congress Cataloging in Publication Data

Glendenning, Norman K.
Direct nuclear reactions.

Includes index.
1. Direct reactions (Nuclear physics) I. Title.
QC794.8.D57G578 1983 539.7'54 82-24365
ISBN 0-12-286320-8

PRINTED IN THE UNITED STATES OF AMERICA

83 84 85 86 9 8 7 6 5 4 3 2 1

In Memory of Florence and Norman,
Mother and Father

Contents

11. *Effective Interactions and the Free Nucleon–Nucleon Force*

12. *Further Developments in the Theory of Inelastic Scattering*

13. *Scattering from Deformed Rotational Nuclei*

14. *Calculation of Specific Components of the Optical Potential*

15. *Two-Nucleon Transfer Reactions*

Preface

The spontaneous disintegration of long-lived, naturally occurring isotopes provides one source of information on nuclei. However, only a limited number of nuclei are accessible for study by this natural process and then only under a narrow range of circumstances. On the other hand, nuclear reactions can be induced in the myriad of pairwise combinations provided by stable or long-lived nuclei and over the wide range of energies provided by the accelerators in the physics laboratories of the world. Reactions therefore provide the greatest volume and widest range of nuclear data. The energy loss of a beam particle can be directly interpreted as an excitation energy in the target nucleus. Usually, however, the data acquire meaning for the structure of nuclei only after they have been interpreted through a reaction theory. The synthesis of such accumulated information into a coherent theory of the nucleus is the main subject of nuclear physics. There are a number of volumes on nuclear structure that review this ongoing process.

This book is about the theory of direct nuclear reactions. It emphasizes the microscopic aspects of the reactions and their description in terms of the changes induced in the motions of the individual nucleons by the reaction. However, because collective motion can be accurately described by a few collective parameters, we sometimes depart from a strictly microscopic description; any account of direct reactions would otherwise be incomplete.

The book begins essentially at the beginning, assuming only a modest knowledge of quantum mechanics and some acquaintance with angular momentum algebra, and ends by describing some of the most recent topics. The principal results of the theory are described with emphasis on the approximations involved to provide a guide to how well the theory can be expected to hold under specific experimental conditions and also to suggest areas in which improvements can be made. Applications to the analysis of experiments are also emphasized, not only because reactions are interesting in themselves, but because they can be used to measure nuclear properties. Indeed, most of our detailed knowledge of nuclear properties has been discovered by means of reactions.

The goal of the book is thus to provide the novice with the means to become competent to do research in direct nuclear reactions and to provide the experienced researcher with a detailed discussion of the advanced topics. Topics covered include coupled equations and the distorted-wave Born approximation, form factors and their nuclear structure content, the basis of the optical potential as an effective interaction, reactions such as inelastic single- and two-nucleon transfer reactions, the effect of nuclear correlations, the role of multiple-step reactions, the theory of inelastic scattering and the relationship of the effective interaction to the free one, reactions between heavy ions, the polarizability of nuclear wave functions during a reaction, and the calculation of components of the optical potential arising from specific collective or transfer reactions.

Acknowledgments

I began thinking of this project in 1969, at the suggestion of the editors at Academic Press. However, it was not until early 1978 that I actually began writing, in response to an invitation from S. K. Bhattacherjee and B. Banerjee of the Tata Institute of Fundamental Research, Bombay, to give a course at a summer school in Bangalore, India. I very much enjoyed the summer school and the hospitality of the organizers and participants. When I had completed the lecture notes I realized that I had in hand a part of the project that had been suggested almost 10 years earlier. More years have elapsed, and it is now much later than my too optimistic estimate of when I would complete the manuscript. The delay did, however, provide the opportunity of including some very recent and quite important developments.

The field of direct nuclear reactions is now more than 30 years old. Many researchers throughout the world have contributed to the development of the theory and its applications to the interpretation of experimental data, the prediction of new phenomena, and of course the design of accelerators, detection equipment, and the experiments that actually obtained the data. Of the voluminous scientific literature that documents the achievements in this field, reference here is directed mainly to the theory of reactions and accordingly does not do justice to all of those whose research inspired the development of the theory and made it a fruitful endeavor.

I especially thank B. G. Harvey of the Lawrence Berkeley Laboratory for nurturing my early interest in this field through his own keen interest and provocative experiments. I am also grateful to M. A. Nagarajan of the Daresbury Laboratory for reading parts of the manuscript and for offering many useful suggestions from his wide knowledge.

Two of the chapters, those having to do with the recent advances in relating the free interaction and the effective interaction in the theory of inelastic scattering, could not have been written in time for this publication were it not for extensive help from F. Petrovich of Florida State University. He contributed theoretical developments before their publication in the journals. I am very grateful to him for his generosity.

Notational Conventions

$\mathscr{A}$	antisymmetrization operator
A, a	*intrinsic* coordinates of nucleus A and a, respectively, of which there are 3(A − 1) and 3(a − 1)
$\alpha, \alpha', \alpha''$	different *channels* in the *partition* α, with α denoting either a typical channel or the ground state channel and α', α'' channels in which one, the other, or both of nuclei a or A is excited
$\alpha, \beta, \ldots$	frequently used to denote different two-body *partitions* a + A, b + B, . . . , of a many-nucleon system
B, b	similar to **A**, **a**, but for nuclei B and b
$\chi_\alpha^{(\pm)}$	optical model wave function in channel α in which the relative motion is distorted by an optical potential (the asymptotic condition of outgoing (+) or incoming (−) waves is indicated)
E	eigenvalue of H, eigenfunction Ψ, or $\Psi_\alpha^{(\pm)}$
E_α	kinetic energy in channel α, eigenvalue of T_α, eigenfunction of $\exp(i\mathbf{k}_\alpha \cdot \mathbf{r}_\alpha)$ ($E_\alpha = E - \varepsilon_\alpha$)
ε_α	eigenvalue of H_α, eigenfunction Φ_α, and where ε_{A} and ε_{a} are eigenvalues of H_{A} and H_{a} with eigenfunction Φ_{A} and Φ_{a} ($\varepsilon_\alpha = \varepsilon_{\mathrm{A}} + \varepsilon_{\mathrm{a}}$)
H	total Hamiltonian of the A + a = B + b . . . many-nucleon system, ordered in terms of various partitions ($H = H_\alpha + T_\alpha + V_\alpha = H_\beta + T_\beta + V_\beta = \cdots$)
$H_{\mathrm{A}}, H_{\mathrm{a}}$	nuclear Hamiltonians for nuclei A and a

H_α	$H_A + H_a$
k_α	wave number of relative motion in channel α, $(k_\alpha^2 = 2m_\alpha E_\alpha/\hbar)$
$\hat{\ell}$	$2\ell + 1$
m_α	reduced mass in channel α
$\binom{N}{n}$	combinatorial symbol is $N!/[(N-n)!n!]$
Φ_α	product $(\Phi_A\Phi_a)$ of nuclear wave functions in channel α and therefore a function of **A** and **a** intrinsic coordinates and of spin and isospin
ϕ_α	the many-nucleon wave function in channel α when the relative motion is a plane (or Coulomb) wave; i.e., $\Phi_\alpha \exp(i\mathbf{k}_\alpha \cdot \mathbf{r}_\alpha)$
Φ_A, Φ_a	nuclear state vectors or wave functions (if coordinate space representation is implied or specified) for nucleus A and a, respectively they are functions of **A** and **a**, respectively, and of spin and isospin coordinates in general (a sometimes denotes a light nucleus like ^{3}He)
$\phi_{n\ell j}$	single-particle spin-orbit wave function
Ψ	state vector or wave function of the A + a many-nucleon system.
$\Psi_\alpha^{(+)}$	state vector or wave function of many-nucleon system as above, for which the boundary conditions are specified as follows: there is a plane (or Coulomb) wave in partition α representing the incident beam in a collision, and asymptotically at infinity there are outgoing waves in all open channels of all partitions; the wave function is a function of the coordinates of A + a nucleons, which in the center of mass frame are taken as **A**, **a**, $\mathbf{r}_\alpha$
$\langle\Psi_\alpha\|\Psi'_\alpha\rangle$	the integration $(\int \Psi_\alpha^*\Psi'_\alpha)$ carried out over all position coordinates and the summation carried out over all spin and isospin coordinates
$(\Psi_\alpha\|\Psi'_\alpha)$	as $\langle\Psi_\alpha\|\Psi'_\alpha\rangle$ but the relative coordinate is *not* integrated (unless otherwise defined); the bracket, therefore, is a function of $\mathbf{r}_\alpha$
$\{\Psi_\alpha\|\Psi'_\alpha\}$	a spin and isospin summation $(\sum \Psi_\alpha^*\Psi'_\alpha)$ unless otherwise defined
$\hat{\mathbf{r}}$	θ, ϕ polar angles of **r**
$\mathbf{r}_\alpha$	relative coordinate between nuclei A and a
T_α	kinetic energy $(-(\hbar^2/2m_\alpha)\nabla_\alpha^2)$ of relative motion in channel α
U_α	optical potential in channel α
U_i	shell-model potential acting on nucleon labeled i
$u_{n\ell}$	radial part of above wave function $\phi_{n\ell j}$ (this can also depend on j if the shell-model potential depends on spin)
V_α	interaction other than optical potential in channel α, often a sum of two-body effective interactions
W	statistical weight of direct integrals contributing to the amplitude of a nucleon transfer reaction see (5.32)
$X_{1/2}^\mu(\sigma)$	nucleon spin eigenfunction (or isospin if τ replaces σ)
$X_S^M(\sigma_1, \sigma_2)$	two-particle spin eigenfunction with eigenvalues $S^2 = S(S+1)$, $S = 0$ or 1 and $S_z = M$

Chapter 1

Introduction: Direct and Compound Nuclear Reactions

A. THE OBSERVABLES

The nature of the experimental observation is the following: A beam of particles, such as protons, deuterons, alpha particles, or heavier nuclei (referred to as heavy ions), is accelerated to a desired energy and then deflected so as to strike a target of known isotopic composition. The energy and scattering angle of some of the products of the collision are measured. For example, in a reaction initiated by protons, the proton itself will sometimes emerge, deflected in angle but having the same energy in the center-of-mass system. This is *elastic scattering*, and (as we shall discuss) the measurement of this cross section is important because its analysis yields the parameters of the *optical potential*.

Sometimes the proton will excite the target nucleus from its ground to some higher-energy state, thus losing some energy and at the same time being deflected in angle. This is an *inelastic event*. The cross section for such an event yields information on the spin and parity of the nuclear transition and, in a way that will be described more fully, is sensitive to the wave functions of the nucleons that are excited. When the detector angle

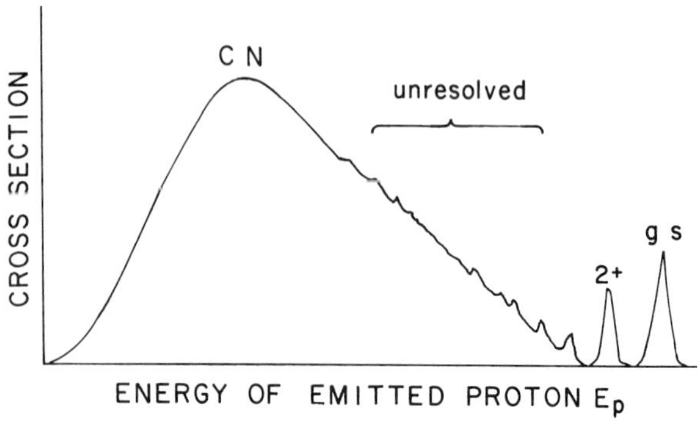

Fig. 1.1. Schematic proton spectrum from proton–nucleus scattering showing a few resolved levels near the beam energy, unresolved, but discrete, levels, and the compound-nucleus continuum.

is fixed and the energy of the scattered protons plotted, the result is typically a spectrum such as that schematically illustrated in Fig. 1.1.

One observes discrete groups of protons. The group that has the same energy as the incident proton constitutes the elastically scattered particles. At somewhat lower energy are discrete groups corresponding to the excitation of low-lying levels in the target nucleus. These are labeled by the spin and parity of the excited nuclear state. At still lower energy, corresponding to a region of closely spaced and higher-lying nuclear states, the protons cannot be resolved into discrete lines because of the imprecision in the measurement of the detected proton's energy (counter resolution) and the spread in energy of the accelerated protons (beam resolution). At still lower energy is a continuous spectrum of energies that corresponds not to the direct reactions but to compound nucleus reactions.

Finally, in addition to the incident proton, other outgoing particles (such as a neutron, deuteron, or triton) may emerge from the collision. These are, respectively, *charge-exchange*, or *knockout*, reactions and *single-* and *two-neutron pickup* reactions. The charge-exchange and single-neutron pickup reactions yield information similar to inelastic collisions, whereas the two-neutron reactions yield, in addition, information on the correlation of neutrons in the nucleus.

All of these matters will be discussed fully later in this book.

B. DIRECT AND COMPOUND NUCLEAR REACTIONS

Each state of a nucleus corresponds to a particular state of motion of all of the nucleons. The shell model of the nucleus provides a useful language.

In lowest approximation, the ground state of a nucleus is a degenerate state having all of the lowest single-particle levels filled. The low-lying excited states are very similar to the ground state in the sense that most of the nucleons are in the same state of motion. Only one or relatively few nucleons are in different states. An ideal example is ^{209}Pb, which has a doubly magic number of neutrons and protons. The lowest few levels differ from each other, to good approximation, only in the state of the odd neutron. A deformed nucleus like ^{154}Sm is an example in which the intrinsic structure of the lowest levels are, to good approximation, identical. The different energy states correspond merely to different rates of rotation of the whole nucleus. Excited states such as these, which differ so little from the ground state, can easily be excited by a projectile. In the case of ^{209}Pb, the projectible has only to interact with the odd neutron, changing its quantum numbers, to produce one of the low excited states. In the case of the deformed ^{154}Sm nucleus, the projectile can scatter *coherently* from all the nucleons, setting it into rotation. Such reactions, in which there is very little change between initial and final states of the nucleus, are referred to as direct nuclear reactions. They occur as a result of the interaction of the projectile with a specific degree of freedom, either of a single nucleon or a collective coordinate, causing that single degree of freedom to change. The projectile thereupon escapes without further change to itself or the target.

At the opposite extreme are compound nuclear reactions. These are reactions in which the incident particle interacts successively with a number of nucleons until most of its energy has been shared among many nucleons. In this extreme, the incident particle initiates a cascade of collisions. The first nucleon struck successively strikes others, and so on, until the randomness introduced by many collisions involving many nucleons causes loss of memory of what type of particle brought in the energy that is now shared throughout the nucleus.

The compound nucleus has too much energy to be stable. However, because the energy of the projectile is now shared among many nucleons, it will survive for a relatively long time compared with the duration of a direct reaction, which is approximately the transit time of the projectile $\sim 2R/v_{\mathrm{p}}$. Eventually, however, in the random course of events, one or more nucleons will acquire sufficient energy to escape. Low-energy neutrons are the most probable decay mode because, unlike charged particles, they experience no Coulomb barrier. The state in which the nucleus is left after this will probably be quite different from the ground state because of the fact that many nucleons shared the projectile energy. In fact, the randomness of the process suggests that one factor involved in predicting the energy of the residual nucleus is the level density in the residual nucleus. The other factor is the penetration factor or the probability that a particle of given energy can escape from the compound system. This is clearly an increasing function of decay energy. No single state will be strongly populated by the reaction because of the competition with other states, but the energy region populated will be at higher excitation where the energy levels are dense. The two factors taken together lead to the broad peaked distributions shown in Fig. 1.1.

The large number of collisions and the randomization of the energy over many nucleons, imply that the energy of the compound nucleus is its main characterization. The manner in which it was formed is largely immaterial. For example, all reactions such as

$$p + {}^{Z-1}A^{N} \to {}^{Z}C^{N}, \qquad d + {}^{Z-1}(A-1)^{N-1} \to {}^{Z}C^{N}, \qquad \text{etc.,} \tag{1.1}$$

at energies that lead to the same energy of C produce essentially the same distribution of compound states in C. The subsequent decay of C is thus seen to be independent of the particular way in which it was formed.

$$\begin{aligned} {}^{Z}C^{N} &\to p + {}^{Z-1}A^{N} \\ &\to n + {}^{Z}A^{N-1} \\ &\to d + {}^{Z-1}(A-1)^{N-1}, \qquad \text{etc.} \end{aligned} \tag{1.2}$$

The initial and final states of the system are decoupled by the multiplicity of collisions that produced the intermediate compound nucleus C. The two phases of reaction (1) and (2) are independent.

This decoupling of the formation and decay of a compound system is in stark contrast with the direct reactions, in which the final state is produced in a single interaction of the projectile with a degree of freedom describing the target nucleus. The cross section is governed, therefore, by a matrix element between the initial and final states and, consequently, involves an overlap between their wave functions. This is what makes direct reactions so useful to the nuclear spectroscopist. The cross section is *directly* related to the *difference* between the final and initial nuclear states and does not involve the states of all the other nucleons.

The two extremes of nuclear reactions have been described. Clearly, there are intermediate situations. A direct reaction is a "doorway" leading toward compound nucleus formation. The compound nucleus is formed by a sequence of collisions leading to increasingly complicated rearrangements of the target nucleus. But if the projectile escapes early in the sequence, then the reaction can be described as a direct one. The difference in the time scales has already been mentioned. There is also a difference in the region of the nucleus where the two extremes dominate. Clearly, a peripheral collision (one in which the projectile passes through the diffuse surface of the nucleus) will most frequently lead to a simple collision, whereas when the projectile passes into the interior of the nucleus, it is more likely to suffer a number of collisions. We therefore expect that direct reactions are localized dominantly in the nuclear surface, whereas compound nuclear reactions are localized in the interior.

The spatial localization can also be stated in terms of an angular momentum localization. Obviously, for a beam of projectiles of given energy, those with high angular momentum relative to the target have large impact parameters. Thus, very roughly speaking, particles with angular momentum ℓ less than some critical value ℓ_c will more often than not form a compound nucleus, whereas those with ℓ greater than ℓ_c will usually scatter elastically or directly.

C. COMPETITION BETWEEN DIRECT AND COMPOUND NUCLEAR REACTIONS

In principle every nuclear state populated in a reaction can be produced by both of the extreme reaction mechanisms previously discussed, as illustrated in Fig. 1.2. Therefore the extreme pictures of direct and compound nuclear reactions would not be so useful were it not for the fact that one or the other mode is usually much stronger than the other.

The low-lying levels are expected to be populated dominantly by the direct reactions for several reasons. The low-lying levels are most closely related to the structure of the ground state, having only some minor rearrangement

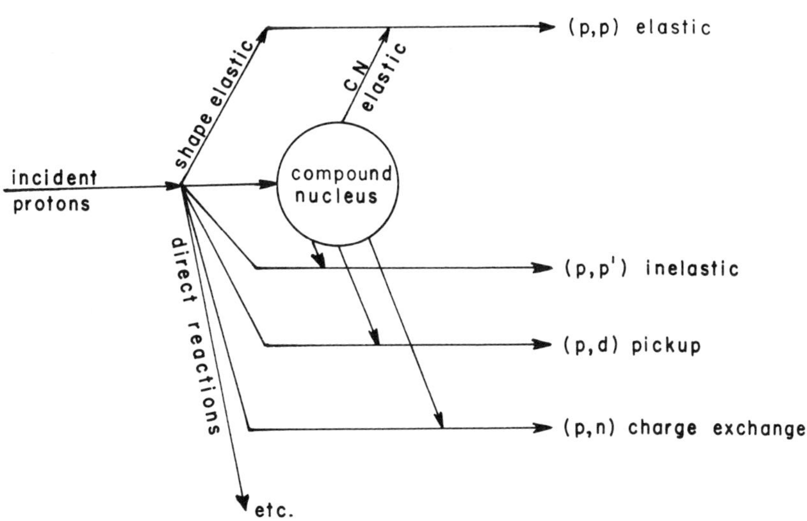

Fig. 1.2. Depiction of the processes that are typical of proton–nucleus interactions. (Adapted from P. E. Hodgson, 1971.)

in the state of motion of the nucleons. That is why they *are* lying in energy close to the ground state. In contrast, most levels at high excitation involve major rearrangement of nucleons. Therefore the former can be populated strongly by direct reactions, whereas the latter can be populated only through small components in the wave function by this reaction and are therefore weakly populated by the direct reaction.

The compound nuclear reaction is not expected to populate any state very strongly, but it is expected to populate the region of high excitation. The reason is simply that at typical reaction energies of tens of million-electron-volts (MeV) there are many open channels, implying that none of them individually is strongly populated. The total cross section cannot exceed the "geometric" cross section. Therefore, the more channels that are open, the smaller (on the average) will be the cross section to any particular one of them. The compound nucleus, because it is itself a highly excited state with the energy distributed over many nucleons, is more likely to decay to a complicated and therefore high-lying state of the residual nucleus. Moreover, because the density of such levels grows rapidly with excitation energy, the decay of the compound nucleus will populate the high-energy region more strongly than the low. Thus, by choosing the energy to be sufficiently high, one can assure that the compound nuclear reaction to low-lying states will be small.

D. HISTORICAL NOTE

The largest part of the reaction cross section is taken up by compound nuclear reactions (Fig. 1.1). The nature of this type of reaction was clarified through the work of Niels Bohr (Bohr, 1936; Kalckar and Bohr, 1937) and was elucidated in subsequent papers by Feshback and Weisskopf (1949) (continuum theory) and Breit and Wigner (1936), Kapur and Peierls (1938), and Feshback, Peaslee, and Weisskopf (1947) (resonance reactions). It was much after Bohr's clarification of the compound nuclear reaction that S. T. Butler (1950, 1951) recognized clearly that a distinctly different mechanism came into play in accounting for the forward-peaked structure observed in the data of Burrows, Gibson, and Rotblat (1950) and of Holt and Young (1950). Subsequently, Bethe and Butler recognized the immense importance that this type of reaction would play in nuclear spectroscopy (Bethe and Butler, 1952; Butler, 1952).

Chapter 2

The Plane-Wave Theory

The difference between direct and compound nuclear reactions has been discussed. In the direct reaction, the incident particle interacts with the target nucleus only several times at most. Therefore the change produced in the target must be a simple rearrangement of one or several nucleons or collective degrees of freedom. Because of the conservation laws, the change produced in the nucleus must be reflected in the energy and angular momentum carried by the outgoing particle.

The spectrum of protons emitted at a fixed angle was illustrated in Fig. 1.1. If the counter is moved successively to various angles, and the cross section to each energy group is recorded, then the angular distribution or differential cross section to specific states is thereby measured. The angular distribution of particles that have produced a specific nuclear level reflects the angular momentum and parity that was exchanged between the scattered particle and the nucleus in making the transition. This is because the particle must scatter in such a way as to obey the conservation laws, including angular momentum, and this imposes a restriction on the directions which it can scatter.

To illustrate how this comes about in a more quantitative way, let us consider the plane-wave Born approximation. The use of plane waves to describe low-energy reactions is not considered accurate enough to be reliable for most direct reactions, but the corrections introduce quantitative rather than qualitative changes.

Consider a direct reaction such as the (p, d) reaction, referred to as pickup, and the inverse stripping reaction. Let us denote the wave function for the initial and final states of the system in the center-of-mass (c.m.) frame by

$$\begin{aligned} |\text{initial}\rangle &= \exp(i\mathbf{k}_p \cdot \mathbf{r}_p)\Phi_{A+1}, \\ |\text{final}\rangle &= \exp(i\mathbf{k}_d \cdot \mathbf{r}_d)\Phi_A \Phi_d(r), \end{aligned} \tag{2.1}$$

corresponding to the reaction

$$p + (A + 1) \to d + A.$$

Here Φ denotes a nuclear wave function, the subscript denotes the atomic number, and $\mathbf{k}_p$ and $\mathbf{k}_d$ denote the relative momenta.

In the Born approximation, the amplitude for the reaction is given by the matrix element of the interaction responsible for the transition between these two states. This must be $V(\mathbf{r}_n - \mathbf{r}_p)$, and so

$$\begin{aligned} T &\propto \langle\text{final}|V|\text{initial}\rangle \\ &= \int \exp(-i\mathbf{k}_d \cdot \mathbf{r}_d)\Phi_d^*(r)\Phi_A^* V(r) \exp(i\mathbf{k}_p \cdot \mathbf{r}_p)\Phi_{A+1}\, d^3(A + 1) \end{aligned} \tag{2.2}$$

(see Fig. 2.1 for coordinates). Most of the integrations are trivial because only the proton and neutron change their state of motion. In the case of the ideal shell model description, the integral is very simple. In that case all nucleons fill the lowest shell model states, and the only difference between Φ_A and Φ_{A+1} is that the latter has one more neutron, say, in the shell-model state $\phi_{n\ell j}$. So by the orthonormality of the commonly occupied states,

$$\int \Phi_A^* \Phi_{A+1}\, d^3A = \phi_{n\ell j}(r_n). \tag{2.3}$$

Now we have for the amplitude

$$T = \int \exp(-i\mathbf{k}_d \cdot \mathbf{r}_d)\Phi_d^*(r)V(r)\phi_{n\ell j}(r_n) \exp(i\mathbf{k}_p \cdot \mathbf{r}_p)\, d\mathbf{r}_n\, d\mathbf{r}_p. \tag{2.4}$$

Now observe the following vector relations between coordinates from Fig. 2.1:

$$\begin{aligned} \mathbf{r}_d &= \mathbf{r}_n - \tfrac{1}{2}\mathbf{r}, \\ \mathbf{r}_p &= \frac{A}{A + 1}\mathbf{r}_n - \mathbf{r}, \end{aligned} \tag{2.5}$$

from which

$$\begin{aligned} \mathbf{k}_p \cdot \mathbf{r}_p - \mathbf{k}_d \cdot \mathbf{r}_d &= -\left(\mathbf{k}_d - \frac{A}{A + 1}\mathbf{k}_p\right) \cdot \mathbf{r}_n - (\mathbf{k}_p - \tfrac{1}{2}\mathbf{k}_d) \cdot \mathbf{r} \\ &\equiv -\mathbf{q} \cdot \mathbf{r}_n - \mathbf{K} \cdot \mathbf{r}, \end{aligned} \tag{2.6}$$

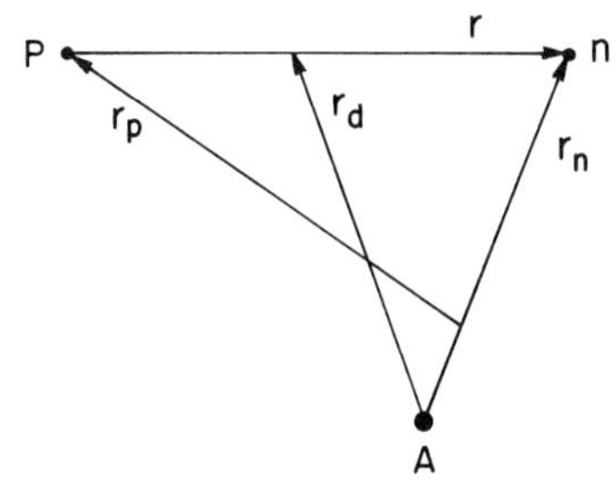

Fig. 2.1. A coordinate for the reaction A + 1 (p, d)A or its inverse $\mathbf{r}_d$ connects the cm of A and deuteron, and $\mathbf{r}_p$ connects cm of nucleus A + n and proton.

so

$$T = \int \exp(-i\mathbf{K} \cdot \mathbf{r})\phi_d^*(r)V(\mathbf{r})\,d\mathbf{r} \times \int \exp(-i\mathbf{q} \cdot \mathbf{r}_n)\phi_{n\ell j}(\mathbf{r}_n)\,d\mathbf{r}_n. \tag{2.7}$$

Thus the amplitude in this approximation is given by the product of two Fourier transforms; one of the deuteron wave function and the interaction, and one of the nuclear wave function of the neutron. It is the second one that interests us most because the first one would be the same no matter which neutron was picked up from the nucleus A.

The angle integrals can be performed easily by noting that the expansion for a plane wave is

$$\exp(i\mathbf{q} \cdot \mathbf{r}) = 4\pi \sum_{LM} i^L j_L(qr) Y_L^{M*}(\hat{\mathbf{q}}) Y_L^M(\hat{\mathbf{r}}), \tag{2.8}$$

where $\hat{\mathbf{q}}$ denotes the direction of the vector $\mathbf{q}$ and j_L is a spherical Bessel function. Also, we note that the shell-model wave function is the product of a radial function and a spherical harmonic

$$\phi = \frac{1}{r_n} u_{n\ell}(r_n) Y_\ell^m(\hat{\mathbf{r}}_n). \tag{2.9}$$

(Because the aim of this discussion is not to calculate the cross section down to the last factor, but to illustrate the physical ideas, the intrinsic spin of the nucleons has been ignored.)

The angle integrals in the second factor of T now can be done because of the orthonormality of the spherical harmonics:

$$T = 4\pi i^{-\ell} G(K) Y_\ell^m(q) \int_0^\infty j_\ell(qr) u_{n\ell}(r)\,dr, \tag{2.10}$$

where $G(K)$, denotes the Fourier transform of the deuteron wave function, that is, the first integral in (2.7).

At this point we introduce a physical approximation corresponding to the fact that we believe that direct reactions occur on the periphery of the nucleus; that is to say, when the projectile penetrates the nucleus, it is more likely to interact many times, forming a compound state, than to undergo a single interaction. Therefore in the preceding integral we reject the contribution from the interior and evaluate

$$\int_R^\infty j_\ell(qr)u_{n\ell}(r)r\,dr. \tag{2.11}$$

This can be done by recalling the radial Schrödinger equations satisfied by j and u:‡

$$\begin{aligned}\left(-\frac{d^2}{dr^2}+\frac{\ell(\ell+1)}{r^2}-q^2\right)rj_\ell(qr)&=0,\\ \left(-\frac{d^2}{dr^2}+\frac{\ell(\ell+1)}{r^2}+t^2\right)u_{n\ell}(r)&=0 \qquad (\text{if } r>R),\end{aligned} \tag{2.12}$$

where $\hbar^2t^2/2m^*=B$ is the binding energy of the neutron in the nucleus A. If we multiply the first of these equations by u and the second by rj and subtract them, we obtain

$$\begin{aligned}(t^2+q^2)\int_R^\infty j_\ell(qr)u_{n\ell}(r)r\,dr&=-\int_R^\infty \frac{d}{dr}\left[u_{n\ell}\frac{d}{dr}(rj_\ell)-(rj_\ell)\frac{d}{dr}u_{n\ell}\right]dr\\ &=u_{n\ell}(R)\frac{d}{dR}[Rj_\ell(qR)]-Rj_\ell(qR)\frac{d}{dR}u_{n\ell}(R).\end{aligned} \tag{2.13}$$

The contribution at $r=\infty$ vanishes because of the vanishing of the bound state wave function u.

Recall the definition of $\mathbf{q}$, which is the momentum carried by the picked-up neutron. The magnitude of $\mathbf{q}$ depends on the scattering angle. Thus the argument qR of the Bessel functions varies rapidly as θ changes (Fig 2.2). Moreover, the Bessel functions $j_\ell(qR)$ have different oscillatory patterns, depending on the value of ℓ. For example,

$$j_0(\rho)=\frac{\sin\rho}{\rho}, \qquad j_1(\rho)=\frac{\sin\rho}{\rho^2}-\frac{\cos\rho}{\rho}. \tag{2.14}$$

This is in agreement with our earlier observation that the outgoing particle must scatter in such a way as to reflect the conserved quantities, including angular momentum and parity. We find, in fact, a quite sensitive dependence of the angular distribution. The cross section of this reaction is proportional to $|T|^2$ and is illustrated in Fig. 2.3 for several values of ℓ. This shows

‡ Recall the method of separation of variables, which in this case are the radial and angular coordinates, see the appendix or Messiah (1962).

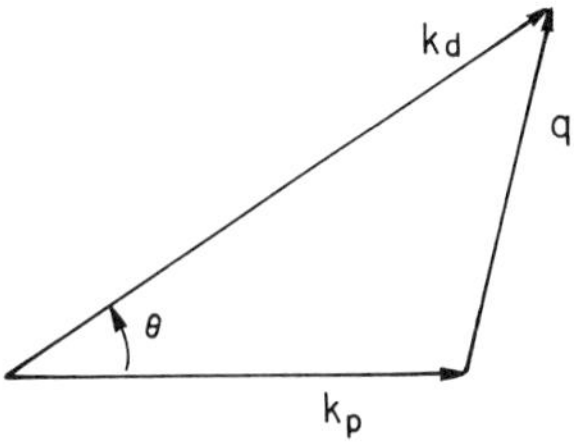

Fig. 2.2. The initial $\mathbf{k}_p$ and final $\mathbf{k}_d$ momenta, and the momentum transfer $\mathbf{q} = \mathbf{k}_d - \mathbf{k}_p$.

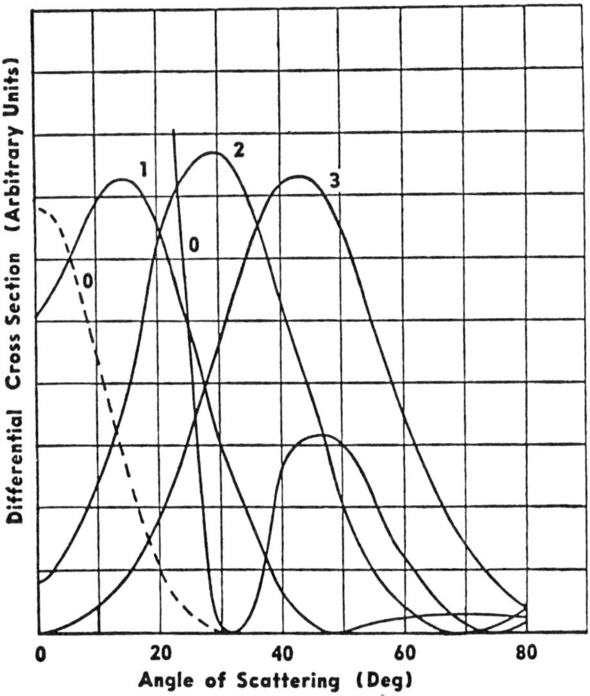

Fig. 2.3. Plane-wave calculation of the cross section by S. T. Butler (1957), showing the strong dependence on the angular momentum of the transferred neutron in the (d, p) reaction.

the dependence of the angular distribution on the angular momentum ℓ that is transferred in the reaction.

It is instructive to consider the following argument. Let the momenta of the incident proton be $A\mathbf{k}_p/(A + 1)$, the outgoing deuteron $\hbar\mathbf{k}_d$, and the picked-up neutron $\hbar q$. Energy conservation specifies the magnitude of $\mathbf{k}_d$ as

$$\frac{\hbar^2 k_d^2}{2m_d} + Q = \frac{\hbar^2 k_p^2}{2m_p},$$

where Q is the energy removed from the nucleus. Momentum conservation requires that

$$\mathbf{q} = \mathbf{k}_d - \frac{A}{A+1}\mathbf{k}_p.$$

Finally, if the neutron is removed from the orbital angular momentum state ℓ at the point R, then

$$\mathbf{R} \times \mathbf{q} = \ell \qquad \text{or} \qquad \left|\mathbf{k}_d - \frac{A}{A+1}\mathbf{k}_p\right| R \geq \ell.$$

This classical argument clearly imposes a condition on the scattering angle. For example, if $\ell > 0$, the scattering angle must be greater than $\theta = 0$. In fact, the minimum scattering angle increases with ℓ. This qualitative argument agrees with our previous calculation of the Born amplitude. However, it also serves to clarify the fact that, because there is no fixed radius R at which the pickup must occur, there is at this point a degree of uncertainty in our theory that needs to be addressed. Our assumption in the Born calculation, that the stripping occurs only for $r > R$, must be replaced by a more realistic treatment in which this condition would be fulfilled in a natural way, a way that is governed in part by the actual radial distribution of the particle in the nucleus.

A second shortcoming of the plane-wave Born approximation is the assumption of straight-line trajectories before and after the single collision event. For example, the Coulomb field can cause an elastic deflection of the charged particles, permitting the pickup event to occur on such a deflected orbit. These defects are remedied in what is called the distorted-wave Born approximation.

Notes to Chapter 2

S. T. Butler, in 1950–1954, in his theory of stripping reactions was the first to recognize direct reactions. A number of people contributed to the early understanding of direct reactions in the plane wave Born approximation including J. A. Spiers, G. R. Satchler, A. B. Bhatia, K. Huang, R. Huby, H. C. Newns, B. B. Daitch, J. B. French, N. Austern, D. M. Brink, J. S. Blair and H. McManus. References to the early literature can be found in the works of Austern (1970), Glendenning (1963), Hodgson (1971), and Jackson (1970).

Chapter 3

Scattering Theory and General Results

A. MOTIVATION

The discussion of the plane-wave Born matrix element for a direct reaction taught us that these reactions are sensitive to the spins and parities of the nuclear levels and to the radial distribution of particles in the nucleus. Although the point has not been stressed so far, the whole discussion also implies that a careful analysis of these reactions must yield information on the degree to which particular levels are simple shell-model states and the degree to which they have more complicated structure. In addition, we shall later discuss reactions in which several nucleons are transferred and see how they depend very sensitively on the correlation of nucleons in the nucleus. For these reasons, direct reactions have proven to be the principal means by which the detailed microscopic structure of nuclei can be discovered. This provides us with the motivation for improving and developing the theory in the many ways that will be discussed in later chapters. In this chapter we begin by considering some of the useful but more formal results of scattering theory.

and the corresponding kinetic energy operator by

$$T_\alpha = -\frac{\hbar^2}{2m_\alpha}\nabla_\alpha^2, \tag{3.4}$$

where m_α is the reduced mass

$$m_\alpha = \frac{m_a m_A}{m_a + m_A} \simeq \frac{aA}{a + A}. \tag{3.5}$$

The two nuclei interact with each other through the two-nucleon interaction, which will be denoted by V_{ij}. It may or may not correspond to the shell-model residual interaction. In any case it is, in practice, an effective interaction for many reasons, which will become clearer as we proceed. Thus the potential energy of the two nuclei is

$$V_\alpha = \sum_{i \in a,\, j \in A} V_{ij}. \tag{3.6}$$

The total Hamiltonian, excluding the c.m. energy, is thus

$$H = H_a + H_A + T_\alpha + V_\alpha, \tag{3.7}$$

where $H_a + H_A$ refers only to the intrinsic structure of the nuclei, which is written as

$$H_\alpha = H_a + H_A. \tag{3.8}$$

For brevity in notation, it will often be convenient to denote the intrinsic state of the two nuclei by a single function, namely,

$$\Phi_\alpha = \Phi_a \Phi_A, \tag{3.9}$$

which obviously satisfies

$$(H_\alpha - \varepsilon_\alpha)\Phi_\alpha = 0, \qquad \varepsilon_\alpha = \varepsilon_a + \varepsilon_A. \tag{3.10}$$

Finally, $T_\alpha + V_\alpha$ refers only to their relative motion.

As result of the interaction V_α between the two nuclei, if they come close enough in the collision, one or the other or both nuclei may be excited to an energy level above the ground state. This channel will be denoted by $\Phi_{\alpha'}$ with energy $\varepsilon_{\alpha'}$. Alternatively, one or more nucleons may be transferred between the nuclei, producing a different *partition* of the system of a + A nucleons. Let these new nuclei be denoted by b and B and denote this channel by β. The *different* nuclear combinations, a + A, b + B, etc., of the original system are referred to as partitions or channels, whereas different excited states within the partition are referred to as different channels. If the kinetic energy of relative motion E_α between the original nuclei a + A, called the entrance channel, is large enough to produce a given partition c + C, that is, if

$$E > \varepsilon_c, \tag{3.11}$$

then this partition is said to be open. In general, at energies of typical tandem or cyclotron accelerators, many channels of both types (inelastic and transfer) are open; and the complete wave function of the system, call it Ψ, must carry the information on all of these possibilities. A moment's reflection reveals that an accurate calculation of the complete wave function Ψ at energies where many channels are open is a horrendous task. It involves solving the many-body scattering problem. Our first task is to find a way of calculating the parts of Ψ that are needed to describe the direct reactions in which we are interested. The general methods for achieving this goal are now outlined.

D. INTEGRAL EQUATIONS AND THE SCATTERING AMPLITUDE

The Schrödinger equation of the entire system of a + A nucleons can be written as

$$(H - E)\Psi = 0, \tag{3.12}$$

where H can be written in terms of any of the partitions of the original nucleons

$$H = H_\alpha + T_\alpha + V_\alpha = H_\beta + T_\beta + V_\beta, \qquad \text{etc.}, \tag{3.13}$$

corresponding merely to a regrouping of terms in H. If the entrance channel is α, then we label Ψ by this to remind ourselves of the initial state of the system. Equation (3.12) is then reordered as

$$(E - H_\alpha - T_\alpha)\Psi_\alpha = V_\alpha\Psi_\alpha. \tag{3.14}$$

If there were no interactions V_α between the nuclei, then the Schrödinger equation would be

$$(E - H_\alpha - T_\alpha)\phi_\alpha = 0. \tag{3.15}$$

Its solution is easy to write in terms of the nuclear wave functions (3.2), because the Hamiltonian is separable into the coordinates a, A, and the relative coordinate r_α. The eigenfunctions of the kinetic energy operator T_α are just plane waves. Therefore

$$\phi_\alpha = \exp(i\mathbf{k}_\alpha \cdot \mathbf{r}_\alpha)\Phi_\alpha, \tag{3.16}$$

where

$$\hbar^2 k_\alpha^2/2m_\alpha = E_\alpha = E - \varepsilon_\alpha \tag{3.17}$$

is the kinetic energy of relative motion.

Consistent with our goal of seeking the parts of Ψ that describe the reactions of interest, we project onto one such channel α by multiplying (3.14) by Φ_α and then integrating the *internal* coordinates. Because T_α does not act on the internal coordinates, the result of this may be written as

$$(E_\alpha - T_\alpha)(\Phi_\alpha, \Psi_\alpha) = (\Phi_\alpha, V_\alpha \Psi_\alpha) \tag{3.18}$$

where we use parentheses to denote that the integration is only over all internal coordinates. The bracket notation will be used frequently, with the convention

$$\begin{aligned}(\Phi_\alpha, \Psi_\alpha) \equiv (\Phi_\alpha | \Psi_\alpha) &= \int \Phi_\alpha^* \Psi_\alpha \, d(\text{internal coordinates}),\\ \langle \phi_\alpha, \Psi_\alpha \rangle \equiv \langle \phi_\alpha | \Psi_\alpha \rangle &= \int \phi_\alpha^* \Psi_\alpha \, d(\text{all coordinates}).\end{aligned} \tag{$*$}$$

Both brackets are thus functions only of the remaining coordinate $\mathbf{r}_\alpha$, and we denote

$$\psi_\alpha(\mathbf{r}_\alpha) \equiv (\Phi_\alpha, \Psi_\alpha), \tag{3.19}$$

which is a wave function that describes the relative motion when the system is in the channel α. Then,

$$(E_\alpha - T_\alpha)\psi_\alpha = (\Phi_\alpha, V_\alpha \Psi_\alpha). \tag{3.20}$$

A useful transformation of this equation can be achieved by using the Green's function. It is a solution of the corresponding equation where the source is a delta function:

$$(E_\alpha - T_\alpha) G_\alpha^0(\mathbf{r}_\alpha, \mathbf{r}'_\alpha) = \delta(\mathbf{r}_\alpha - \mathbf{r}'_\alpha). \tag{3.21}$$

By direct substitution one can see that a solution to (3.20) can be written

$$\psi_\alpha(\mathbf{r}_\alpha) = \int G_\alpha^0(\mathbf{r}_\alpha, \mathbf{r}'_\alpha)(\Phi_\alpha, V_\alpha \Psi_\alpha) \, d\mathbf{r}'_\alpha. \tag{3.22}$$

This is, however, not the most general solution, because we can add a solution of

$$(E_\alpha - T_\alpha) \exp(i\mathbf{k}_\alpha \cdot \mathbf{r}_\alpha) = 0, \tag{3.23}$$

and still have a solution, which is obvious. Thus

$$\psi_\alpha(\mathbf{r}_\alpha) = \exp(i\mathbf{k}_\alpha \cdot \mathbf{r}_\alpha) + \int G_\alpha^0(\mathbf{r}_\alpha, \mathbf{r}'_\alpha)(\Phi_\alpha, V_\alpha \Psi_\alpha) \, d\mathbf{r}_\alpha. \tag{3.24}$$

The situation we wish to describe is one in which the two nuclei a and A are initially in a plane-wave state. As a result of the interaction, the nuclei are scattered, causing spherical waves to emanate from the scattering site.

A solution to (3.21) is

$$G_\alpha^{0(+)}(\mathbf{r}, \mathbf{r}') = -\frac{2m_\alpha}{4\pi\hbar^2} \frac{\exp(ik_\alpha|\mathbf{r} - \mathbf{r}'|)}{|\mathbf{r} - \mathbf{r}'|} \xrightarrow[r\to\infty]{} -\left(\frac{2m_\alpha}{4\pi\hbar^2}\right) \frac{\exp(ik_\alpha r)}{r} \exp(-i\mathbf{k}'_\alpha \cdot \mathbf{r}'), \tag{3.25}$$

where $\mathbf{k}'_\alpha$ is a vector directed along $\mathbf{r}$ with length k_α.‡ This result for the free outgoing Green's function may be well known to the reader. Therefore, we shall not interrupt the main line of argument to derive it. We will derive it in Chapter 6, while in Chapter 5, we shall derive the Green's function in the more general case of a potential, of which (3.25) is a special case. The Green's function (3.25) is referred to as outgoing because the quantum-mechanical flux associated with $\exp(ik_\alpha r)/r$ is directed radially outward. The incoming Green's function has $i \to -i$ everywhere, that is, it is the adjoint of the outgoing one.

Introducing the value of the Green's function (3.25) into (3.24) we have the desired solution, possessing the properties required to describe the physical situation:

$$\psi_{\mathbf{k}_\alpha}^{(+)}(\mathbf{r}_\alpha) \to \exp(i\mathbf{k}_\alpha \cdot \mathbf{r}_\alpha) - \frac{m_\alpha}{2\pi\hbar^2} \frac{\exp(ik_\alpha r_\alpha)}{r_\alpha} \int \exp(-i\mathbf{k}'_\alpha \cdot \mathbf{r})(\Phi_\alpha, V_\alpha \Psi_\alpha^{(+)})\, d\mathbf{r}. \tag{3.26}$$

The amplitude of the scattered wave in the direction $\mathbf{k}'_\alpha$ is called the scattering amplitude,

$$f_{\alpha\alpha}(\theta) = -\frac{m_\alpha}{2\pi\hbar^2} \int \exp(-i\mathbf{k}'_\alpha \cdot \mathbf{r})(\Phi_\alpha, V_\alpha \Psi_\alpha^{(+)})\, d\mathbf{r}. \tag{3.27}$$

The angle between $\mathbf{k}_\alpha$ and $\mathbf{k}'_\alpha$ is denoted by θ. The $(+)$ sign that appears on the Green's function (3.25) and on our desired solution (3.26) is used to denote the outgoing wave condition at $r \to \infty$, whereas a $(-)$ sign would be used to denote an incoming wave. Two subscripts $\alpha\alpha$ appear on $f(\theta)$ to indicate that the entrance channel is α *and* that we have looked

‡ The limit in (3.25) can be evaluated as follows:

$$|\mathbf{r} - \mathbf{r}'| = [(\mathbf{r} - \mathbf{r}') \cdot (\mathbf{r} - \mathbf{r}')]^{1/2} = r\left(1 - \frac{2\mathbf{r} \cdot \mathbf{r}'}{r} + \frac{r'^2}{r^2}\right)^{1/2}$$
$$\to r - \mathbf{r} \cdot \mathbf{r}'/r, \qquad \text{for} \quad r \gg r'$$

Define $\mathbf{k}'_\alpha = k_\alpha \mathbf{r}/r$. Then

$$\exp(ik_\alpha|\mathbf{r} - \mathbf{r}'|)/r \to \frac{1}{r} \exp(ik_\alpha r) \exp(-i\mathbf{k}'_\alpha \cdot \mathbf{r}')$$

at the same channel as exit channel. So far we have calculated nothing more than the elastic scattering amplitude.

The generalization to an inelastic or a reaction channel is straightforward. Clearly, instead of the projection (3.19), we want, for a reaction, the projection

$$\psi_\beta(\mathbf{r}_\beta) = (\Phi_\beta, \Psi_\alpha), \tag{3.28}$$

which, with reference to (3.14), is governed by the equation

$$(E_\beta - T_\beta)\psi_\beta = (\Phi_\beta, V_\beta \Psi_\alpha). \tag{3.29}$$

We leave the label α on Ψ because the initial partition is still a + A. Now we must compute an expression for the amplitude of

$$a + A \to b + B. \tag{3.30}$$

The Green's function for this problem is similar to (3.25). Now, however, there is no incoming wave in the channel β, only outgoing ones. Consequently, our desired solution, for large r_β, has the form

$$\psi_\beta^{(+)}(\mathbf{r}_\beta) \to -\frac{m_\beta}{2\pi\hbar^2}\frac{\exp(ik_\beta r_\beta)}{r_\beta}\int \exp(-i\mathbf{k}_\beta \cdot \mathbf{r})(\Phi_\beta, V_\beta \Psi_\alpha^{(+)})\, d\mathbf{r}, \tag{3.31}$$

from which the scattering amplitude is

$$f_{\beta\alpha}(\theta) = -\frac{m_\beta}{2\pi\hbar^2}\int \exp(-i\mathbf{k}_\beta \cdot \mathbf{r})(\Phi_\beta, V_\beta \Psi_\alpha^{(+)})\, d\mathbf{r}. \tag{3.32}$$

Again, the scattering angle is denoted by θ, being in this case the angle between $\mathbf{k}_\beta$ and $\mathbf{k}_\alpha$.

Frequently the $\mathscr{T}$ matrix is used in place of the scattering amplitude, which, under the physical conditions of equal energies in all channels, is simply the integral in (3.32). Referring to the notation (3.16), it is

$$\mathscr{T}_{\beta\alpha} = \langle \phi_\beta | V_\beta | \Psi_\alpha^{(+)} \rangle, \tag{3.33}$$

where the angle brackets are used to denote integration over all coordinates. This is sometimes referred to as the transition amplitude, and the whole collection of such amplitudes for various α, β is referred to as the transition matrix. Frequent usage employs the terms interchangeably.

E. ASYMPTOTIC FORM OF THE COMPLETE WAVE FUNCTION

The complete wave function Ψ_α, a solution of (3.12), must contain the description of all possible events associated with the collision of a and A. We have performed several projections of the intrinsic nuclear states onto this function. The results can be summarized by the following expression, which gives the behavior of Ψ_α when it is examined in the limit in which the relative coordinate in any two-body partition becomes very large:

$$\Psi_\alpha^{(+)} \to \begin{cases} \Phi_\alpha\left[\exp(i\mathbf{k}_\alpha\cdot\mathbf{r}_\alpha) + f_{\alpha\alpha}(\theta)\dfrac{\exp(ik_\alpha r_\alpha)}{r_\alpha}\right] & \\ \qquad + \displaystyle\sum_{\alpha'\neq\alpha}\Phi_{\alpha'}f_{\alpha'\alpha}(\theta)\frac{\exp(ik_{\alpha'}r_\alpha)}{r_\alpha}, & r_\alpha\to\infty, \\ \displaystyle\sum_\beta \Phi_\beta f_{\beta\alpha}(\theta)\frac{\exp(ik_\beta r_\beta)}{r_\beta}, & r_\beta\to\infty. \end{cases} \tag{3.34}$$

The three terms correspond to the elastic, inelastic, and transfer channels, respectively. These manipulations have given us a concise expression for the scattering amplitude, or, equivalently, the $\mathscr{T}$ matrix (3.33). We have not solved the scattering problem in the sense that our expression for $\mathscr{T}$ contains the unknown complete wave function Ψ. However, as soon as we make an approximation of Ψ, (3.33) tells us how to calculate the corresponding amplitude, from which, as will be shown shortly, the cross section can be calculated.

F. THE FIRST BORN APPROXIMATION

The crudest possible approximation for Ψ in (3.33) is to neglect the effect of the scattering altogether; that is, drop the right side of (3.14). Then we obtain (3.15). Hence, we are taking

$$\Psi_\alpha \simeq \phi_\alpha.$$

Then, from (3.33),

$$\mathscr{T}_{\beta\alpha} \simeq \langle\phi_\beta|V_\beta|\phi_\alpha\rangle = \langle\exp(i\mathbf{k}_\beta\cdot\mathbf{r}_\beta)\Phi_\beta|V_\beta|\exp(i\mathbf{k}_\alpha\cdot\mathbf{r}_\alpha)\Phi_\alpha\rangle. \tag{3.35}$$

This is known as the first Born approximation and corresponds to calculating $\mathscr{T}$ as the matrix element of the interaction between unperturbed wave functions—obviously the lowest-order approximation that we can make. This is the approximation that we used in Chapter 2 for the (p, d) reaction.

G. CROSS SECTION

The incident beam of particles is represented in (3.34) by the plane wave, and the outgoing spherical waves in the various channels represent the scattered waves arising from the interaction of the incident beam with the target located at the origin. The differential cross section of particles scattered in the direction θ is defined as the flux of scattered particles through the area $dA = r^2\,d\Omega$ in the direction θ, per unit incident flux. The quantum-mechanical current associated with a wave function ψ is

$$\mathbf{j} = \frac{\hbar}{2mi}(\psi^*\nabla\psi - \psi\nabla\psi^*). \tag{3.36}$$

From (3.34) the scattered flux passing through dA in the channel β, for example, is obtained from the radial component of $\mathbf{j}$. (The other two components vanish at large distance from the scattering site.) The radial component is

$$(\hat{\mathbf{r}}_\beta \cdot \mathbf{j}_\beta)\,dA = \frac{\hbar k_\beta}{m_\beta}|f_{\beta\alpha}(\theta)|^2\,d\Omega + O\left(\frac{1}{r_\beta}\right), \tag{3.37}$$

where the second term vanishes as $r_\beta \to \infty$.‡ The incident flux passing through unit area, which is along the axis, is obtained from the plane wave part of (3.34) as $\hbar k_\alpha/m_\alpha$. Thus the differential cross section is

$$d\sigma = \frac{m_\alpha}{m_\beta}\frac{k_\beta}{k_\alpha}|f_{\beta\alpha}(\theta)|^2\,d\Omega, \tag{3.38}$$

or, written in terms of the $\mathscr{T}$ matrix,

$$\left(\frac{d\sigma}{d\Omega}\right)_{\beta\alpha} = \frac{m_\alpha m_\beta}{(2\pi\hbar^2)^2}\frac{k_\beta}{k_\alpha}|\mathscr{T}_{\beta\alpha}|^2. \tag{3.39}$$

This is the differential cross section for the transition $\alpha \to \beta$. We have not specifically spoken of spin angular momentum as yet. Naturally the spins of the nuclear states a, A and b, B are included in the labels α and β in addition to their z components. Usually, in an experimental measurement, the projectile a is unpolarized, and the target A is unaligned. Typically, the final spin orientations of b and B also are not measured. The cross section corresponding to this type of measurement is therefore an average of the preceding one over the initial angular momentum projections of J_a and J_A, and a sum over the final projections of J_b and J_B. Thus

$$\frac{d\sigma}{d\Omega} = \frac{1}{(2J_a+1)(2J_A+1)}\sum_{M_a M_A M_b M_B}\left(\frac{d\sigma}{d\Omega}\right)_{\beta\alpha}. \tag{3.40}$$

The Q value of a reaction is defined as the change in kinetic energy between final and initial partitions. The convention universally adopted is

‡ Throughout this book, a vector with a caret, e.g., $\hat{\mathbf{r}}$, is used to denote a unit vector in the direction $\mathbf{r}$.

that a positive Q denotes an increase of kinetic energy in the final state. From (3.17) it can be written as

$$Q = E_\beta - E_\alpha = \varepsilon_\beta - \varepsilon_\alpha = \varepsilon_{\mathrm{b}} + \varepsilon_{\mathrm{B}} - \varepsilon_{\mathrm{a}} - \varepsilon_{\mathrm{A}} \tag{3.41}$$

Notes to Chapter 3

The nuclear shell model is discussed by Bohr and Mottelson (1969, 1975), de Shalit and Feshback (1974), de-Shalit and Talmi (1963), Eisenberg and Greiner (1976), Preston (1962), and Preston and Bhaduri (1975). An account of scattering theory can be found in many textbooks. Two that are readily accessible are "Quantum Mechanics" by Messiah (1962) and "Introduction to the Quantum Theory of Scattering" by Rodberg and Thaler (1967).

Chapter 4

The Phenomenological Optical Potential

A. RATIONALE FOR THE OPTICAL POTENTIAL

In the last chapter the amplitude for a transition of the system from one partition to another,

$$a + A \rightarrow b + B,$$

was derived. The general result is

$$\mathscr{T}_{\beta\alpha} = \langle \phi_\beta | V_\beta | \Psi_\alpha^{(+)} \rangle.$$

This depends on the exact wave function $\Psi_\alpha^{(+)}$ that, being exact, contains a description of all processes that can occur at the given bombarding energy. Even if it were possible to compute Ψ_α, and hence to compute the overlap for $\mathscr{T}_{\beta\alpha}$, this would be an inefficient procedure. However, we cannot compute Ψ_α in any case. For these practical reasons, an approximation of the amplitude would be useful. Equation (3.35) showed the crudest possible approximation, which formed the basis for the earliest work on direct reactions. In this first Born approximation, the relative motion of the two parts of the system is represented by a plane wave. We now seek to lay the groundwork for improving the calculation in this respect.

The central potential U of the shell model (3.1) is used to represent the average field of all the nucleons on a particular one, and the corrections to this average are calculated by taking into account the residual interaction V_{ij}. In much the same sense, we understand intuitively that two nuclei, as they approach each other, are deflected by the field generated by the mutual interactions of all nucleons in one nucleus with those of the other, both before and after any specific interaction exchanges energy or nucleons between them.

Thus we introduce a potential $U(r_\alpha)$ that depends only on the relative coordinate between two nuclei in the partition α. This potential is therefore intended to describe the relative motion in this channel. Intuitively, we would be motivated to choose U in such a way that it would describe the elastic scattering as accurately as possible. The direct reactions are then thought of as a perturbation on the elastic scattering. Indeed, most reaction cross sections are small compared with the elastic cross section. Because the same function U cannot be expected to account for the elastic cross section in every partition, we are prepared to define potentials U_α, $U_\beta, \ldots$ for each partition that we wish to consider.

Although the motivation for the optical potential has been presented from the point of view of improving the description of direct reactions, as an historical note, it was first introduced for the purpose of describing the elastic scattering of nucleons from nuclei; and it was from this perspective that the experimental data motivated and justified the concept of an optical potential (Feshback *et al.*, 1954; Levier and Saxon, 1952; Melkanoff *et al.*, 1956; Woods and Saxon, 1954).

The Schrödinger equation, if U_α alone were the whole interaction between a and A, would be

$$(H_\alpha + T_\alpha + U_\alpha - E)\Psi = 0. \tag{4.1}$$

In contrast to (3.14), this Schrödinger equation is separable into the nuclear coordinates contained in H_α and the relative coordinate contained in U_α. The solution can be written as a product of the nuclear wave functions (3.9) and a relative wave function $\chi_\alpha(\mathbf{r}_\alpha)$, which satisfies the Schrödinger equation

$$(T_\alpha + U_\alpha - E_\alpha)\chi_\alpha(\mathbf{r}_\alpha) = 0, \tag{4.2}$$

as direct substitution of $\Psi = \Phi_\alpha\chi_\alpha$ into (4.1) shows. We shall often refer to this as the optical-model Schrödinger equation. Because U depends only on the relative coordinate, it can produce no change in the nuclei and so can only describe elastic scattering.

An important remark should be made at this point. At very low bombarding energy, charged nuclei can only scatter elastically. As the energy is raised, inelastic and reaction channels open. Thus some flux is removed from the elastic channel. Indeed, to assure that the direct reaction mechanism dominates the production of low-lying levels of the product nucleus, a sufficiently high bombarding energy must be chosen so that the compound nucleus formed at that energy will, as a result of the competition of the many open channels, decay to such levels with negligible probability. At such energies there is a very significant reaction cross section and, consequently, *a large removal of flux* from the initial or elastic channel.

This discussion implies that U ought to be complex, because a real potential conserves flux. This can be proven with reference to the Schrödinger equation (Levier and Saxon, 1952). The one governing χ^* is obviously

$$(T_\alpha + U_\alpha^* - E_\alpha)\chi_\alpha^* = 0. \tag{4.3}$$

Multiplying (4.2) by χ_α^* and (4.3) by χ_α and subtracting leads to

$$-\frac{\hbar^2}{2m}(\chi_\alpha^* \nabla^2 \chi_\alpha - \chi_\alpha \nabla^2 \chi_\alpha^*) = (U_\alpha^* - U_\alpha)\chi_\alpha^* \chi_\alpha.$$

Recalling the definition of the quantum-mechanical current, this equation reads

$$-i\hbar \nabla \cdot \mathbf{j} = (U_\alpha^* - U_\alpha)\chi_\alpha^* \chi_\alpha \qquad \text{or} \qquad \hbar \nabla \cdot \mathbf{j} = 2\rho \,\mathrm{Im}\, U_\alpha. \tag{4.4}$$

Integrating this over a volume containing the scattering center, we find that if U_α is real there is no net change of flux, whereas if Im U_α is negative there is a loss.

Consequently, for U to describe accurately the elastic cross section when nuclear reactions can also occur, it must be complex with a negative imaginary part. Later in this chapter we shall describe the parametrization of the optical potential, and in Chapter 8 we shall describe its theoretical structure. The explicit calculation of certain components from its theoretical structure will be discussed in Chapter 14.

B. PARTIAL WAVE EXPANSION, THE RADIAL WAVE FUNCTION, ITS ASYMPTOTIC BEHAVIOR

The Schrödinger equation (4.2) for the optical model represents one of the simpler scattering problems. However, except for several special cases such as a square potential and a pure Coulomb potential, analytic solutions are not known, and the Schrödinger equation must be solved on a computer. The equation is in three dimensions. However, because the optical potential is usually regarded as a function only of the separation $r_\alpha = |\mathbf{r}_\alpha|$ between the centers of mass of the two nuclei, the equation can be separated into radial and angular parts. The separation is still possible for a potential that depends on the vector $\mathbf{r}_\alpha$, but the radial equations are more complicated. A consideration of this situation, pertinent to deformed nuclei, is deferred to Chapter 13.

The kinetic energy operator (3.4) in (4.2) can be written in spherical coordinates using the results stated in the Appendix for the Laplace operator and the spherical harmonics $Y_\ell^m(\theta, \phi)$. It is then straightforward to show that one solution of (4.2) is

$$\psi_{\ell m}(\mathbf{r}_\alpha) = \frac{1}{k_\alpha r_\alpha} f_\ell(k_\alpha, r_\alpha) Y_\ell^m(\theta, \phi), \tag{4.5}$$

provided that the radial function $f_\ell(k_\alpha, r_\alpha)$ satisfies the one-dimensional differential equation

$$(E_\alpha - T_{\alpha\ell} - U_\alpha) f_\ell(k_\alpha, r_\alpha) = 0, \qquad T_{\alpha\ell} = \frac{\hbar^2}{2m_\alpha}\left(-\frac{d^2}{dr^2} + \frac{\ell(\ell+1)}{r^2}\right). \tag{4.6}$$

Also, any linear combination is a solution. Because of the azimuthal symmetry of the scattering about the direction $\mathbf{k}_\alpha$ of the incident particle, the desired solution will depend on the angle θ between $\mathbf{k}_\alpha$ and $\mathbf{r}_\alpha$ but not on ϕ. We write this solution as

$$\begin{aligned}
\chi_\alpha^{(+)}(\mathbf{k}_\alpha, \mathbf{r}_\alpha) &= \frac{1}{k_\alpha r_\alpha} \sum_\ell [4\pi(2\ell+1)]^{1/2} i^\ell e^{i\sigma_\ell} f_\ell(k_\alpha, r_\alpha) Y_\ell^0(\theta) \\
&= \frac{1}{k_\alpha r_\alpha} \sum_\ell (2\ell+1) i^\ell e^{i\sigma_\ell} f_\ell(k_\alpha, r_\alpha) P_\ell(\cos\theta) \\
&= \frac{4\pi}{k_\alpha r_\alpha} \sum_{\ell, m} i^\ell e^{i\sigma_\ell} f_\ell(k_\alpha, r_\alpha) Y_\ell^m(\hat{\mathbf{r}}_\alpha) Y_\ell^{m*}(\hat{\mathbf{k}}_\alpha),
\end{aligned} \tag{4.7}$$

where in the last two lines relations from the Appendix were used. This is the partial wave expansion of the wave function. The particular factors with which the solutions (4.5) arecombined were chosen for convenience, as will be seen shortly, and, of course, merely effect the normalization of the radial functions f_ℓ. Although (4.7) is a general form of the solution, in the placement of $(+)$ as a superscript on χ, we have already anticipated the particular physical situation that is to be described. This denotes that the radial solutions f_ℓ will be chosen in such a way that χ has outgoing spherical waves at infinity, as must be the case if we are to describe a scattering state. The factor $\exp i\sigma_\ell$ may seem mysterious at this point. It is absent for uncharged particles but otherwise is inserted in accord with the convention used for Coulomb wave functions. This will be discussed more fully at a later point in the chapter.

It is worth noting at this point that if the potential U_α were zero, then eqs. (4.6) would be satisfied by the spherical Bessel functions j_ℓ and n_ℓ:

$$f_\ell(k, r) = \begin{cases} F_\ell(kr) = krj_\ell(kr), \\ G_\ell(kr) = -krn_\ell(kr), \end{cases} \quad \text{if} \quad U = 0. \tag{4.8}$$

In this case, along with the regular solutions j_ℓ, we find from (2.8) the plane wave

$$\chi_\alpha(r_\alpha) = \exp(i\mathbf{k}_\alpha \cdot \mathbf{r}_\alpha), \qquad \text{if} \quad U = 0. \tag{4.9}$$

If, on the other hand, U is the Coulomb potential alone, then two independent solutions of (4.6) are the regular and irregular Coulomb functions

$$f_\ell(kr) = \begin{cases} F_\ell(\eta, kr), \\ G_\ell(\eta, kr), \end{cases} \tag{4.10}$$

$$\eta = m_\alpha Z_A Z_a e^2/(\hbar^2 k),$$

if $U = Z_A Z_a e^2/r$; and σ_ℓ in (4.7) is the Coulomb phase shift

$$\sigma_\ell = \arg \Gamma(\ell + 1 + i\eta). \tag{4.11}$$

It is also helpful to note the asymptotic behavior of these solutions as $r \to \infty$:

$$\begin{aligned} F_\ell(kr) &\to \sin \theta_\ell, \\ G_\ell(kr) &\to \cos \theta_\ell, \\ \theta_\ell &= kr - \eta \ln(2kr) - \ell\pi/2 + \sigma_\ell \end{aligned} \tag{4.12}$$

which also gives the behavior of krj_ℓ and krn_ℓ by letting $\eta = \sigma_\ell = 0$.

Of course any convenient combinations of solutions are also solutions. Such combinations are

$$\begin{aligned} I_\ell &= G_\ell - iF_\ell \to e^{-i\theta_\ell}, \\ O_\ell &= G_\ell + iF_\ell \to e^{i\theta_\ell}, \end{aligned} \tag{4.13}$$

because they represent incoming and outgoing waves. In the absence of the Coulomb field, the same notation of I and O will be used.

Returning to a consideration of the partial wave expansion and the equation satisfied by f, we note that, in the region of large r where U has become zero (or purely Coulombic), f must become a linear combination of solutions to the Bessel (or Coulomb) equation:

$$f_\ell(k_\alpha, r_\alpha) \to AF_\ell + BG_\ell, \tag{4.14}$$

where A and B are as yet undetermined constants. They can be determined by the physical circumstances to be described: there ought to be a plane wave representing the incident beam of particles and an outgoing spherical wave representing the effects of the collision. We have just learned that F_ℓ and O_ℓ are the radial functions that correspond to plane wave and spherical outgoing wave, respectively. The physical situation demands, therefore, that f_ℓ behave like

$$f_\ell(k_\alpha, r_\alpha) \to F_\ell + T_\ell(G_\ell + iF_\ell). \tag{4.15}$$

When this is used in (4.7), one sees that the first term reproduces the plane wave $\exp(ik_\alpha \cdot r_\alpha)$ (if the Coulomb field is absent), whose expansion was written in (2.8). The second term has the desired outgoing wave character with an as yet undetermined amplitude. It is usual to write T_ℓ in the particular form

$$T_\ell = e^{i\delta_\ell} \sin \delta_\ell, \tag{4.16}$$

because, referring to the asymptotic forms of (4.12),

$$f_\ell(k_\alpha, r_\alpha) \to e^{i\delta_\ell} \sin(k_\alpha r_\alpha - \ell\pi/2 + \delta_\ell). \tag{4.17}$$

Here δ_ℓ is referred to as the phase shift. Its value is determined by the requirement that the wave function f_ℓ be regular at the origin. In practice this can be achieved by numerically integrating the Schrödinger equation from the origin with initial value zero and finite slope, to R, a point beyond which the nuclear potentials have become negligible, and there matching it to the asymptotic form (4.17) to determine δ_ℓ.

C. ELASTIC SCATTERING AMPLITUDE

We have found the asymptotic form of the radial function f. Upon substitution into (4.7), it leads to the result

$$\chi_\alpha^{(+)}(\mathbf{k}_\alpha, \mathbf{r}_\alpha) \to \exp(i\mathbf{k}_\alpha \cdot \mathbf{r}_\alpha) + \frac{\exp(ik_\alpha r_\alpha)}{k_\alpha r_\alpha} \sum_\ell (2\ell + 1)\, e^{i\delta_\ell} \sin \delta_\ell\, P_\ell(\cos\theta). \tag{4.18}$$

The coefficient of the outgoing wave e^{ikr}/r is called the scattering amplitude

$$f(\theta) = \frac{1}{k_\alpha} \sum_\ell (2\ell + 1)\, e^{i\delta_\ell} \sin \delta_\ell\, P_\ell(\cos\theta). \tag{4.19}$$

The outgoing scattered current is obtained from the second term as

$$\frac{1}{r^2} \frac{\hbar k_\alpha}{m_\alpha} |f|^2 + O\left(\frac{1}{r^3}\right), \tag{4.20}$$

compared with the incident current of the plane wave, which is $\hbar k_\alpha/m_\alpha$. Comparing the ratio of flux through $r^2\,d\Omega$ and the incident flux, we find for the cross section

$$d\sigma/d\Omega = |f(\theta)|^2. \tag{4.21}$$

D. COULOMB AND NUCLEAR POTENTIALS

In all but the case of neutron scattering, the Coulomb potential is present in addition to the specifically nuclear potential. This does not present any real problem. The mathematical treatment of the Coulomb potential has been carried out carefully in various texts (see Messiah, 1962 or Rodberg and Thaler, 1967), so we shall not reproduce that material but rather quote the needed results. However, it is important to understand the peculiar modification by the Coulomb potential of the preceding discussion of a short-range potential. What puts the Coulomb potential in a class by itself is the slow $1/r$ fall off. As a consequence, there is no solution of the Schrödinger equation that asymptotically has a plane-wave part, as in (4.18), when the Coulomb potential is present. Instead, it can be shown (preceding references) that to leading order, the regular solution behaves asymptotically for large r as

$$\chi_c(\mathbf{r}) \to e^{i[kz + \eta \ln k(r - z)]} + \frac{1}{r} e^{i(kr - \eta \ln 2kr)} f_c(\theta), \tag{4.22}$$

where the incident beam direction $\mathbf{k}$ is chosen to coincide with the z axis. The Coulomb scattering amplitude $f_c(\theta)$ has the explicit form (i.e., the partial wave expansion can be summed)

$$f_c(\theta) = -\frac{\eta}{2k} \frac{1}{\sin^2(\theta/2)} e^{-i[\eta \ln \sin^2(\theta/2) - 2\sigma_0]}. \tag{4.23}$$

Comparing (4.22) with (4.18) we see that both the incident and scattered waves are affected even in the asymptotic region. The Coulomb potential causes scattering, however small, even at infinity. However, the current of the scattered wave is still directed radially outward so that the interpretation of the second term of (4.22) as a scattered wave is still valid. Indeed, the cross section, defined in the usual way, is

$$d\sigma_c/d\Omega = |f_c(\theta)|^2, \tag{4.24}$$

to leading order for large r.

As implied earlier, with the introduction of the regular and irregular Coulomb functions $F_\ell(\eta, kr)$ and $G_\ell(\eta, kr)$, it can be proven that the regular solution of the Schrödinger equation with a Coulomb potential has the partial wave expansion

$$\chi_c(\mathbf{r}) = \frac{1}{kr} \sum_\ell i^\ell (2\ell + 1) e^{i\sigma_\ell} F_\ell(\eta, kr) P_\ell(\cos\theta). \tag{4.25}$$

We are now in a position to discuss the combined scattering of a Coulomb and nuclear (or other short-ranged) potential. In the asymptotic region, beyond the range of the nuclear potential, the Schrödinger equation reduces to the Coulomb case previously discussed. Therefore, as in the discussion of the pure nuclear potential, the regular solution (4.5) can be expressed in the form (4.14), *in* the asymptotic region. The particular constants in (4.14), however, must be chosen so that the solution differs from (4.22) only in the outgoing spherical wave part. To allow the modification of the incoming wave part of (4.22) would violate the situation to be described, namely, an incident wave from the accelerator impinges on the target, giving rise to scattered outgoing waves. These conditions, as before, are expressed by (4.15). Another way of expressing the same thing is to note that $F_\ell(\eta, kr)$ in (4.25) can also be written in terms of outgoing and incoming waves as

$$F_\ell = \frac{1}{2i}(O_\ell - I_\ell). \tag{4.26}$$

The only modification allowed by the physical situation is that the amplitude of O_ℓ is altered by the nuclear potential, that is, f_ℓ in (4.5) asymptotically behaves like

$$f_\ell \to \frac{1}{2i}(S_\ell O_\ell - I_\ell), \tag{4.27}$$

instead of like (4.26). The connection between (4.15) and (4.27) is

$$S_\ell = 2iT_\ell + 1 = e^{2i\delta_\ell}. \tag{4.28}$$

Returning to the main line of argument, with the recognition that in the asymptotic region the regular solution must behave as in (4.15) with F and G understood to be the Coulomb functions, then in view of (4.25) we find immediately that

$$\chi_\alpha^{(+)}(\mathbf{r}) \to \chi_c(\mathbf{r}) + \frac{1}{kr} \sum_\ell i^\ell (2\ell + 1) e^{i\sigma_\ell} T_l O_l(\eta, kr) P_l(\cos\theta) \tag{4.29}$$

Referring to the asymptotic behavior of χ_c and O_ℓ given by (4.22) and (4.13), we have, finally,

$$\chi_\alpha^{(+)}(\mathbf{r}) \to e^{i[kz + \eta \ln k(r-z)]} + \frac{1}{r} e^{i(kr - \eta \ln 2kr)} f(\theta), \tag{4.30}$$

$$f(\theta) = f_c(\theta) + \frac{1}{k} \sum_\ell (2\ell + 1) e^{i(2\sigma_\ell + \delta_\ell)} \sin \delta_\ell P_\ell(\cos \theta). \tag{4.31}$$

The scattering amplitude now consists of a pure Coulomb amplitude and an amplitude in which the nuclear and Coulomb potentials are inextricably intertwined.

The Coulomb phase shifts are given by the closed form expression (4.11). The additional phase shift δ_ℓ can be found, in general, only by numerical integration of the Schrödinger equation, as discussed earlier. The boundary condition at large r, (4.15) or equivalently (4.27), when expressed in terms of the phase shift, reads

$$f_\ell(k, r) \to e^{i\delta_\ell} \sin(kr - \eta \ln 2kr - \ell\pi/2 + \sigma_\ell + \delta_\ell), \tag{4.32}$$

as follows from (4.13) and (4.28). Because the nuclear optical potential is a complex function, the phase shifts δ_ℓ are also complex.

E. PARAMETERIZATION OF THE OPTICAL POTENTIAL

Consider the scattering of a light particle like a proton or alpha particle from a nucleus. Because the density ρ of nucleons in the nucleus is fairly constant in the interior and falls smoothly to zero beyond the nuclear radius R (see Fig. 4.1), it is reasonable to adopt the same shape for the potential U acting between particle and nucleus because the nucleon–nucleon potential is short range. A convenient parameterization is the Fermi distribution, which, in the context of the optical model, is referred to as the Woods–Saxon form (1954). Thus we suppose that U has a radial dependence similar to

$$f(r) = \frac{1}{1 + \exp[(r - R)/a]}. \tag{4.33}$$

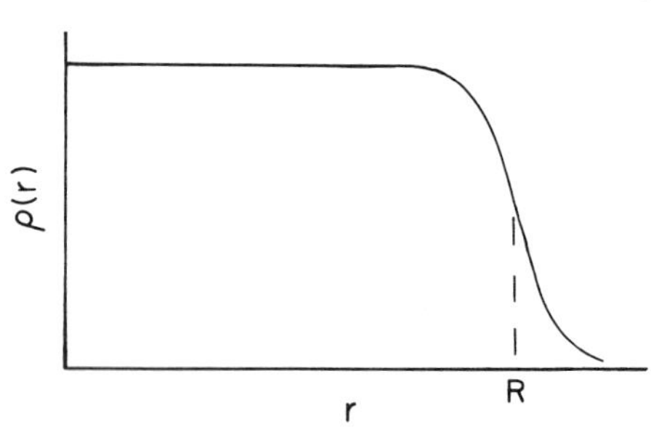

Fig. 4.1. Nuclear density as a function of distance from center.

As already discussed, U ought to be complex. Thus a suitable form is

$$U(r) = \frac{V}{1 + e^{(r - R_V)/a_V}} + \frac{iW}{1 + e^{(r - R_W)/a_W}}. \tag{4.34}$$

Usually, the nuclear radius is expressed in the form

$$R_v = r_v A^{1/3}, \qquad R_w = r_w A^{1/3}.$$

Some authors prefer to introduce an imaginary potential that is peaked in the nuclear surface by using the derivative of f as the form factor for the imaginary part, instead of f itself.

In the case of the charged particles, a Coulomb potential is added to this. The potential corresponding to a uniform distribution of charge for $r < R_c$ and zero for $r > R_c$ is

$$V_c(r) = \begin{cases} \dfrac{Z_1 Z_2 e^2}{2R_c}\left[3 - \left(\dfrac{r}{R_c}\right)^2\right], & r < R_c, \\ \dfrac{Z_1 Z_2 e^2}{r}, & r > R_c. \end{cases} \tag{4.35}$$

F. ELASTIC SCATTERING OF ALPHA PARTICLES

As a example of the elastic scattering of spinless particles for which the optical potential described so far would be suitable, consider the scattering of 50-MeV alpha particles from the four isotopes of samarium shown in Fig. 4.2 (O. N. Jarvis, personal communication). The corresponding optical model parameters are shown in Table I. This sequence of cross sections is unusual in the sense that there is quite a rapid evolution among neighboring nuclei, because of the increasing collectivity of the excited states of these nuclei, which span a transition region from spherical to deformed nuclei. This matter will be taken up again in Chapter 13.

G. SPIN–ORBIT INTERACTION AND NUCLEON ELASTIC SCATTERING

Just as the central potential in the shell model is believed to have a spin–orbit interaction because of the energy splitting of spin–orbit partners $j = \ell \pm \frac{1}{2}$ observed in odd-mass nuclei, so too one might expect that the optical potential of particles having spin and, in particular, nucleons should have a spin–orbit part. Agreement with data is improved thereby. The form of this

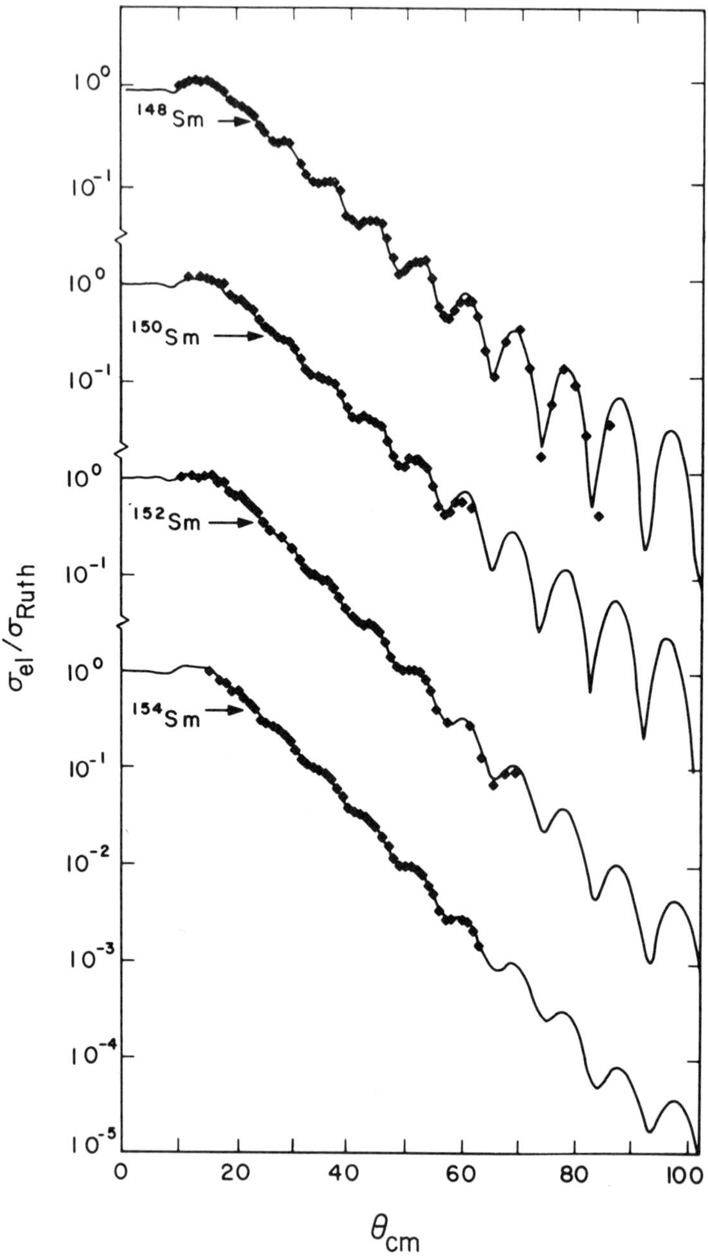

Fig. 4.2. The elastic scattering of 50-MeV alpha particles from samarium isotopes that span the spherical (A = 148) to deformed (A = 154) region. Note the systematic trend to weaker oscillations and steeper slope of the envelope of maxima with increasing collectivity. Solid lines are elastic optical-model calculations of the cross section by Jarvis *et al.* (1967).

TABLE I

Optical-Model Parameters for 50-MeV Alpha Particles

Isotope	V	W	$r_v = r_w$	$a_v = a_w$	r_c
^{148}Sm	−65.5	−29.8	1.427	0.671	1.4
^{154}Sm	−34.6	−29.4	1.404	0.819	1.4

part of the optical potential is usually taken to be

$$U_{so}(r)(2\ell \cdot \mathbf{s}) = (V_{so} + iW_{so})\left(\frac{\hbar}{m_\pi c}\right)^2 \frac{1}{r}\frac{df_{so}(r)}{dr}(2\ell \cdot \mathbf{s}), \tag{4.36}$$

where $f_{so}(r)$ is given by (4.34) with a radius and diffuseness parameter that may be different from those of the central part. In (4.36), ℓ is the orbital angular momentum of relative motion of the scattered particle and the nucleus, and $\mathbf{s}$ is its spin.

In the presence of a term like (4.36) in the Schrödinger equation (4.2), the partial wave expansion has to be generalized. Because the potential now depends on the spin orientation, the radial function must also be sensitive to this. It is most convenient to work with eigenfunctions of the total angular momentum j of the scattered particle and to use, instead of (4.7), the partial wave expansion

$$\chi_\alpha^{m(+)}(\mathbf{k}_\alpha, \mathbf{r}_\alpha) = \frac{1}{k_\alpha r_\alpha}\sum_{\ell j}\{4\pi(2\ell+1)\}^{1/2} A_{\ell(1/2)jm} i^\ell e^{i\sigma_\ell} f_{\ell j}(k_\alpha, r_\alpha)\,[Y_\ell(\theta)\chi_{1/2}(\boldsymbol{\sigma})]_j^m, \tag{4.37}$$

where the square bracket denotes the vector coupling of orbital angular momentum ℓ and spin $\frac{1}{2}$ to j and projection m. Some of the elements of angular momentum algebra are given in the Appendix, and a familiarity with that material will ultimately be needed to follow the derivations in this book. The preceding bracketed term can be written in terms of the Clebsch–Gordan coefficients that produce an eigenfunction of $\mathbf{j} = \ell + \mathbf{s}$ out of a sum of products of eigenfunctions of ℓ and s:

$$[Y_\ell(\hat{\mathbf{r}})X_{1/2}(\boldsymbol{\sigma})]_j^m = \sum_\mu C_{\mu(m-\mu)m}^{l\,(1/2)\,j}\, Y_l^\mu(\theta) X_{1/2}^{m-\mu}(\boldsymbol{\sigma}). \tag{4.38}$$

To determine the expansion coefficients A in (4.37) we recall that beyond the region where the potential is finite, $f_{\ell j}$ should equal the spherical Bessel function $(kr)j_\ell(kr)$ as in (4.8) if there is no Coulomb potential, and should equal the regular Coulomb function (4.10) otherwise. In any case,

$$\chi_\alpha^{m(+)} \to \sum_j [4\pi(2l+1)]^{1/2} j_\ell i^\ell A_{\ell 1/2 jm} \sum_\mu C_{\mu(m-\mu)m}^{\ell\ (1/2)\ j} Y_\ell^\mu X_{1/2}^{m-\mu}. \tag{4.39}$$

Now, outside the potential region we want this to be $X_{1/2}^m \exp(i\mathbf{k}_\alpha \cdot \mathbf{r}_\alpha)$ in the absence of the Coulomb potential and to be the product of the spin function and the Coulomb generalization otherwise. The desired asymptotic form of χ_α will be realized if the linear combination is taken with the coefficients A chosen as

$$A_{\ell(1/2)jm} = C_{0\ m\ m}^{\ell(1/2)j}, \tag{4.40}$$

because

$$\sum_j C_{0\ m\ m}^{\ell(1/2)j} C_{\mu(m-\mu)m}^{\ell\ (1/2)\ j} = \delta_{\mu 0}. \tag{4.41}$$

To find the differential equation that the radial functions $f_{\ell j}$ satisfy, we substitute

$$\frac{1}{kr} f_{\ell j}[Y_\ell(\theta) X_{1/2}(\boldsymbol{\sigma})]_j^m$$

into (4.2) and using (4.6) find

$$[E_\alpha - T_{\alpha\ell} - U(r) - U_{\text{so}}(r)\, 2\ell \cdot \mathbf{s}] f_{\ell j}(k_\alpha, r)\, [Y_\ell(\theta) X_{1/2}(\boldsymbol{\sigma})]_j^m = 0.$$

We can evaluate the action of $\ell \cdot \mathbf{s}$ on the bracket by noting

$$2\ell \cdot \mathbf{s} = \mathbf{j}^2 - \ell^2 - \mathbf{s}^2 = j(j+1) - \ell(\ell+1) - 3/4$$

$$= \begin{cases} \ell, & j = \ell + \frac{1}{2}, \\ -(\ell+1), & j = \ell - \frac{1}{2}, \end{cases} \tag{4.42}$$

so that the $j = \ell \pm \frac{1}{2}$ functions obey the following differential equations:

$$\begin{aligned} \{E_\alpha - T_{\alpha l} - U(r) - lU_{\text{so}}(r)\} f_{l+}(k_\alpha, r) &= 0, \\ \{E_\alpha - T_{\alpha l} - U(r) + (l+1)U_{\text{so}}(r)\} f_{l-}(k_\alpha, r) &= 0. \end{aligned} \tag{4.43}$$

As an example of the elastic scattering of nucleons calculated with an optical potential having a spin–orbit interaction, see Fig. 4.3 (taken from an analysis by Satchler, 1967a). This figure shows remarkable agreement with the experimental data over the entire angular range for several nuclei.

Generally speaking, the optical model is not as successful in accounting for the elastic scattering of more massive spin-$\frac{1}{2}$ particles like He^3 and the triton, although Fig. 4.4 shows He^3 scattering at several energies from ^{58}Ni, and the agreement is quite good over the range of the data (see Gibson *et al.*, 1967). For the mass-3 particles the spin–orbit interaction is weaker than for nucleons, whereas to rough approximation, the central part is three times stronger (Abul-Magd and El-Nadi, 1966).

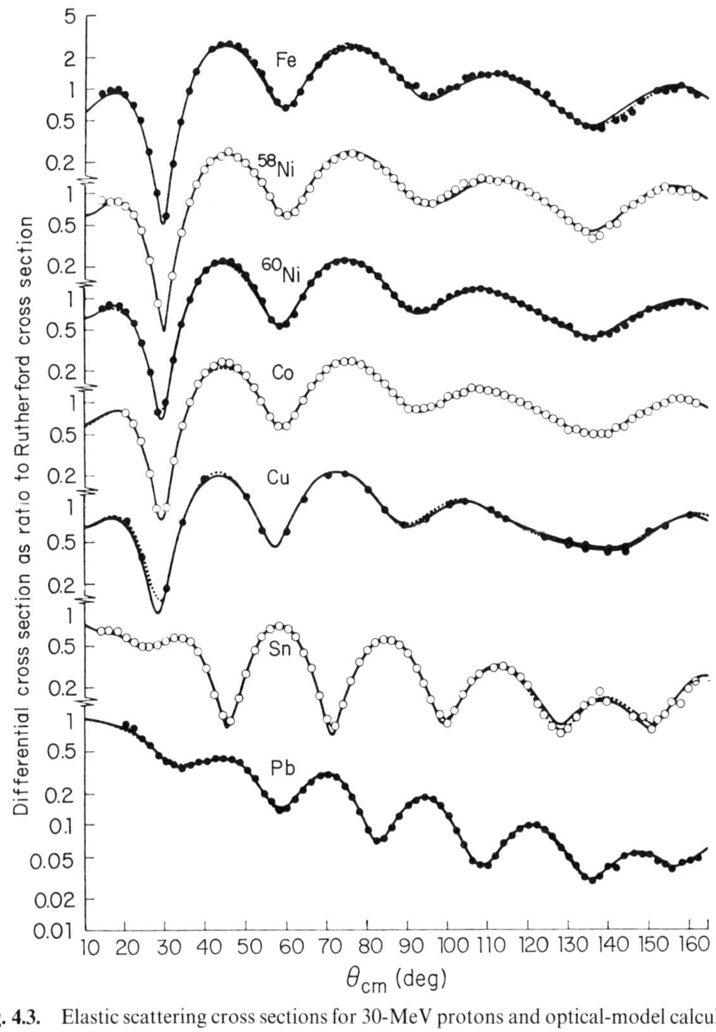

Fig. 4.3. Elastic scattering cross sections for 30-MeV protons and optical-model calculations from Satchler (1967).

H. ELASTIC SCATTERING OF HEAVY IONS

The optical potential can also be used to parameterize the elastic scattering of one nucleus by another. It is plausible that as two nuclei approach, each of whose densities is such as that depicted in Fig. 4.1, their mutual interaction can again be represented by a nuclear potential that becomes effective as the

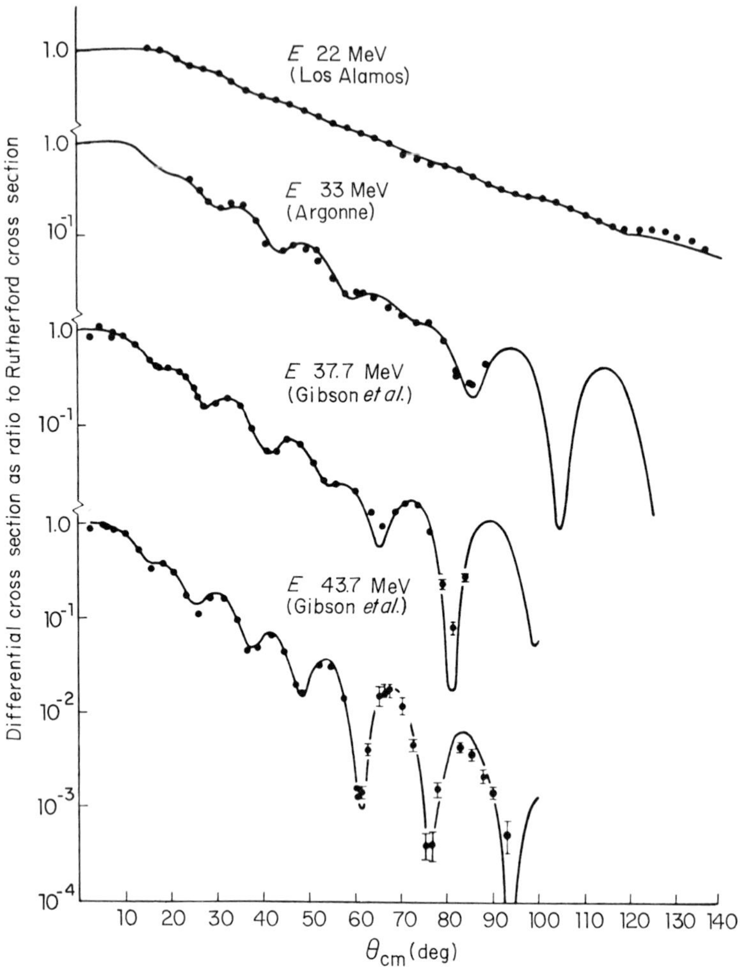

Fig. 4.4. Elastic-scattering cross sections for tritons ^{58}Ni (*h*, *h*) at several energies. Optical-model calculations by Gibson *et al.*, (1967).

distance *r* between their centers approaches the sum of their radii. Generally, the same form as (4.23) is used except that the radius of the potential should now be taken as the sum of the radii.

Of course, it would not be physically reasonable to assume that the interaction of two complex nuclei could be described solely in terms of a potential acting between their centers once the nuclear surfaces overlap appreciably. In this case new degrees of freedom must come into play in describing their

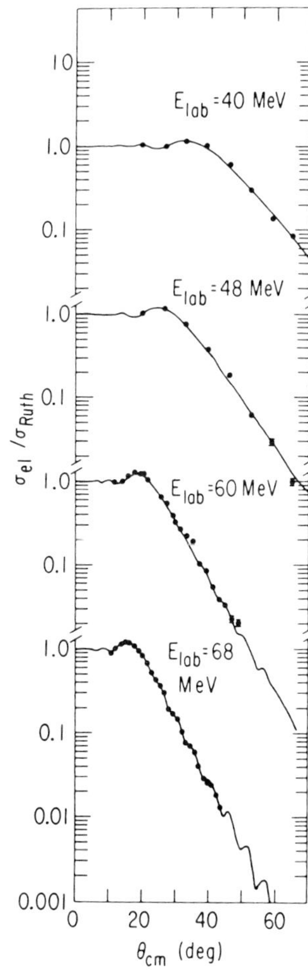

Fig. 4.5. Elastic-scattering cross sections of ^{13}C and ^{40}Ca. Optical model calculations by Bond *et al.*, (1973).

relative motion. Indeed, we cannot easily imagine that two nuclei could overlap strongly without destroying each other. Consequently, we can anticipate (and it is true) that the phenomenologically derived optical-model parameters for nucleus–nucleus scattering will be such that no flux corresponding to appreciable interpenetration remains in the elastic channel (Glendenning, 1974; Satchler, 1967b).

The cross section for elastic scattering of two nuclei from each other is shown in Fig. 4.5 for a range of energies (Bond *et al.*, 1973). The cross section becomes more forward-peaked at higher energy, as is easily understood.

I. THE IMAGINARY POTENTIAL AND MEAN FREE PATH

At this point it is appropriate to relate the magnitude for the imaginary part of the optical potential to the decay length (or mean free path) of a wave propagating in the presence of such a potential. For simplicity consider a particle of mass m incident on a semi-infinite slab of nuclear material whose effects on the particle are described by an optical potential

$$U = -(V + iW), \tag{4.44}$$

where V and W are positive real numbers. As expressed in (4.4), the imaginary part of U has to be negative to account for absorption. Intuitively, we expect that as W increases the decay length will decrease. Let us see the exact relationship in the idealization of a slab (which therefore neglects the curvature and diffuseness of the nuclear surface). In the presence of a constant potential, the wave function in the medium can be written down immediately as

$$\psi = e^{iKz}, \tag{4.45}$$

where K is the wave number in the medium. If E is the incident energy, the wave number is evidently given by

$$K^2 = \frac{2m}{\hbar^2}(E + V + iW). \tag{4.46}$$

Assuming W to be small compared with $E + V$, we evaluate K as

$$\begin{aligned} K &= \left(\frac{2m(E+V)}{\hbar^2}\right)^{1/2}\left(1 + i\frac{W}{E+V}\right)^{1/2} \\ &\simeq \left(\frac{2m(E+V)}{\hbar^2}\right)^{1/2}\left(1 + \frac{i}{2}\frac{W}{E+V}\right). \end{aligned} \tag{4.47}$$

Therefore, the amplitude of $|\psi|^2$ decreases as the particle penetrates the medium according to $e^{-z/l}$, where

$$\ell = \left(\frac{\hbar^2}{2m}\right)^{1/2}\frac{(E+V)^{1/2}}{W}. \tag{4.48}$$

This shows an inverse relationship to W, but the scale, so to speak, is set by $[(E + V)/m]^{1/2}$. Thus a value $W = 10$ MeV corresponds to twice the absorption of an alpha particle compared with a nucleon, for the same $E + V$.

Using the optical-model parameters of Table I for ^{148}Sm, we calculate a mean free path for 50-MeV alpha particles of $\sim$0.8 fm. On the other hand, for nucleons of the same energy, using the average parameters for nucleons shown in Fig. 4.6, we find a mean free path of $\sim$4.5 fm. Not unreasonably,

Fig. 4.6. Schematic behavior of real and imaginary parts of optical potential as a function of energy.

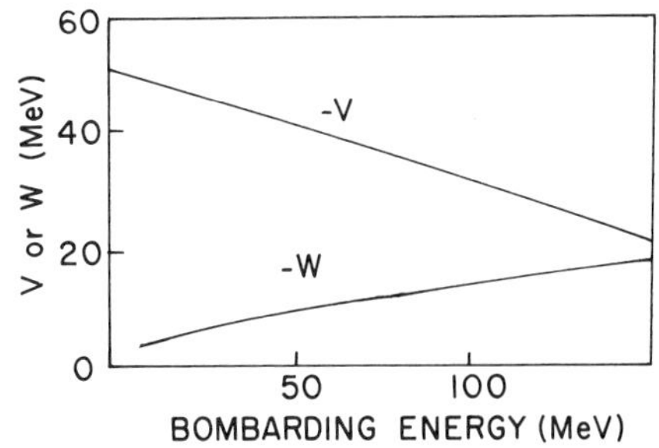

alpha particles are more strongly absorbed from the elastic beam than are neutrons. Consequently, direct reactions involving alpha particles are more strongly localized on the nuclear surface than in the case of nucleons. This is also true for nucleus–nucleus collisions.

J. SYSTEMATICS OF THE PARAMETERS

The parameters of the optical potential usually vary smoothly with energy and are similar for neighboring nuclei. The origin of any rapid evolution or erratic variation of the parameters for different nuclei can be traced to differences in their structure, especially in collective states, as will become clear in Chapter 8. Figure 4.6 shows the energy dependence of nucleon optical-model parameters (averaged) (Bowen *et al.*, 1961). The abscissa is the bombarding energy for neutrons. For protons it should be understood to be the bombarding energy in excess of the Coulomb barrier energy

$$B = \frac{Ze^2}{R} \simeq \frac{Ze^2}{r_0 \mathrm{A}^{1/3}} \approx \tfrac{1}{2}\mathrm{A}^{2/3} \quad \mathrm{MeV}. \tag{4.49}$$

The radius and diffuseness parameters are

$$R = r_0 \mathrm{A}^{1/3}, \qquad r_0 = 1.25\ \mathrm{fm}, \qquad \mathrm{a} = 0.65\ \mathrm{fm}. \tag{4.50}$$

For a best fit to individual nuclei, all parameters can be varied.

The fact that $|W|$ increases with energy corresponds to the opening of more channels that drain flux from the elastic channel. That $|V|$ decreases slowly with energy can probably be attributed, in part, to the properties of the nucleon–nucleon interaction. This interaction is repulsive at short distance. At low energy it is the long-range attraction that is mainly contributing to the optical potential. As the energy increases, the repulsion at short distance in the primary nucleon–nucleon collisions partially overcomes the attractive contribution.

K. NONLOCALITY OF THE OPTICAL POTENTIAL

In addition to the previously mentioned dependencies on energy, the theoretical structure of the optical potential (which will be discussed in Chapter 8) exhibits an explicit energy dependence *and* a nonlocal character. A nonlocal potential can be replaced, at least phenomenologically, by a local but energy-dependent one. To say that U is nonlocal implies that the Schrödinger equation is

$$(H_\alpha + T_\alpha - E)\Psi(\mathbf{r}) + \int U_\alpha(\mathbf{r}, \mathbf{r}')\Psi(\mathbf{r}')\,d\mathbf{r}' = 0, \tag{4.51}$$

instead of (4.1). If the "range" of the nonlocality is not too long, then we can expand $\Psi(\mathbf{r}')$ via a Taylor series about $\mathbf{r}$:

$$\Psi(\mathbf{r}') = \Psi(\mathbf{r}) + (\mathbf{r}' - \mathbf{r}) \cdot \nabla\Psi(\mathbf{r}) + \frac{1}{2}(\mathbf{r}' - \mathbf{r})^2\nabla^2\Psi(\mathbf{r}) + \cdots. \tag{4.52}$$

Replacing $\nabla = \mathbf{p}/i\hbar$ and noting that the odd terms in $\mathbf{r}' - \mathbf{r}$ will average to zero, we see that the nonlocal term reduces to the form

$$[U_0(\mathbf{r}) + U_2(\mathbf{r})p^2 + \cdots]\Psi(\mathbf{r}). \tag{4.53}$$

This exhibits the energy dependence of which we spoke. So, in fact, we fully expect (and it is indeed so) that the local phenomenological potential depends on the collision energy.

Phenomenological representations of a nonlocal optical potential have been investigated (Perey and Buck, 1962), but no special features in the calculated elastic cross sections are attributable to the nonlocality (Perey, 1974). There is one possible effect of significance, however. The wave functions of a local and nonlocal optical potential, which are arranged to yield the same elastic cross section, differ from each other inside the potential region, although, of course, they must be asymptotically the same if they yield the same cross section (Perey and Saxon, 1964). Although the nonlocality has no particular importance for elastic scattering, it could affect cross sections for other processes such as stripping and inelastic scattering. This is because these processes also depend on the scattering of the particles in the optical potential of the nucleus and, as well, on an additional transition density, which depends on the nucleus and the particular reaction under discussion. The wave function of relative motion, which depends on the optical model, will in principle overlap differently with the transition density, depending on whether a local or nonlocal optical potential is employed. When the nonlocality is taken into account, an approximate scheme is usually employed (Perey and Saxon, 1964).

L. CONVERGENCE OF THE PARTIAL WAVE SUM

Although the sum on ℓ in the scattering amplitude extends from $\ell = 0$ to ∞, convergence of the cross section is obtained for a finite number of terms, which depend on the collision energy and the range of the interaction. This can be understood in terms of the behavior of the solutions of (4.6). When this is written as

$$\frac{1}{f_\ell}\frac{d^2 f_\ell}{dr^2} = \frac{\ell(\ell+1)}{r^2} + \frac{2m_\alpha}{\hbar^2}[U(r) - E], \tag{4.54}$$

it is readily seen that f_ℓ at first increases as $r^{\ell+1}$ for r very small, because everything on the right side can be neglected except the $1/r^2$ term. It has an inflection point at the value of $r = r_c$, for which the right side is zero (called the classical turning point). Thereafter, it reaches a maximum and then begins to oscillate. Beyond the range of the interaction the wavelength of the oscillations is $\lambda = 1/k$, $k = (2m_\alpha E/\hbar^2)^{1/2}$. The solution is sketched in Fig. 4.7. The inflection point r_c, and hence the maximum, moves to larger values as ℓ increases. Therefore, for some sufficiently large value of ℓ, the wave function f_ℓ is unaffected by the interaction, that is, when $r_c(\ell)$ is larger than the range of the nuclear part of $U(r)$. Of course, it is always affected by the Coulomb force, but the solutions are known in the region where the force is purely Coulombic. Thus if R is a point where

$$U_N(R) \simeq 0, \tag{4.55}$$

then the maximum needed ℓ value is given roughly by the zero of the right side of (4.64):

$$\frac{\ell(\ell+1)}{R^2} + \frac{2m_\alpha Z_a Z_A e^2}{\hbar^2 R} - k^2 = 0, \tag{4.56}$$

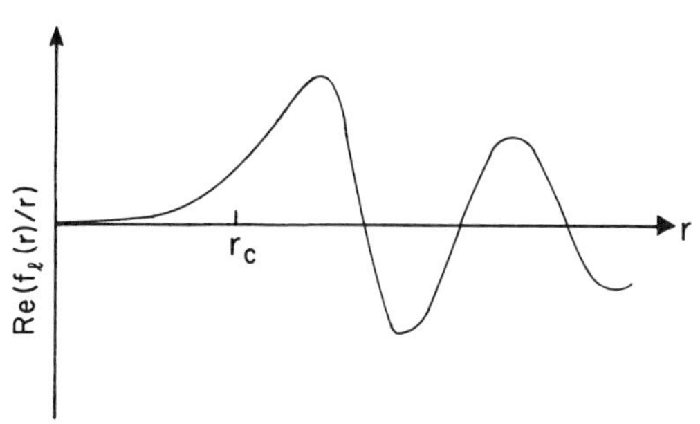

Fig. 4.7. Schematic depiction of radial wave function.

and R is the maximum radius to which the solutions need to be integrated. If the Coulomb force is unimportant (e.g., neutrons), then $l \simeq kR$. In numerical calculations, one can check whether a large enough value of R has been chosen by varying it. If the answer is insensitive to R, then it is large enough.

Notes to Chapter 4

A discussion of the optical potential and its parameters can be found in the work of Bohr and Mottelson (1969, 1975) and in Hodgson (1963, 1971) and Perey (1974). The theoretical justification for the optical model is reviewed by Feshback (1958). A familiarity with some of the results and theory of angular momentum algebra is necessary in nuclear physics. Some of the essential results can be found in Appendix A. Two excellent monographs are by Brink and Satchler (1968) and by Edmonds (1957). Brink and Satchler use a different convention for reduced matrix elements than Edmonds. In this book we use the convenction of Edmonds, which is the original 1942 convention of Racah. They differ simply by a factor.

Chapter 5

Distorted-Wave Born Approximation

A. INTRODUCTION

In the last chapter, the introduction of the optical potential was motivated as a means of describing the average field experienced by the colliding nuclei. The direct nuclear reactions are then viewed as being caused by a weak interaction that induces a transition to occur between two channels, in each of which the relative motion is described by an optical potential. In this chapter, we shall derive an approximation of the transition amplitude that corresponds to this interpretation, which is known as the distorted-wave Born approximation (DWBA). This approximation is without doubt the most useful one in direct nuclear reaction theory and is valid as long as the reaction is sufficiently weak. For stronger transitions, the coupled-channel formalism offers the most useful means of going beyond first order in the interaction causing the transition. We shall come to this method in Chapter 8.

We shall derive the DWBA from a general transformation of the transition amplitude due to Gell-Mann and Goldberger, (1953). They introduce an arbitrary auxiliary potential, which is supposed to present a solvable problem. They then transform the transition amplitude so that it depends on the solution of this solvable problem. In our context, the auxiliary potential will be the optical potential.

This chapter will be devoted to the general problem of deriving the DWBA. The application to specific nuclear reactions will be taken up in Chapter 7.

B. DISTORTED-WAVE GREEN'S FUNCTIONS

Recognizing that the scattering of nuclei from each other through the effect of their central field can be accounted for by an optical potential whose parameters can be adjusted to reproduce the elastic and reaction cross section, we are able to isolate for special attention the weaker interaction causing the minor rearrangements called direct reactions. Return, therefore, to Eq. (3.18), and subtract from both sides a common term $(\Phi_\alpha, U_\alpha \Psi_\alpha)$ to obtain

$$(E_\alpha - T_\alpha - U_\alpha)\psi_\alpha = (\Phi_\alpha, [V_\alpha - U_\alpha]\Psi_\alpha), \tag{5.1}$$

where ψ_α, defined by (3.19), describes the relative motion in channel α. Because U_α depends only on the relative coordinate r_α, whereas Φ_α depends only on intrinsic coordinates, then obviously

$$(\Phi_\alpha, U_\alpha \Psi_\alpha) = U_\alpha(\Phi_\alpha, \Psi_\alpha) = U_\alpha \psi_\alpha. \tag{5.2}$$

The formal solution to Eq. (5.1) follows along similar lines to the development in Chapter 3. The corresponding Green's function satisfies

$$(E_\alpha - T_\alpha - U_\alpha)G_\alpha(\mathbf{r}_\alpha, \mathbf{r}'_\alpha) = \delta(\mathbf{r}_\alpha - \mathbf{r}'_\alpha). \tag{5.3}$$

Then a general solution of (5.1) can be written as

$$\psi_\alpha = A\chi_\alpha + \int G_\alpha(\mathbf{r}, \mathbf{r}')\,[\Phi_\alpha, (V_\alpha - U_\alpha)\Psi_\alpha]\,d\mathbf{r}', \tag{5.4}$$

where χ is a solution of (4.2) and A a constant. The physical conditions to be described, of course, are what determines A and the particular Green's function.

The Green's function satisfying (5.3) cannot generally be written in closed form because we are allowing ourselves freedom in the choice of U_α. However, we can construct it from the solutions of (4.2), whatever the choice of U_α, and (4.2) is merely a differential equation, which can be solved on a computer if not in closed form. In the process, we shall obtain the field-free result (3.25) as well. The construction of the Green's function is simple in one dimension, so we seek an expansion of G in spherical harmonics.

First, note that because of the closure and orthonormality of the spherical harmonics, we have[‡]

$$\delta(\hat{\mathbf{r}}' - \hat{\mathbf{r}}) = \sum_{\ell m} Y_\ell^{m*}(\hat{\mathbf{r}}')Y_\ell^m(\hat{\mathbf{r}}), \tag{5.5}$$

where $\hat{\mathbf{r}}$ denotes the polar angles θ and ϕ of the vector $\mathbf{r}$. From this it follows that the three-dimensional δ function is

$$\delta(\mathbf{r}' - \mathbf{r}) = \frac{\delta(r' - r)}{rr'} \sum_{\ell m} Y_\ell^{m*}(\hat{\mathbf{r}}')Y_\ell^m(\hat{\mathbf{r}}). \tag{5.6}$$

Hence, it follows from (5.3) that

$$G(\mathbf{r}, \mathbf{r}') = \frac{1}{rr'} \sum_{\ell m} g_\ell(r, r')Y_\ell^{m*}(\hat{\mathbf{r}}')Y_\ell^m(\hat{\mathbf{r}}), \tag{5.7}$$

where g_ℓ satisfies the equation corresponding to (4.6) and (4.7):

$$(E_\alpha - T_{\alpha\ell} - U_\alpha)g_\ell(r, r') = \delta(r - r'). \tag{5.8}$$

We have already exhibited one solution (4.17) of the homogeneous equation that is a regular solution. Another independent solution is therefore an irregular solution at the origin, which we denote by $h_\ell^{(+)}(k, r)$, with the boundary condition that

$$h_\ell^{(+)}(k, r) \rightarrow O_\ell = G_\ell + iF_\ell \rightarrow \exp[i(kr - \tfrac{1}{2}\ell\pi)]. \tag{5.9}$$

Clearly, from these two solutions, g can be constructed as

$$\begin{aligned} g_\ell(r, r') &= N_\ell f_\ell(r_<)h_\ell^{(+)}(r_>), \\ r_< &= \text{lesser of} \quad r, r'. \end{aligned} \tag{5.10}$$

This satisfies (5.8) for $r \neq r'$, as can be seen by direct substitution. It has the desirable property of being the outgoing-wave Green's function. From the singular point we learn the normalization. Integrate (5.8) over an infinitesimal interval around r' and find

$$\begin{aligned} \frac{2m}{\hbar^2} &= \int_{r'-\delta}^{r'+\delta} dr \frac{d^2}{dr^2} g_\ell(r, r') = \frac{d}{dr} g_\ell(r, r')\bigg|_{r'-\delta}^{r'+\delta} \\ &= N_\ell[f_\ell(r')h_\ell^{(+)\prime}(r' + \delta) - f_\ell'(r' - \delta)h_\ell^{(+)}(r')] \\ &\xrightarrow[\delta \to 0]{} N_\ell W[f_\ell(r'), h_\ell^{(+)}(r')]. \end{aligned} \tag{5.11}$$

The prime on h or f denotes differentiation, and W denotes a Wronskian. The Wronskian[§] is independent of the value of r at which it is evaluated. This can be easily proven with reference to the Schrödinger equation satisfied

[‡] Simply check that $\int \delta(\hat{\mathbf{r}}' - \hat{\mathbf{r}})Y_\lambda^\mu(\hat{\mathbf{r}}')\, d\hat{\mathbf{r}}'$, evaluated with the right side of (5.5) substituted for the δ function, yields $Y_\lambda^\mu(\hat{\mathbf{r}})$, as it should.

[§] The Wronskian of two functions f and h is simply $fh' - hf'$.

by f and h. We find

$$fT_\ell h - hT_\ell f = 0.$$

Integrate between two arbitrary points, a to b, to obtain

$$\begin{aligned} 0 &= \int_a^b dr[fT_\ell h - hT_\ell f] \\ &= -\int_a^b \frac{d}{dr}\left[f\frac{dh}{dr} - h\frac{df}{dr}\right]dr \\ &= -W(f, h)\Big|_a^b, \end{aligned}$$

which proves that W has the same value no matter where it is evaluated. In particular, we can evaluate it for large r using (5.9) and (4.17):

$$W[f_\ell, h_\ell^{(+)}] = -k,$$

which shows that the one-dimensional Green's function is

$$g_\ell(r, r') = -\frac{2m}{\hbar^2}\frac{1}{k}f_\ell(r_<)h_\ell^{(+)}(r_>). \tag{5.12}$$

Now the desired three-dimensional Green's function is formed:

$$\begin{aligned} G_\alpha^{(+)}(\mathbf{r}, \mathbf{r}') &= -\frac{2m_\alpha}{\hbar^2}\frac{1}{k_\alpha rr'}\sum_\ell f_\ell(r_<)h_\ell^{(+)}(r_>)\sum_m Y_\ell^{m*}(\hat{\mathbf{r}}')Y_l^m(\hat{\mathbf{r}}) \\ &\xrightarrow[r\to\infty]{} -\frac{2m_\alpha}{\hbar^2}\frac{e^{ik_\alpha r}}{r}\sum_\ell i^{-\ell}\frac{1}{k_\alpha r'}f_\ell(k_\alpha, r')\sum_m Y_\ell^{m*}(\hat{\mathbf{r}}')Y_l^m(\hat{\mathbf{r}}) \\ &= -\frac{m_\alpha}{2\pi\hbar^2}\frac{e^{ik_\alpha r}}{r}\chi_\alpha^{(-)*}(\mathbf{k}_\alpha', \mathbf{r}'), \end{aligned} \tag{5.13}$$

where

$$\chi_\alpha^{(-)*}(\mathbf{k}_\alpha', \mathbf{r}') \equiv \frac{4\pi}{k_\alpha r'}\sum i^{-\ell}f_\ell(k_\alpha, r')Y_\ell^{m*}(\hat{\mathbf{r}}')Y_\ell^m(\hat{\mathbf{k}}_\alpha'). \tag{5.14}$$

This is the partial wave expansion, analogous to (4.7), for the wave function with incoming spherical waves. Here $\mathbf{k}_\alpha'$ has the magnitude of $\mathbf{k}_\alpha$ and the direction of $\mathbf{r}$ in (5.4). The $(-)$ sign on (5.14) indicates, as is clear from the asymptotic form of (4.15), that $\chi^{(-)}$ has incoming spherical waves, in contrast to $\chi^{(+)}$ and $\chi^{(-)*}$. [If there is a Coulomb potential present, then the phase $e^{i\sigma_\ell}$ should be inserted into (5.14).] If $U \equiv 0$, then χ becomes a plane wave, and (5.14) becomes (3.25).

C. THE GELL-MANN–GOLDBERGER TRANSFORMATION

Employing the Green's function derived above we can write the particular solution of (5.1) that describes an incident wave in channel α and outgoing spherical waves. It is

$$\psi_\alpha = \chi_\alpha^{(+)}(\mathbf{k}_\alpha, \mathbf{r}_\alpha) + \int G^{(+)}(\mathbf{r}_\alpha, \mathbf{r}'_\alpha)[\Phi_\alpha, (V_\alpha - U_\alpha)\Psi_\alpha^{(+)}]\, d\mathbf{r}'_\alpha$$

$$\to \chi_\alpha^{(+)}(\mathbf{k}_\alpha, \mathbf{r}_\alpha) - \frac{m_\alpha}{2\pi\hbar^2}\frac{e^{ik_\alpha r_\alpha}}{r_\alpha}\langle \chi_\alpha^{(-)}\Phi_\alpha | U_\alpha - V_\alpha | \Psi_\alpha^{(+)}\rangle. \tag{5.15}$$

In an entirely analogous development, if we look at a projection of the wave function Ψ_α onto a different channel β that is not the entrance channel, we find

$$\psi_\beta \to -\frac{m_\beta}{2\pi\hbar^2}\frac{e^{ik_\beta r_\beta}}{r_\beta}\langle \chi_\beta^{(-)}\Phi_\beta | V_\beta - U_\beta | \Psi_\alpha^{(+)}\rangle. \tag{5.16}$$

Note that $\chi_\alpha^{(+)}$ has an outgoing wave, so that the scattering amplitude, or T matrix, has two parts. The amplitude in $\chi_\alpha^{(+)}$ can be written immediately from the rule (3.32) or (3.33), namely,

$$\langle \exp(i\mathbf{k}'_\alpha \cdot \mathbf{r}_\alpha) | U_\alpha | \chi_\alpha^{(+)}\rangle.$$

Thus we may write the entire T matrix as

$$\mathscr{T}_{\beta\alpha} = \langle \exp(i\mathbf{k}'_\alpha \cdot \mathbf{r}_\alpha) | U_\alpha | \chi_\alpha^{(+)}\rangle \delta_{\alpha\beta} + \langle \chi_\beta^{(-)}\Phi_\beta | V_\beta - U_\beta | \Psi_\alpha^{(+)}\rangle. \tag{5.17}$$

No approximations were made in obtaining this expression. Therefore it is an alternative expression to (3.33), known as the Gell-Mann–Goldberger transformation (see Gell-Mann and Goldberger, 1953). (Note that $\mathbf{k}'_\alpha$ differs from $\mathbf{k}_\alpha$ in direction, as in (3.25).)

From definitions (4.7) and (5.14) one can find the connection between distorted waves having incoming and outgoing spherical waves, namely,

$$\chi^{(-)*}(\mathbf{k}, \mathbf{r}) = \chi^{(+)}(-\mathbf{k}, \mathbf{r}), \tag{5.18}$$

where use is made of

$$Y_\ell^m(\hat{\mathbf{k}}) = (-)^\ell Y_\ell^m(-\hat{\mathbf{k}}),$$

and $-\hat{\mathbf{k}}$ means the reflected direction of $\hat{\mathbf{k}}$, that is, $(\theta, \phi \to \pi - \theta, \phi + \pi)$. Relation (5.18) establishes that $\chi^{(-)}(\mathbf{k}, \mathbf{r})$ is the time reversal of the wave function $\chi^{(+)}(\mathbf{k}, \mathbf{r})$.

D. TWO-POTENTIAL FORMULA

The preceding results can also be used to obtain the two-potential formula. Suppose

$$H = T + V_1 + V_2, \tag{5.19}$$

where $T + V_1$ is a Hamiltonian for which we can obtain the solution (as for the optical potential, for example), whereas V_2 poses too difficult a problem and we would be content with an approximate solution. The Schrödinger equation is

$$(E - T - V_1)\Psi = V_2\Psi. \tag{5.20}$$

As supposed, we can solve the simpler problem

$$(E - T - V_1)\chi = 0. \tag{5.21}$$

From the results of this chapter it is clear that

$$\mathscr{T} = \langle \exp(i\mathbf{k}' \cdot \mathbf{r})|V_1|\chi^{(+)}\rangle + \langle \chi^{(-)}|V_2|\Psi^{(+)}\rangle. \tag{5.22}$$

By supposition, the first term is known and the second must be approximated. For example, $\Psi^{(+)} \simeq \chi^{(+)}$, which follows from (5.15).

E. THE DWBA TRANSITION AMPLITUDE

The preceding results can be used to obtain the distorted-wave Born approximation (DWBA) of the amplitude, exact expressions of which are given by (3.33) and (5.17). As previously noted, the exact expressions cannot be solved because the exact scattering solution to the many-body problem Ψ appears in them. However, in (5.15) we have a valuable result that incorporates at least the effects of the scattering of the optical potential U through the first term χ. This is a solution of the solvable optical-model problem, which *is* solvable because it is a simple one-body problem.

Recalling the definition of ψ as a projection onto Ψ, we learn from (5.15) that an approximation of Ψ is

$$\Psi_\alpha^{(+)} \simeq \Phi_\alpha \chi_\alpha^{(+)}(\mathbf{k}_\alpha, \mathbf{r}_\alpha). \tag{5.23}$$

This is an approximation that describes the motion in the entrance channel due to the optical potential U_α. It does not have all the other components of other channels that are actually there.

Inserting the approximation of Ψ into (5.17),

$$\mathscr{T}_{\beta\alpha} \simeq \langle \exp(i\mathbf{k}'_\alpha \cdot \mathbf{r}_\alpha)|U_\alpha|\chi_\alpha^{(+)}\rangle\delta_{\alpha\beta} + \langle \chi_\beta^{(-)}\Phi_\beta|V_\beta - U_\beta|\Phi_\alpha\chi_\alpha^{(+)}\rangle. \tag{5.24}$$

We are interested in this expression for the nonelastic amplitudes for which the first term is absent. The DWBA is therefore

$$\mathscr{T}_{\beta\alpha}^{\mathrm{DWBA}} = \langle \chi_\beta^{(-)}\Phi_\beta|V_\beta - U_\beta|\Phi_\alpha\chi_\alpha^{(+)}\rangle, \tag{5.25}$$

where χ_α and χ_β are obtained from solvable optical-model problems in channels α and β. This is analogous to the first Born approximation (3.35), but it contains more of the physics—the description of the relative motion in channels α and β, as effected by the overall potential scattering and absorption. Figure 5.1 compares the plane-wave theory of Chapter 2 with

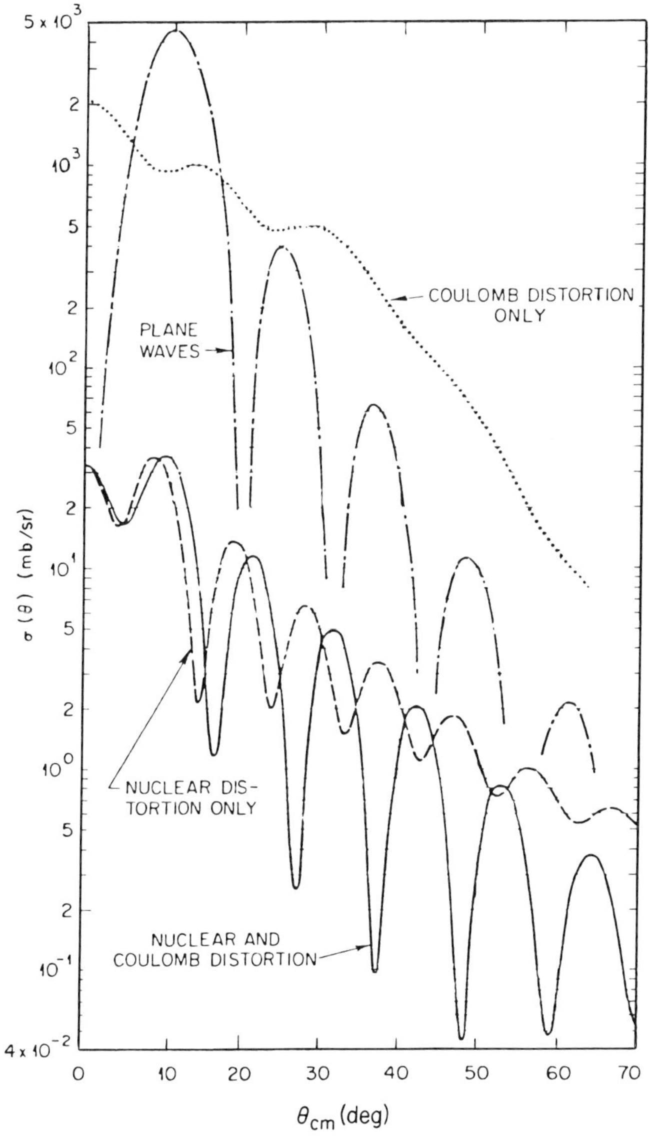

Fig. 5.1. The plane-wave theory for inelastic scattering of 43-MeV alpha particle that excites the $2^+ : Q = -1.45$ state in ^{58}Ni is compared with several DWBA calculations with various distortion potentials U. [From Rost, (1962).]

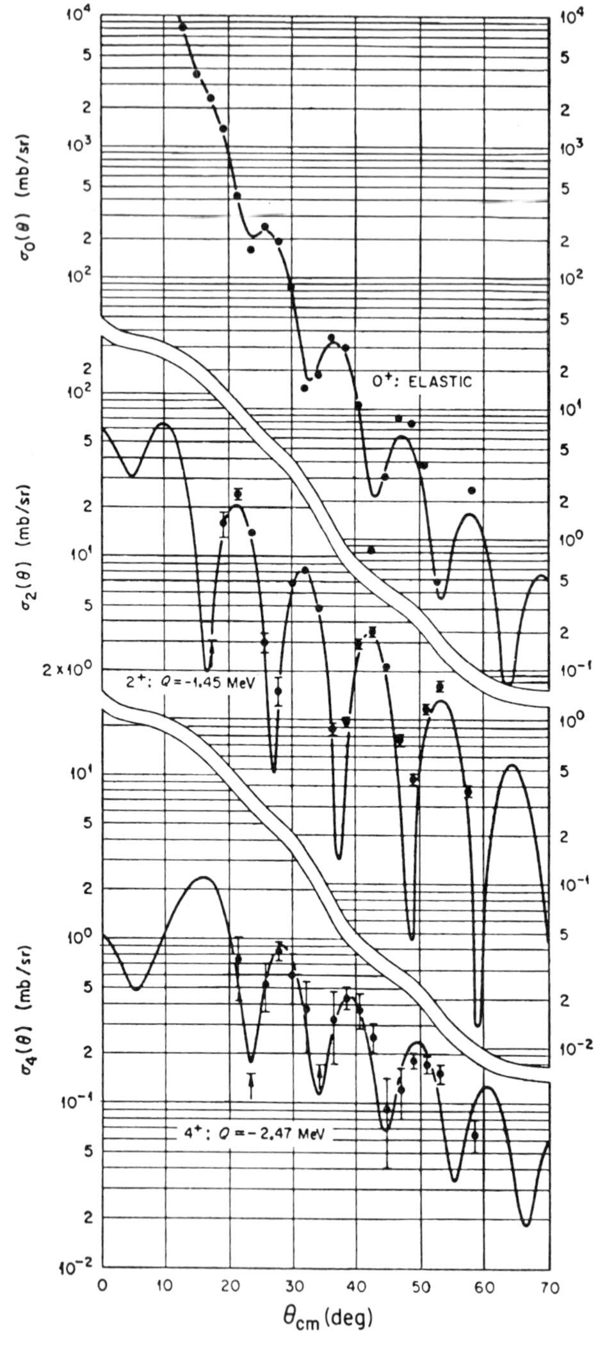

Fig. 5.2. Cross sections calculated for the 0^+, 2^+, and 4^+ states, showing distinct dependence on multipolarity. In this case, higher order contributions are also included. The complex Saxon potential parameters here are: $V = 47.6$ MeV, $W = 13.8$ MeV, $r_0 = 1.6$ fm, $a = 0.55$ fm, and $\beta = \{\langle\sum_m |a_m|^2\rangle\}^{1/2} = 0.2$ (Buck, 1962).

distorted-wave Born approximation calculations (Rost, 1962). The calculation makes it clear that the distortion in the wave functions of relative motion, caused by the interaction between the colliding nuclei, is very strong. However, the dependence on the multipolarity is very distinct, as Fig. 5.2 makes clear (Buck, 1962). In this case the calculation is more complicated than described so far because it includes higher-order contributions than the first-order Born approximation, and this emphasizes the point being made.

F. DISCUSSION OF THE APPROXIMATIONS

Two exact expressions for the T matrix or scattering amplitude have been derived: (3.33) and (5.17). Both contain Ψ_α, the exact solution to the many-body scattering problem. As such, Ψ, if it were available, contains information on all the processes that can take place; elastic, inelastic, particle transfer, and many-body channels. It is not available, but its presence in the exact amplitude emphasizes the approximate nature of any practical application of the theory. Let us emphasize this point by noting that the T matrix for nonelastic events (5.17) can be written as

$$\mathscr{T}_{\beta\alpha} = \langle \chi_\beta^{(-)} \Phi_\beta | V_\beta - U_\beta | \Psi_\alpha^{(+)} \rangle = \sum_\gamma \langle \chi_\beta^{(-)} \Phi_\beta | V_\beta - U_\beta | \Phi_\gamma \psi_\gamma^{(+)} \rangle, \quad (5.26)$$

where the completeness relation has been used as

$$\sum_\gamma |\Phi_\gamma)(\Phi_\gamma| = 1. \quad (5.27)$$

This tells us clearly that the transition $\alpha \to \beta$ is affected by all the other channels. In going to the DWBA (5.25), we are making two approximations immediately. Of all the terms in the sum on γ, we select just the entrance channel, because it is most important generally; and having made that restriction, we approximate ψ_α by the first term in (5.15).

Therefore the DWBA assumes, first of all, that the final channel β is reached directly from the entrance channel α. Processes that, for example, excite the entrance channel to α' and then undergo the final reaction to β are neglected, as are intermediate transfer channels γ. Intuitively one feels that this neglect may be frequently but not always, justified. Later we shall extend the treatment to include such multiple-step processes by the coupled-channels method.

The DWBA additionally assumes that the distorted waves χ_α and χ_β are used to represent the relative motion in these channels. Although the optical potentials may be adjusted so that the elastic and total reaction cross sections are accurately reproduced, this only assures us that χ_α and χ_β are asymptotically correct, that is, they have the correct phase shifts as seen at infinity.

Hence there are ambiguities in these wave functions in the region of the nucleus, which is just from where the contribution to the matrix element comes. This ambiguity is further compounded when α represents a channel in which both a and A are complex nuclei.

Finally, the approximation $\psi_\alpha \simeq \chi_\alpha$ implies that the reaction is treated in first order only in the interaction causing $\alpha \to \beta$. This is often justified by the fact that many other reactions have small cross sections compared with elastic reactions.

G. ANTISYMMETRIZATION

So far we have not mentioned the fact that the wave function for the system ought to be antisymmetric upon the interchange of any pair of like particles. Of course we may assume that the wave functions Φ_a and Φ_A for the separated parts of the system, which are stationary solutions to a shell-model problem, are already antisymmetrized. All that remains is to take account of antisymmetrization *between* them. In this case an antisymmetric function is

$$\tilde{\Phi}_\alpha = \frac{1}{\sqrt{N_\alpha}} \sum_P (-)^P P \Phi_a(a) \Phi_A(A), \tag{5.28}$$

where P is a permutation operator. For example, if there are only three nucleons,

$$\tilde{\Phi} = \frac{1}{\sqrt{3}} [\phi(1)\phi(2, 3) - \phi(3)\phi(1, 2) + \phi(2)\phi(3, 1)]$$

is totally antisymmetric. If the isospin formalism is used in constructing Φ_a and Φ_A, then we do not distinguish neutrons from protons. Otherwise the antisymmetrization is done separately for neutrons and protons.

The normalization constant is

$$N_\alpha = \binom{A + a}{a} = \frac{(A + a)!}{A!\, a!}, \tag{5.29}$$

that is, the number of ways of selecting the a nucleons from A + a. Now, to examine the consequence of antisymmetrization for (5.25),

$$\mathscr{T}_{\text{anti}} = \frac{1}{\sqrt{N_\alpha N_\beta}} \sum_{P_\alpha P_\beta} (-)^{P_\alpha + P_\beta} P_\alpha P_\beta \langle \chi_\beta^{(-)} \Phi_b \Phi_B | V_\beta - U_\beta | \Phi_a \Phi_A \chi_\alpha^{(+)} \rangle. \tag{5.30}$$

In the sum there will be direct integrals and exchange integrals. We illustrate this for a four-particle system and a one-particle exchange reaction. Then $+ \langle \phi_b(1)\phi_B(2, 3, 4) V \phi_a(1, 2)\phi_A(3, 4) \rangle$ is a typical direct term and

$-\langle\phi_b(4)\phi_B(1,2,3)V\phi_a(1,2)\phi_A(3,4)\rangle$ is a typical exchange term. In the former, particle 2 has been removed from a and added to A to form b, B. In the exchange integral, the same was done to particle 2, but in addition, particle 1 was interchanged with particle 4. The exchange integrals are expected to be smaller than the direct ones because of poorer wave-function overlap. We defer discussion of an approximate method of taking exchange into account until Chapter 12.

Therefore we shall consider only the direct integrals, noting, however, that all of them must be counted. This is easy. Returning to the general example, suppose we are dealing with a stripping reaction (a > b, B > A), which we denote by

$$\mathrm{a} + \mathrm{A} \to (\mathrm{a} - x) + (\mathrm{A} + x) \equiv \mathrm{b} + \mathrm{B}, \tag{5.31}$$

where x is the number of transferred particles. *Each* of the N_α terms in the sum over permutations of $\Phi_a\Phi_A$ will form $\binom{a}{x}$ direct integrals, all of which are equal. Therefore there are $N_\alpha\binom{a}{x}$ equal direct integrals. Together with the normalization in (5.28), this means

$$\begin{aligned}\mathscr{T}_{\text{anti}} &= N_\alpha \binom{\mathrm{a}}{x} \frac{1}{\sqrt{N_\alpha N_\beta}} \mathscr{T} + \text{exchange} \\ &= \left[\binom{\mathrm{a}}{x}\binom{\mathrm{B}}{x}\right]^{1/2} \mathscr{T} + \text{exchange}.\end{aligned} \tag{5.32}$$

The statistical weight in front of $\mathscr{T}$ will sometimes be denoted by W.

As implied in the derivation of (5.25), $\mathscr{T}$ denotes a single amplitude computed without antisymmetrization *between* nuclei. The meaning of the statistical weight is fairly obvious. There are $\binom{a}{x}$ ways of selecting x nucleons from the projectile, and each such x nucleons can be distributed in $\binom{B}{x}$ ways among the nucleons in B. Therefore, there are altogether $\binom{a}{x}$ $\binom{B}{x}$ indistinguishable ways of *directly* transferring the nucleons, all of them with amplitude equal to $\mathscr{T}$. There are, in addition, the exchange type of integrals. They are usually not computed on the grounds that the greater the degree of rearrangement the smaller the integral.

Austern (1970) has written a more general expression than (5.32), arrived at by a somewhat different argument, which keeps track of both direct and exchange integrals. His result can be written as

$$\mathscr{T}_{\text{anti}} = \left(\frac{N_\beta}{N_\alpha}\right)^{1/2} \sum_{P_\alpha} (-)^{P_\alpha} \langle \chi_\beta^{(-)}\Phi_\beta | V_\beta - U_\beta | P_\alpha \Psi_\alpha^{(+)} \rangle. \tag{5.33}$$

The number of direct integrals in this expression is given precisely by W. To see this, note that of the N_α terms in the sum over permutations, $\binom{B}{x}$ of them are equal direct integrals. For example, in a schematic notation for the sum, involving for simplicity four nucleons,

$$\begin{aligned}\sum_{P\alpha} = \int \phi_b(1)\phi_B(2, 3, 4)V\{&\phi_a(1, 2)\phi_A(3, 4) \\ &- \phi_a(1, 3)\phi_A(2, 4) + \phi_a(1, 4)\phi_A(2, 3) \\ &+ \phi_a(2, 3)\phi_A(1, 4) - \phi_a(2, 4)\phi_A(1, 3) \\ &+ \phi_a(3, 4)\phi_A(1, 2)\}.\end{aligned} \tag{5.34}$$

The second term is found to be identical to the first by interchanging 2 and 3 in ϕ_B in the second term and relabeling these two coordinates of integration. Similarly, the third term is equal to the first. The last three terms are the exchange integrals. The first three are direct integrals. There are $\binom{3}{1}$ such terms in $\sum_{P\alpha}$, or, in general, $\binom{B}{x}$, corresponding to the number of ways x particles can be selected from B to form a and A. Thus in (5.33) the weight of the direct terms is

$$\left(\frac{N_\beta}{N_\alpha}\right)^{1/2}\binom{B}{x} \equiv \binom{a}{x}^{1/2}\binom{B}{x}^{1/2}, \tag{5.35}$$

as was to be proven.

If the isospin formalism is not employed in the construction of the nuclear wave functions, then the product $\Phi_a\Phi_A$ must be separately antisymmetrized with respect to neutrons and protons. The corresponding weight W is the obvious generalization of the preceding, being a product of two such weights, one for neutrons and one for protons. If the calculations are all carried through correctly, the end results should be identical.

H. MULTIPOLE EXPANSION OF THE TRANSITION AMPLITUDE

In applications of the distorted-wave Born approximation of the transition amplitude [(5.25) and (5.32)], a certain pattern having to do with the angular momentum couplings will emerge each time. The particular factors that go into this, of course, depend on the type of reaction under consideration, some examples of which we shall examine in Chapter 7. At this point we can anticipate the general form that will be taken by the amplitude (5.25) and at the same time make it a little more explicit. The brackets in (5.25) denotes an integration over the position coordinates of all nucleons and a sum over spin (and isospin if used) coordinates. There are (A + a) nucleons, so there are 3(A + a) spatial coordinates. We choose three of them to be the c.m. coordinates and work in this sytem. The remainder are relative coordinates. For a rearrangement collision (5.31) a convenient set of such

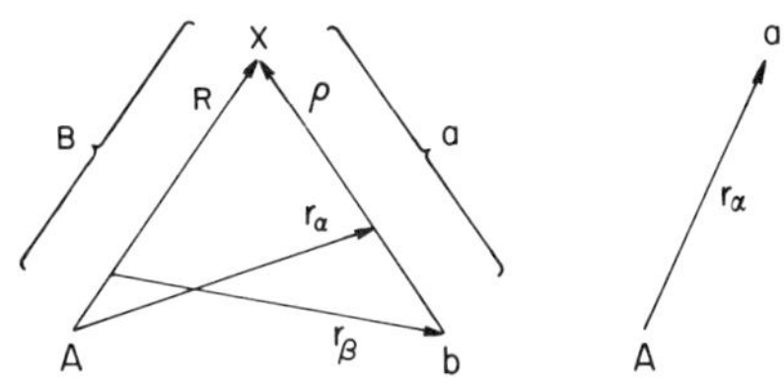

Fig. 5.3. For a transfer reaction of x nucleons, part (a) shows coordinate relations. The channel coordinates $\mathbf{r}_\alpha$ and $\mathbf{r}_\beta$ connect the cm's of A and a, and B and b, respectively, whereas $\mathbf{R}$ and $\boldsymbol{\rho}$ connect the c.m.'s of A and x, and b and x, respectively. For elastic and inelastic scattering part (b) shows the channel coordinate r_α connecting the cm's of A and a.

relative coordinates consists of the 3(A − 1), 3(x − 1), and 3(b − 1) intrinsic coordinates of A, x, and b together with $\mathbf{R}$ and $\boldsymbol{\rho}$, as shown in Fig. 5.3, which comprise the correct number 3(A + a) − 3 of relative coordinates. For a reaction in which no nucleons are transferred, such as inelastic or charge-exchange reactions, the intrinsic coordinates of the two nuclei of the partition, together with the channel coordinate r_α, comprise a convenient set of relative coordinates. A zero-range approximation is sometimes employed in nucleon-transfer reactions, especially when a is a light nucleus like a deuteron or triton. The interaction in this case is taken as a delta function on the coordinate $\boldsymbol{\rho}$. This is convenient because it reduces the number of integration variables. Moreover, the distorted waves depend on the channel coordinates $\mathbf{r}_\alpha$ and $\mathbf{r}_\beta$, which in terms of $\mathbf{R}$ and $\boldsymbol{\rho}$ are given by

$$\mathbf{r}_\alpha = \mathbf{R} - (\mathrm{b}/\mathrm{a})\boldsymbol{\rho}, \qquad \mathbf{r}_\beta = (\mathrm{A}/\mathrm{B})\mathbf{R} - \boldsymbol{\rho}. \tag{5.36}$$

In the zero-range approximation,

$$\mathbf{r}_\alpha \to \mathbf{R}, \qquad \mathbf{r}_\beta \to (\mathrm{A}/\mathrm{B})\mathbf{R}, \qquad \text{if} \quad \boldsymbol{\rho} \to 0, \tag{5.37}$$

so that the channel coordinates are conveniently proportional. If the zero-range approximation is not made, then it is most convenient to transform $\mathbf{R}$ and $\boldsymbol{\rho}$ to $\mathbf{r}_\alpha$ and $\mathbf{r}_\beta$ because the distorted waves are only known numerically as solutions to the optical-potential Schrödinger equation of Chapter 4. The transformation involves a nonunit Jacobian

$$d\boldsymbol{\rho}\, d\mathbf{R} = J\, d\mathbf{r}_\alpha\, d\mathbf{r}_\beta, \tag{5.38}$$

where J is a 6 × 6 matrix, indicated symbolically by

$$J = \frac{\partial(\boldsymbol{\rho}, \mathbf{R})}{\partial(\mathbf{r}_\alpha, \mathbf{r}_\beta)} = \left(\frac{\mathrm{a}}{x}\frac{\mathrm{B}}{\mathrm{A} + \mathrm{a}}\right)^3, \tag{5.39}$$

which relates the volume elements in the two coordinate systems. Let us denote by

$$(\Phi_\mathrm{B}\Phi_\mathrm{b}|V_\beta - U_\beta|\Phi_\mathrm{A}\Phi_\mathrm{a}) = W\int d\mathbf{A}\, d\mathbf{b}\, d\mathbf{x}\, \{\Phi_\mathrm{B}\Phi_\mathrm{b}|V_\beta - U_\beta|\Phi_\mathrm{A}\Phi_\mathrm{a}\} \tag{5.40}$$

the integrations over the intrinsic coordinates, where $d\mathbf{A}$ denotes a 3(A − 1)-dimensional integration, etc., and the curly brackets denote a spin integration. Then (5.25) and (5.32) can be written as

$$\mathscr{T}_{\beta\alpha} = \int \chi_\beta^{(-)*}(\mathbf{k}_\beta, \mathbf{r}_\beta)(\Phi_B\Phi_b|V_\beta - U_\beta|\Phi_A\Phi_a)\chi_\alpha^{(+)}(\mathbf{k}_\alpha, \mathbf{r}_\alpha)\, d\mathbf{R}\, d\rho \tag{5.41a}$$

$$= J\int \chi_\beta^{(-)*}(\mathbf{k}_\beta, \mathbf{r}_\beta)(\Phi_B\Phi_b|V_\beta - U_\beta|\Phi_A\Phi_a)\chi_\alpha^{(+)}(\mathbf{k}_\alpha, \mathbf{r}_\alpha)\, d\mathbf{r}_\alpha\, d\mathbf{r}_\beta. \tag{5.41b}$$

In some of the examples the intrinsic matrix element appearing here will be evaluated, and then one can see the following expansion emerge in an explicit way. At this point we can make a general argument about the form that these explicit evaluations will take (Satchler, 1964 and Austern *et al.* 1964). The reaction, either by virtue of particle transfer or particle excitation, will generally exchange angular momentum and parity between the relative motion and the internal states of the system.

Let us denote the following angular momenta transfer:

$$\mathbf{j} = \mathbf{J}_B - \mathbf{J}_A, \qquad \mathbf{s} = \mathbf{J}_a - \mathbf{J}_b; \tag{5.42a}$$

and

$$\boldsymbol{\ell} = \mathbf{j} - \mathbf{s}. \tag{5.42b}$$

The symbols chosen are arbitrary. Denote the angular momentum of a partial wave in the expansion of the distorted wave in channel α by ℓ_α and in channel β by ℓ_β. Conservation of angular momentum reads

$$\mathbf{J}_a + \mathbf{J}_A + \boldsymbol{\ell}_\alpha = \mathbf{J}_b + \mathbf{J}_B + \boldsymbol{\ell}_\beta \tag{5.43}$$

or, rearranging,

$$\boldsymbol{\ell}_\alpha - \boldsymbol{\ell}_\beta = \mathbf{J}_B - \mathbf{J}_A + \mathbf{J}_b - \mathbf{J}_a = \mathbf{j} - \mathbf{s} = \boldsymbol{\ell}, \tag{5.44}$$

which shows us that ℓ denotes the angular momentum transferred to the relative motion. These angular momentum relations can be declared by expanding the bracket in (5.40) in a multipole series. Thus

$$J(\Phi_B\Phi_b|V_\beta - U_\beta|\Phi_A\Phi_a) = \sum_{\ell sj,\, m_\ell m_s m_j} f_{\ell sj}^{*m_\ell}(\mathbf{r}_\beta, \mathbf{r}_\alpha) C^{J_A\ j\ J_B}_{M_A m_j M_B} C^{J_b\ s\ J_a}_{M_b m_s M_a} C^{\ell\ s\ j}_{m_\ell m_s m_j}. \tag{5.45}$$

Note that the summations $m_\ell m_s m_j$ have only one term each because the first two z projections in C must equal the third.

Remember that Φ_B is an abbreviated notation, which carries the labels $J_B M_B$, and similarly for the other wave functions. This expression simply declares by construction the vector relations previously stated. The factor G depends either as shown on r_β and r_α or on R and ρ, which are related by

(5.36). The inverse relationship can be derived by the orthonormal properties of the Clebsch–Gordan coefficients. Thus

$$f^{*m_l}_{lsj}(r_\beta, r_\alpha) = \frac{\hat{s}\,\hat{\ell}}{\hat{J}_a \hat{J}_B} \sum_{M_B M_A M_b M_a,\, m_s m_j} J(\Phi_B \Phi_b | V_\beta - U_\beta | \Phi_A \Phi_a) C^{J_A\ j\ J_B}_{M_A m_j M_B} C^{J_b\ s\ J_a}_{M_b m_s M_a} C^{l\ s\ j}_{m_l m_s m_j}, \tag{5.46}$$

where $\hat{\ell} = 2l + 1$.

One may wonder what advantage results from having made the preceding definitions, for indeed nothing has been calculated. In our examples in later chapters, the form of (5.45) will emerge in the calculation quite naturally. How could it be otherwise? It is simply an expression of the conservation of angular momentum and their z projections. These expressions do, however, underline the *common* features of all direct reactions.

Let us insert (5.45) into (5.41b). We then find

$$\mathscr{T}_{\beta\alpha} = \sum_{\ell s j, m_\ell m_s m_j} C^{J_A\ j\ J_B}_{M_A m_j M_B} C^{J_b\ s\ J_a}_{M_b m_s M_a} C^{\ell\ s\ j}_{m_\ell m_s m_j} \sqrt{\hat{\ell}}\, i^\ell B^{m_\ell}_{\ell sj}, \tag{5.47}$$

where

$$B^{m_\ell}_{\ell sj} = \frac{i^{-\ell}}{\sqrt{\hat{\ell}}} \int \chi^{(-)*}_\beta(\mathbf{k}_b, \mathbf{r}_\beta) f^{*m_\ell}_{\ell sj}(\mathbf{r}_\beta, \mathbf{r}_\alpha) \chi^{(+)}_\alpha(\mathbf{k}_\alpha, \mathbf{r}_\alpha)\, d\mathbf{r}_\alpha\, d\mathbf{r}_\beta. \tag{5.48}$$

The cross section is given by (3.39) for the transition

$$J_A M_A J_a M_a \to J_B M_B J_b M_b. \tag{5.49}$$

Generally, the incident beam is unpolarized and the target unaligned, and one counts the outgoing particles irrespective of orientations. Therefore we want to average over the orientations M_A, M_a and sum over M_B, M_b. The ordinary cross section therefore is

$$\begin{aligned} \frac{d\sigma}{d\Omega} &= \frac{m_\alpha m_\beta}{(2\pi\hbar^2)^2} \frac{k_\beta}{k_\alpha} \frac{1}{\hat{J}_A \hat{J}_a} \sum_{M_A M_a M_B M_b} |\mathscr{T}_{\beta\alpha}|^2 \\ &= \frac{m_\alpha m_\beta}{(2\pi\hbar^2)^2} \frac{k_\beta}{k_\alpha} \frac{\hat{J}_B}{\hat{J}_A} \sum_{\ell s j m_\ell} \frac{1}{\hat{s}} |B^{m_\ell}_{\ell sj}|^2. \end{aligned} \tag{5.50}$$

Note that the sum over ℓ, s, j is incoherent. However, if there were a spin–orbit interaction in the optical potential, then, as discussed in Chapter 4, the partial waves would become dependent on spin orientation and its coupling to the orbital angular momentum. In this event, only j would be incoherent, and there would be interference terms between different ℓ and s.

In the final expression of (5.50), the sums over magnetic quantum numbers have, through the properties of the Clebsch–Gordan coefficients, eliminated all cross terms among *lsj* that otherwise would occur in the square. It would be instructive to the reader who is not already practiced in evaluating sums on the geometrical factors of angular momentum algebra to confirm this result, referring to the Appendix for the relevant orthogonality and symmetry properties of the Clebsch–Gordan coefficients. It is usually easier in carrying out such evaluations to substitute the 3-*j* symbols for the Clebsch–Gordan coefficients according to their relationship, because the 3-*j* symbols have simpler symmetries. Both symbols are used in this book, with the Clebsch–Gordan coefficients usually employed for vector coupling, for convenience.

Notes to Chapter 5

The DWBA was first discussed by Mott and Massey (1933) in their famous book, "The Theory of Atomic Collisions." Other early developments of the theory are due to Gibbs and Tobocman (1961), Glendenning (1959), Horowitz and Messiah (1953, 1954), Levinson and Banerjee (1957a, 1958). Tobocman (1954, 1959), Tobocman and Gibbs (1962). and Tobocman and Kalos (1955). A review of much of the literature prior to 1970 is provided by Austern (1970).

Chapter 6

Operator Formalism

A. INTRODUCTION

In the preceding chapters, we have used the coordinate-space representation for wave functions and operators. It is a familiar representation and one in which the scattering amplitude can be extracted by examining the asymptotic behavior of the wave function.

Although the results of Chapter 5 were obtained straightforwardly, in some situations the coordinate-space representation can be awkward. There is another symbolic, or operator, formalism that is very powerful when applied to the derivation of general results and relationships. This formalism dispenses with many of the labels that are needed to specify the wave function in coordinate space and that would tend to obscure the meaning and hinder certain manipulations. Once having derived a new result in the operator formalism, in which the manipulations are very easy and transparent, one typically takes the coordinate-space representation to extract the scattering amplitude and to carry out numerical calculations.

In this chapter we shall rederive some of the earlier results to establish a familiarity with the formalism and derive some new results that will be useful and will also deepen the understanding of some of the material encountered in later chapters. Among the new results is a generalization to rearrangement collisions of the Gell-Mann–Goldberger transformation that we encountered in the simpler case of an optical potential or a one-channel problem in the last chapter. The other new result of note is Watson's multiple-scattering series.

We shall not rigorously justify the manipulations of this chapter nor review in any detail the representation theory of quantum mechanics, which is discussed very elegantly by Dirac in his textbook on quantum mechanics. (Dirac, 1947) Simply recall that the Schrödinger equation

$$(E - H_\alpha - T_\alpha - V_\alpha)\Psi = 0, \tag{6.1}$$

which we have tacitly assumed to be given in a coordinate representation, has meaning independent of any particular representation. A particular representation can be achieved by selecting a maximal set of commuting observables for the system under discussion, whose simultaneous eigenfunctions we can denote by $|n\rangle$, where n denotes collectively a set of eigenvalues of the observables. The complete set allows us to write

$$|\Psi\rangle = \sum_n |n\rangle\langle n|\Psi\rangle. \tag{6.2}$$

This is analogous to a vector with the numbers $\langle n|\Psi\rangle$ representing the coordinates on the axis $|n\rangle$. The sum can be discrete or continuous or both, depending on the eigenvalues. We refer to $|\Psi\rangle$ as the state vector and to its coordinate space representation $\langle \mathbf{r}_1, \mathbf{r}_2, \ldots |\Psi\rangle$ as the wave function, where Ψ will be used to denote either one for notational simplicity and the content will determine which is meant.

The operator multiplying Ψ in (6.1) can be described as a matrix once a particular representation is chosen and is otherwise an operator. Thus we may write

$$\sum_{n'} \langle n|(E - H_\alpha - T_\alpha - V_\alpha)|n'\rangle\langle n'|\Psi\rangle = 0, \tag{6.3}$$

wherein we see the matrix elements of the operator. We can, therefore, think of the manipulations in the next section as matrix manipulations, in which, however, we shall not always specify the basis.

B. LIPPMANN–SCHWINGER EQUATION

Let us rewrite (6.1) by taking $V\Psi$ to the right side:

$$(E - H_\alpha - T_\alpha)\Psi = V_\alpha\Psi. \tag{6.4}$$

Multiplying through on the left by the inverse operator $(E - H_\alpha - T_\alpha)^{-1}$, which is a matrix operation if we choose a representation, we have

$$\Psi = \frac{1}{E - H_\alpha - T_\alpha} V_\alpha\Psi. \tag{6.5}$$

This is analogous to (3.22). We can add to this a solution of

$$(E - H_\alpha - T_\alpha)\phi_\alpha = 0, \qquad \langle \mathbf{r}_\alpha | \phi_\alpha \rangle = \Phi_\alpha \exp(i\mathbf{k}_\alpha \cdot \mathbf{r}_\alpha), \tag{6.6}$$

and still have a valid equation. [Recall the notation established in Chapter 3 (3.2)–(3.13)]. Thus, more generally,

$$\Psi_\alpha = \phi_\alpha + \frac{1}{E - H_\alpha - T_\alpha} V_\alpha \Psi_\alpha \tag{6.7}$$

still satisfies (6.1). However, there are other solutions. The one of interest in scattering problems has outgoing waves at infinity. This is prescribed by adding an infinitesimal positive imaginary part to the energy. Thus

$$\Psi_\alpha^{(+)} = \phi_\alpha + \frac{1}{E - H_\alpha - T_\alpha + i\varepsilon} V_\alpha \Psi_\alpha^{(+)}, \tag{6.8}$$

in which it is understood that the limit $\varepsilon \to 0$ is taken after a particular representation has been chosen and the calculation made. To prove that this prescription produces the desired boundary condition, one can examine the coordinate-space representation of the inverse operator in (6.8). We are considering here a many-particle system, that consists of two separate bound systems (two-body channel), and the eigenfunctions of H_α can be used immediately to reduce the problem to one of looking at the behavior with respect to the relative coordinate r. Consider the following matrix element.

$$\left\langle \mathbf{r}, \Phi_\alpha \left| \frac{1}{E - H_\alpha - T_\alpha + i\varepsilon} \right| \mathbf{r}', \Phi_{\alpha'} \right\rangle = \left\langle \mathbf{r} \left| \frac{1}{E_\alpha - T_\alpha + i\varepsilon} \right| \mathbf{r}' \right\rangle \delta_{\alpha\alpha'}. \tag{6.9}$$

This is now most easily evaluated by going to the momentum-space representation, because T_α is known there.

We shall use the normalization conventions

$$\langle \mathbf{r} | \mathbf{r}' \rangle = \delta(\mathbf{r} - \mathbf{r}'), \qquad \langle \mathbf{k} | \mathbf{k}' \rangle = (2\pi)^3 \delta(\mathbf{k} - \mathbf{k}'), \tag{6.10a}$$

with corresponding completeness relations

$$\int |\mathbf{r}\rangle \, d\mathbf{r} \, \langle \mathbf{r}| = 1 = (2\pi)^{-3} \int |\mathbf{k}\rangle \, d\mathbf{k} \, \langle \mathbf{k}|, \tag{6.10b}$$

having the consequence that the transformation function has normalization unity:

$$\langle \mathbf{r} | \mathbf{k} \rangle = \exp(i\mathbf{k} \cdot \mathbf{r}). \tag{6.11}$$

We can now evaluate (6.9):

$$\left\langle \mathbf{r} \left| \frac{1}{E_\alpha - T_\alpha + i\varepsilon} \right| \mathbf{r}' \right\rangle = (2\pi)^{-6} \iint \left\langle \mathbf{r} \middle| \mathbf{k} \right\rangle \left\langle \mathbf{k} \left| \frac{1}{E_\alpha - T_\alpha + i\varepsilon} \right| \mathbf{k}' \right\rangle \left\langle \mathbf{k}' \middle| \mathbf{r}' \right\rangle d\mathbf{k} \, d\mathbf{k}'. \tag{6.12}$$

Now

$$T_\alpha|\mathbf{k}\rangle = \frac{\hbar^2 k^2}{2m_\alpha}|\mathbf{k}\rangle, \tag{6.13}$$

$$\left\langle k \left| \frac{1}{E_\alpha - T_\alpha + i\varepsilon} \right| \mathbf{k}' \right\rangle = \frac{2m_\alpha}{\hbar^2}(2\pi)^3 \frac{\delta(\mathbf{k}' - \mathbf{k})}{k_\alpha^2 - k^2 + i\varepsilon}, \tag{6.14}$$

where we define k_α by

$$E_\alpha = \hbar^2 k_\alpha^2/(2m_\alpha). \tag{6.15}$$

We now have, writing $\mathbf{R} = \mathbf{r} - \mathbf{r}'$,

$$\begin{aligned}
\left\langle \mathbf{r} \left| \frac{1}{E_\alpha - T_\alpha + i\varepsilon} \right| \mathbf{r}' \right\rangle &= \frac{2m_\alpha}{\hbar^2}(2\pi)^{-3} \int \frac{\exp(i\mathbf{k}\cdot\mathbf{R})}{k_\alpha^2 - k^2 + i\varepsilon}\, d\mathbf{k} \\
&= \frac{2m_\alpha}{\hbar^2}(2\pi)^{-3} \int \frac{\exp(ikR\cos\theta)}{k_\alpha^2 - k^2 + i\varepsilon}\, k^2\, dk\, d(\cos\theta)\, d\phi \\
&= \frac{2m_\alpha}{\hbar^2}(2\pi)^{-3} \frac{2\pi}{iR} \int_0^\infty \frac{\exp(ikR) - \exp(-ikR)}{k_\alpha^2 - k^2 + i\varepsilon}\, k\, dk \\
&= \frac{2m_\alpha}{\hbar^2}(2\pi)^{-3} \frac{2\pi}{iR} \int_{-\infty}^\infty \frac{\exp(ikR)}{k_\alpha^2 - k^2 + i\varepsilon}\, k\, dk.
\end{aligned} \tag{6.16}$$

Represent $i\varepsilon$ by setting $k_\alpha \to k_\alpha + i\lambda$. Then we see that the integrand has two singularities at

$$k = \pm(k_\alpha + i\lambda).$$

The integral can be evaluated using Cauchy's theorem:

$$f(z_0) = \frac{1}{2\pi i} \oint \frac{f(z)}{z - z_0}\, dz. \tag{6.17}$$

By choosing the contour to be on the real axis and to be closed in the upper plane, we then find, after more algebra,

$$G_\alpha^{0(+)}(\mathbf{r}, \mathbf{r}') \equiv \left\langle \mathbf{r} \left| \frac{1}{E_\alpha - T_\alpha + i\varepsilon} \right| \mathbf{r}' \right\rangle = -\frac{m_\alpha}{2\pi\hbar^2} \frac{\exp(ik_\alpha|\mathbf{r} - \mathbf{r}'|)}{|\mathbf{r} - \mathbf{r}'|}. \tag{6.18}$$

This should not be a surprising result, corresponding as it does to the free Green's function (3.25). Its asymptotic value as $r \to \infty$ does indeed have the desired outgoing wave, indicated in (3.25). This completes the proof that $+i\varepsilon$ in (6.8) yields outgoing waves.

It will now be recognized that the Φ_α projection of (6.8) is our operator way of writing the integral equation (3.24). This equation is known as the

Lippmann–Schwinger equation (see Lippmann and Schwinger, 1950). Its usefulness, as compared with the Schrödinger equation, lies in the fact that it incorporates both boundary conditions, corresponding to the initial state of the system and to the outgoing wave nature of the scattered waves.

C. FORMAL SOLUTION

Now note that we can rewrite the integral equation

$$\Psi_\alpha^{(+)} = \phi_\alpha + \frac{1}{E - H_\alpha - T_\alpha + i\varepsilon} V_\alpha \Psi_\alpha^{(+)}$$

as a series by repeatedly using the definition of $\Psi_\alpha^{(+)}$ in the right-hand side. Thus

$$\Psi_\alpha^{(+)} = \phi_\alpha + \frac{1}{E - H_\alpha - T_\alpha + i\varepsilon} V_\alpha \phi_\alpha + \frac{1}{E - H_\alpha - T_\alpha + i\varepsilon} V_\alpha \frac{1}{E - H_\alpha - T_\alpha + i\varepsilon} V_\alpha \phi_\alpha + \cdots . \quad (6.19)$$

This Born series can also be summed by observing the operator identity

$$\frac{1}{\mathrm{A}} - \frac{1}{\mathrm{B}} = \frac{1}{\mathrm{A}}(\mathrm{B} - \mathrm{A})\frac{1}{\mathrm{B}} = \frac{1}{\mathrm{B}}(\mathrm{B} - \mathrm{A})\frac{1}{\mathrm{A}}. \quad (6.20\mathrm{a})$$

This identity can be iterated to yield,

$$\begin{aligned} \frac{1}{A} &= \frac{1}{B} + \frac{1}{B}(B - A)\frac{1}{A} \\ &= \frac{1}{B} + \frac{1}{B}(B - A)\frac{1}{B} + \frac{1}{B}(B - A)\frac{1}{B}(B - A)\frac{1}{B} + \cdots . \end{aligned} \quad (6.20\mathrm{b})$$

Thus putting

$$A \equiv E - H_\alpha - T_\alpha - V_\alpha + i\varepsilon$$
$$B \equiv E - H_\alpha - T_\alpha + i\varepsilon,$$

we see how the series in (6.19) can be summed to yield

$$\Psi_\alpha^{(+)} = \phi_\alpha + \frac{1}{E - H_\alpha - T_\alpha - V_\alpha + i\varepsilon} V_\alpha \phi_\alpha . \quad (6.21\mathrm{a})$$

Alternatively, the identity can be used directly in (6.8) to obtain the same result by following the steps below:

$$\begin{aligned}\Psi_\alpha^{(+)} &= \phi_\alpha + \frac{1}{B} V\Psi_\alpha^{(+)} \\ &= \phi_\alpha + \left[\frac{1}{A} - \frac{1}{A}(B - A)\frac{1}{B}\right] V_\alpha \Psi_\alpha^{(+)} \\ &= \phi_\alpha + \frac{1}{A} V_\alpha \Psi_\alpha - \frac{1}{A}(B - A)(\Psi_\alpha^{(+)} - \phi_\alpha) \\ &= \phi_\alpha + \frac{1}{A} V_\alpha \phi_\alpha\end{aligned}$$

This is (6.21a).

Whereas (6.8) is the integral equation for Ψ_α, (6.21a) is its solution, or at least a way of writing its solution in operator form. We can also write it as

$$\Psi_\alpha^{(+)} = \left(1 + \frac{1}{E - H + i\varepsilon} V_\alpha\right) \phi_\alpha, \tag{6.21b}$$

where H is the total Hamiltonian (3.13). The operator multiplying ϕ_α is sometimes called the Möller operator and denoted by Ω^+:

$$\Psi_\alpha^{(+)} = \Omega_\alpha^+ \phi_\alpha. \tag{6.21c}$$

The same result could have been obtained differently. From (6.1) and (6.6),

$$(E - H_\alpha - T_\alpha - V_\alpha)\Psi_\alpha = (E - H_\alpha - T_\alpha - V_\alpha)\phi_\alpha + V_\alpha \phi_\alpha,$$

from which (6.21a) follows by multiplying through by the inverse operator on the left and then inserting the boundary condition prescription $i\varepsilon$. This is not a derivation, as was the passage from (6.8) to (6.21a) by means of (6.20). It is simply a fast way of recalling the answer.

D. TRANSITION AMPLITUDE

Recall from Chapter 3 that a general and exact expression for the transition amplitude is

$$\mathcal{T}_{\beta\alpha} = \langle \phi_\beta | V_\beta | \Psi_\alpha^{(+)} \rangle. \tag{6.22}$$

This can be rederived as before. We take a projection on the channel of interest, say α', go to the coordinate-space representation for the relative coordinate $\mathbf{r}_\alpha$ in this channel, and examine the asymptotic behavior as $\mathbf{r}_\alpha$ tends to infinity,

$$\begin{aligned}\langle \mathbf{r}_\alpha, \Phi_{\alpha'} | \Psi_\alpha^{(+)} \rangle &\\ = \delta_{\alpha'\alpha} \exp(i\mathbf{k}_\alpha \cdot \mathbf{r}_\alpha) &\\ + \sum_{\alpha''} \int d\mathbf{r}'_\alpha \left\langle \mathbf{r}_\alpha, \Phi_{\alpha'} \right| \frac{1}{E - H_\alpha - T_\alpha + i\varepsilon} &\left| \mathbf{r}'_\alpha, \Phi_{\alpha''} \right\rangle \langle \mathbf{r}'_\alpha, \Phi_{\alpha''} | V_\alpha \Psi_\alpha^{(+)} \rangle \\ = \delta_{\alpha'\alpha} \exp(i\mathbf{k}_\alpha \cdot \mathbf{r}_\alpha) - \frac{m_\alpha}{2\pi\hbar^2} \int d\mathbf{r}'_\alpha \frac{\exp(ik'_\alpha|\mathbf{r}_\alpha - \mathbf{r}'_\alpha|)}{|\mathbf{r}_\alpha - \mathbf{r}'_\alpha|} &\langle \mathbf{r}'_\alpha, \Phi_{\alpha'} | V_\alpha \Psi_\alpha^{(+)} \rangle. \qquad (6.23)\end{aligned}$$

The limit as $r_\alpha \to \infty$ can be taken as in (3.25), with the same result, namely,

$$\mathscr{T}_{\alpha'\alpha} = \int d\mathbf{r}'_\alpha \exp(-i\mathbf{k}'_\alpha \cdot \mathbf{r}'_\alpha) \langle \mathbf{r}'_\alpha, \Phi_{\alpha'} | V_\alpha \Psi_\alpha^{(+)} \rangle = \langle \phi_{\alpha'} | V_\alpha | \Psi_\alpha^{(+)} \rangle.$$

What we have learned, quite generally, from these manipulations on the operator equation (6.8), is that when the *free* Green's function operator

$$G_\alpha^{0(+)} \equiv \frac{1}{E - H_\alpha - T_\alpha + i\varepsilon} \qquad (6.24a)$$

appears in an expression like (6.8), the transition amplitude to an eigenstate of H_α can be read immediately as $\langle \phi_{\alpha'} |$ acting on whatever vector appears to the right of G_α^0. In the particular case of (6.8), it is $\mathscr{T}_{\alpha'\alpha}$. Symbolically we may write this as

$$\begin{aligned}\langle \mathbf{r}_\alpha, \Phi_{\alpha'} | G_\alpha^{0(+)} X \rangle &\to -\frac{m_\alpha}{2\pi\hbar^2} \frac{\exp(ik'_\alpha r_\alpha)}{r_\alpha} \int \exp(-i\mathbf{k}'_\alpha \cdot \mathbf{r}'_\alpha) \langle \mathbf{r}'_\alpha \Phi_{\alpha'} | X \rangle \, d\mathbf{r}'_\alpha \\ &= -\frac{m_\alpha}{2\pi\hbar^2} \frac{\exp(ik'_\alpha r_\alpha)}{r_\alpha} \langle \phi_{\alpha'} | X \rangle. \qquad (6.24b)\end{aligned}$$

where $\mathbf{k}'_\alpha = k'_\alpha \mathbf{r}_\alpha / r_\alpha$ and X is any vector.

The rearrangement amplitude can also be obtained from (6.8) after some manipulation. The object of such manipulations must, of course, be to bring in the appearance of the free Green's function in channel β:

$$G_\beta^{0(+)} \equiv \frac{1}{E - H_\beta - T_\beta + i\varepsilon}. \qquad (6.24c)$$

To accomplish this, use the operator identity (6.20) to rewrite (6.8) as

$$\begin{aligned}\Psi_\alpha^{(+)} &= \phi_\alpha + \frac{1}{E - H_\beta - T_\beta + i\varepsilon}\left[V_\alpha + (V_\beta - V_\alpha)\frac{1}{E - H_\alpha - T_\alpha + i\varepsilon}V_\alpha\right]\Psi_\alpha^{(+)} \\ &= \phi_\alpha + \frac{1}{E - H_\beta - T_\beta + i\varepsilon}[V_\alpha\Psi_\alpha^{(+)} + (V_\beta - V_\alpha)(\Psi_\alpha^{(+)} - \phi_\alpha)] \\ &= \left[1 - \frac{1}{E - H_\beta - T_\beta + i\varepsilon}(V_\beta - V_\alpha)\right]\phi_\alpha + \frac{1}{E - H_\beta - T_\beta + i\varepsilon}V_\beta\Psi_\alpha^{(+)} \\ &= \frac{i\varepsilon}{E - H_\beta - T_\beta + i\varepsilon}\phi_\alpha + \frac{1}{E - H_\beta - T_\beta + i\varepsilon}V_\beta\Psi_\alpha^{(+)}. \end{aligned} \tag{6.25}$$

We shall now prove that, if $\beta \neq \alpha$, the norm of the first term vanishes in the limit $\varepsilon \to 0$. The norm is

$$\begin{aligned} N(\varepsilon) &= \varepsilon^2\langle\phi_\alpha|\frac{1}{E - H_\beta - T_\beta - i\varepsilon}\frac{1}{E - H_\beta - T_\beta + i\varepsilon}|\phi_\alpha\rangle \\ &= \varepsilon\sum_{\beta'}\int\frac{d\mathbf{k}'}{(2\pi)^3}\frac{\varepsilon}{(E - E')^2 + \varepsilon^2}|\langle\phi_\alpha|\mathbf{k}', \Phi_{\beta'}\rangle|^2, \end{aligned}$$

where $E' = \varepsilon_{\beta'} + \hbar^2 k'^2/2m_\beta$ and we inserted a complete set of energy states in channel β. The first factor in the integrand is one of the many representations of the δ-function (see Messiah, 1962, Vol. I, Appendix A):

$$\lim_{\varepsilon\to 0}\frac{\varepsilon}{(E - E')^2 + \varepsilon^2} = \pi\delta(E - E').$$

Hence, $N(\varepsilon)$ is the product of ε and a factor independent of ε, and it therefore vanishes in the limit. Consequently, in the limit $\varepsilon \to 0$, we find (Lippman, 1956; Gerjuoy, 1958)

$$\Psi_\alpha^{(+)} = \frac{1}{E - H_\beta - T_\beta + i\varepsilon}V_\beta\Psi_\alpha^{(+)}, \qquad \beta \neq \alpha. \tag{6.26}$$

This is to be contrasted with (6.8), which has a homogeneous term representing the boundary condition. Clearly, such a term ϕ_α cannot be added to (6.26) and still have $\Psi_\alpha^{(+)}$ satisfy the Schrödinger equation. If (6.26) were solved, one would have to take care that the boundary condition is satisfied. On the other hand, (6.26) makes physical sense. The inverse operator is the free outgoing Green's function in the partition β. Going to the representation $\mathbf{r}_\beta, \Phi_\beta$, we find, according to the analog of (6.24b), that (6.26) possesses only outgoing waves, as should be the case, and their amplitude is given by (6.22), as was to be proven.

E. TRANSITION OPERATOR T

Use (6.21) in (6.22) to obtain

$$\mathscr{T}_{\beta\alpha} = \left\langle \phi_\beta \middle| V_\beta + V_\beta \frac{1}{E - H + i\varepsilon} V_\alpha \middle| \phi_\alpha \right\rangle$$
$$\equiv \langle \phi_\beta | T_{\beta\alpha} | \phi_\alpha \rangle. \tag{6.27}$$

The operator

$$T_{\beta\alpha} = V_\beta + V_\beta \frac{1}{E - H + i\varepsilon} V_\alpha \tag{6.28a}$$

could be referred to as the transition operator. Its matrix element with respect to the unperturbed or asymptotic states of the system provides the transition amplitude. We shall see that it has another form. Notice that

$$\langle \phi_\beta | V_\beta - V_\alpha | \phi_\alpha \rangle = \langle \phi_\beta | (H - H_\beta - T_\beta) - (H - H_\alpha - T_\alpha) | \phi_\alpha \rangle$$
$$= \langle \phi_\beta | (H_\alpha + T_\alpha) - (H_\beta + T_\beta) | \phi_\alpha \rangle.$$

According to (6.6), because ϕ_α and ϕ_β have the same energy (by conservation) this vanishes, giving

$$\langle \phi_\beta | V_\beta | \phi_\alpha \rangle = \langle \phi_\beta | V_\alpha | \phi_\alpha \rangle. \tag{6.29}$$

We can therefore write the symmetric form

$$T_{\beta\alpha} = \frac{1}{2}(V_\beta + V_\alpha) + V_\beta \frac{1}{E - H + i\varepsilon} V_\alpha \tag{6.28b}$$

or an asymmetrical form in the sense opposite to (6.28a)

$$T_{\beta\alpha} = V_\alpha + V_\beta \frac{1}{E - H + i\varepsilon} V_\alpha. \tag{6.28c}$$

This latter form, when used in (6.27), yields a result of interest:

$$\mathscr{T}_{\beta\alpha} = \left\langle \phi_\beta \middle| \left(1 + V_\beta \frac{1}{E - H + i\varepsilon}\right) V_\alpha \middle| \phi_\alpha \right\rangle$$
$$= \left\langle \left(1 + \frac{1}{E - H - i\varepsilon} V_\beta\right) \phi_\beta \middle| V_\alpha \middle| \phi_\alpha \right\rangle, \tag{6.30}$$

where H and V are assumed to be Hermitian. The bracket in this expression has the same form as (6.21), so that we may write

$$\Psi_\beta^{(-)} = \phi_\beta + \frac{1}{E + H - i\varepsilon} V_\beta \phi_\beta, \tag{6.31}$$

which is the solution to (6.1) having incoming waves at infinity and a plane wave in the channel β. We now have two ways of writing the transition amplitude,

$$\mathscr{T}_{\beta\alpha} = \langle \Psi_\beta^{(-)} | V_\alpha | \phi_\alpha \rangle = \langle \phi_\beta | V_\beta | \Psi_\alpha^{(+)} \rangle. \tag{6.32}$$

These are sometimes referred to as prior and post forms of writing $\mathscr{T}$. According to (6.29), the equality of the two forms also holds in the first Born approximation.

Note that the transition operator T, according to (6.21) allows one to express $\mathscr{T}$ as a matrix element between plane wave states ϕ. That is to say,

$$V_\beta \Psi_\alpha = T_{\beta\alpha} \phi_\alpha. \tag{6.33}$$

Integral equations can also be derived for the T operator. Using the operator identity (6.20) and, variously, (6.28) and (6.29), we can obtain the following results:

$$\begin{aligned} T_{\beta\alpha} &= V_\beta + V_\beta \frac{1}{E - H_\beta - T_\beta + i\varepsilon} \left[1 + V_\beta \frac{1}{E - H + i\varepsilon} \right] V_\alpha \\ &= V_\beta + V_\beta \frac{1}{E - H_\beta - T_\beta + i\varepsilon} T_{\beta\alpha} \\ &= V_\alpha + V_\beta \frac{1}{E - H_\beta - T_\beta + i\varepsilon} T_{\beta\alpha} \end{aligned} \tag{6.34a}$$

or

$$T_{\beta\alpha} = V_\beta + T_{\beta\alpha} \frac{1}{E - H_\alpha - T_\alpha + i\varepsilon} V_\alpha = V_\alpha + T_{\beta\alpha} \frac{1}{E - H_\alpha - T_\alpha + i\varepsilon} V_\alpha. \tag{6.34b}$$

We can also derive several series expansions for T by iterating the integral equation (6.34b). Thus, for example,

$$\begin{aligned} T_{\beta\alpha} &= V_\alpha + V_\alpha \frac{1}{E - H_\alpha - T_\alpha + i\varepsilon} V_\alpha \\ &\quad + V_\alpha \frac{1}{E - H_\alpha - T_\alpha + i\varepsilon} V_\alpha \frac{1}{E - H_\alpha - T_\alpha + i\varepsilon} V_\alpha + \cdots \end{aligned} \tag{6.35}$$

and a similar series with $\alpha \to \beta$ everywhere on the right side. This is an important equation. If V_α is a potential that can be treated in perturbation theory, then the series provides a way of calculating $\mathscr{T}_{\beta\alpha}$ to any desired order in V_α. However, the free nucleon–nucleon interaction is not such a well-behaved potential that this expansion can be used for that purpose. The reason for writing the series is to show explicitly that the scattering operator invovles all orders of the interaction, corresponding to single, double, triple, . . . scattering by the potential V.

F. GELL-MANN–GOLDBERGER TRANSFORMATION

It is frequently the case in scattering problems that an interaction can be divided into two pieces, $V = V_1 + V_2$, such that one of them poses a solvable problem. The solution to this problem can then be used in an approximation scheme for handling the entire interaction. Direct nuclear reactions can be made to fit this scheme by the artifice of writing $V = U + (V - U)$. In the last chapter, this device was employed to introduce the optical potential, a phenomenological interaction whose parameters can be adjusted to yield the observed elastic scattering rather well, as we saw in Chapter 4. The optical potential that we have discussed so far is a one-body potential that is independent of all internal nuclear degrees of freedom. However, many nuclei have certain collective states that are strongly excited in reactions. It is important to develop a scheme that treats such collective inelastic transitions on a comparable basis with the elastic scattering. We shall therefore consider the division $V = U + (V - U)$ as one that separates out the part of the interaction U that is responsible for elastic and inelastic transitions. We suppose that this part of the problem can be solved "exactly." The remaining interaction $(V - U)$ is regarded as weak and is responsible, say, for the nucleon transfer reactions, which typically have smaller cross sections than the elastic or collective inelastic transitions. In a more general context, we simply regard U as an auxiliary potential introduced to represent an important part of the scattering. In the development of this section, which is a generalization of the Gell-Mann–Goldberger transformation to rearrangement collisions, we shall not make any special assumptions about U, other than that it differs from V. In general, then, it can cause elastic, inelastic, and rearrangement scattering. It is important to notice that, whereas the equality

$$H = H_\alpha + T_\alpha + V_\alpha = H_\beta + T_\beta + V_\beta = \cdots$$

corresponds merely to a rearrangement of terms in the Hamiltonian, as discussed in Chapter 3, no such equality exists for the auxiliary Hamiltonians

$$H \neq H_\alpha + T_\alpha + U_\alpha \neq H_\beta + T_\beta + U_\beta.$$

We shall retain the notation of Chapter 5 in denoting the state vector corresponding to U_α as χ_α. However, in Chapter 5, U_α was a one-body potential in the channel coordinate, so the problem separated into one involving the relative motion and one involving the intrinsic nuclear degrees of freedom. In this case factorization does not occur, so that χ_α denotes the state of the entire system. It is solution to

$$(E - H_\alpha - T_\alpha - U_\alpha)\chi_\alpha = 0. \tag{6.36}$$

We can also write (6.1) together with (6.36) as

$$(E - H_\alpha - T_\alpha - U_\alpha)\Psi_\alpha = (V_\alpha - U_\alpha)\Psi_\alpha + (E_\alpha - H_\alpha - T_\alpha - U_\alpha)\chi_\alpha$$

or

$$\Psi_\alpha^{(+)} = \chi_\alpha^{(+)} + \frac{1}{E - H_\alpha - T_\alpha - U_\alpha + i\varepsilon}(V_\alpha - U_\alpha)\Psi_\alpha^{(+)}. \tag{6.37}$$

This is the integral equation (5.15).‡ Its "solution" can be obtained either by using the identity (6.20) or by observing that

$$(E - H_\alpha - T_\alpha - V_\alpha)\Psi_\alpha = (E - H_\alpha - T_\alpha - V_\alpha)\chi_\alpha + (V_\alpha - U_\alpha)\chi_\alpha,$$

from which

$$\begin{aligned}\Psi_\alpha^{(+)} &= \chi_\alpha^{(+)} + \frac{1}{E - H_\alpha - T_\alpha - V_\alpha + i\varepsilon}(V_\alpha - U_\alpha)\chi_\alpha^{(+)} \\ &= \chi_\alpha^{(+)} + \frac{1}{E - H + i\varepsilon}(V_\alpha - U_\alpha)\chi_\alpha^{(+)}.\end{aligned} \tag{6.38}$$

The $\mathscr{T}$ matrix can also be written in terms of the "distorted waves" χ, which are solutions of (6.36). In other words, we can rederive (5.17) using these operator techniques. Use (6.38) for $\Psi_\alpha^{(+)}$ in

$$\begin{aligned}\mathscr{T}_{\beta\alpha} &= \langle \phi_\beta | V_\beta | \Psi_\alpha^{(+)} \rangle \\ &= \langle \phi_\beta | V_\beta | \chi_\alpha^{(+)} \rangle + \left\langle \phi_\beta \middle| V_\beta \frac{1}{E - H + i\varepsilon}(V_\alpha - U_\alpha) \middle| \chi_\alpha^{(+)} \right\rangle \\ &= \langle \phi_\beta | V_\beta - V_\alpha + U_\alpha | \chi_\alpha^{(+)} \rangle + \langle \Psi_\beta^{(-)} | V_\alpha - U_\alpha | \chi_\alpha^{(+)} \rangle,\end{aligned} \tag{6.39}$$

where (6.31) was used in the last step. An alternative derivation of this last result is given by Messiah (1962, Vol. II, Ch. XIX, Sect. 20). Note that in the case of no rearrangement, that is to say, simply elastic or inelastic scattering, the two potentials V_β and V_α are identical and only U_α is left in the first term. It

‡This was not a derivation but a fast way of "guessing" the answer. It can be derived from (6.21) using the identity (6.20) and the solution of (6.36) analogous to (6.21). Alternatively, by direct substitution of (6.37) into (6.1) with the aid of (6.20), we find

$$(E - H)\Psi_\alpha^{(+)} = \frac{-i\varepsilon}{E - H_\alpha - T_\alpha - U_\alpha + i\varepsilon}(V_\alpha - U_\alpha)\Psi_\alpha^{(+)}.$$

In the same way as the $i\varepsilon$ term of (6.25) was studied, the square of the norm of the right side is found to vanish as $\varepsilon \to 0$.

then has the traditional form of (6.22). But for a rearrangement collision $\beta \neq \alpha$, the first term of (6.39) has an unfamiliar form. Why? Before providing the answer, note that (6.39) was obtained by substituting an expression for $\Psi_\alpha^{(+)}$ into the general result (6.22). Of course, we should be able to obtain it by looking at the asymptotic behavior of $\Psi_\alpha^{(+)}$, in the same way that (6.22) was derived. It will prove very instructive to do so. We start this time with (6.38). Using the operator identity (6.20), the standard procedure is to extract the free Green's function. For the rearrangement channel β, we can write

$$\Psi_\alpha^{(+)} = \chi_\alpha^{(+)} + \frac{1}{E - H_\beta - T_\beta + i\varepsilon}\left[1 + V_\beta \frac{1}{E - H + i\varepsilon}\right](V_\alpha - U_\alpha)\chi_\alpha^{(+)}. \tag{6.40}$$

Going to the coordinate representation we find, in the by now familiar way, that the transition amplitude to Φ_β in the second term is given by

$$\langle\phi_\beta|\left[1 + V_\beta \frac{1}{E - H + i\varepsilon}\right](V_\alpha - U_\alpha)\chi_\alpha^{(+)}\rangle = \langle\Psi_\beta^{(-)}|V_\alpha - U_\alpha|\chi_\alpha^{(+)}\rangle.$$

This is the expected term in (6.39). Of course, $\chi_\alpha^{(+)}$ may also have a scattering amplitude. Let us note that the integral equation and solution corresponding to (6.36) are

$$\chi_\alpha^{(+)} = \phi_\alpha + \frac{1}{E - H_\alpha - T_\alpha + i\varepsilon} U_\alpha \chi_\alpha^{(+)} \tag{6.41a}$$

and

$$\chi_\alpha^{(+)} = \phi_\alpha + \frac{1}{E - H_\alpha - T_\alpha - U_\alpha + i\varepsilon} U_\alpha \phi_\alpha, \tag{6.41b}$$

respectively. From the integral equation we recognize the free outgoing Green's function in channel α and can write immediately

$$\mathcal{T}_{\alpha'\alpha}^{U_\alpha} = \langle\phi_{\alpha'}|U_\alpha|\chi_\alpha^{(+)}\rangle. \tag{6.42}$$

This is the alternative derivation of the first term in (6.39). What about the rearrangement amplitude in $\chi_\alpha^{(+)}$? Rewrite the Schrödinger equation for χ_α as

$$(E - H_\beta - T_\beta)\chi_\alpha^{(+)} = (V_\beta - V_\alpha + U_\alpha)\chi_\alpha^{(+)}.$$

Hence,

$$\chi_\alpha^{(+)} = \frac{1}{E - H_\beta - T_\beta + i\varepsilon}(V_\beta - V_\alpha + U_\alpha)\chi_\alpha^{(+)}. \tag{6.41c}$$

The same result could be derived more rigorously from (6.41), just as (6.26) was derived. Also, just as in the case of (6.26), we cannot add a homogeneous solution ϕ_α to this, for then it would not satisfy the Schrödinger equation. Again in the usual way, the amplitude for outgoing scattered waves in the channel Φ_β can be read from (6.41c). It is precisely the first term of (6.39).

Had we been careless, we might have said that the amplitude for the rearrangement transition $\alpha \to \beta$ caused by U is $\langle \phi_\beta | U_\beta | \chi_\alpha^{(+)} \rangle$, in analogy with (6.22). This would be wrong although (6.22) is perfectly correct. The point is that the eigenstates defining the partitions α, β, . . . are eigenstates of the true interactions V_{ij} of the problem and not of the auxiliary potentials U_{ij}.

As can be imagined, there is an alternative form of (6.39) that involves the exact solution Ψ with boundary conditions corresponding to the initial channel, that is, $\Psi_\alpha^{(+)}$, rather than the final channel. However, because in many applications of these results the auxiliary potentials are non-Hermitian optical potentials, we must adopt a convention concerning the use of U and $U^\dagger$. Having written the outgoing wave solution $\Psi_\alpha^{(+)}$ in terms of $\chi_\alpha^{(+)}$ in (6.38), then physically it makes sense that $\chi_\alpha^{(+)}$ should be calculated from an absorptive potential. Because χ_α satisfies (6.36), we shall associate absorption (negative imaginary part) with U_α. We have already derived one expression for the T matrix corresponding to the auxiliary potential U_α, namely (6.42). Substituting (6.41b) in that expression, we have

$$\mathscr{T}_{\alpha'\alpha}^{U_\alpha} = \left\langle \phi_{\alpha'} \middle| U_\alpha + U_\alpha \frac{1}{E - H_\alpha - T_\alpha - U_\alpha + i\varepsilon} U_\alpha \middle| \phi_\alpha \right\rangle$$

$$= \left\langle \left(1 + \frac{1}{E - H_\alpha - T_\alpha - U_\alpha - i\varepsilon} U_\alpha^\dagger\right) \phi_{\alpha'} \middle| U_\alpha \middle| \phi_\alpha \right\rangle. \tag{6.43}$$

Note the appearance of $U_\alpha^\dagger$ because we are not assuming the Hermiticity of U_α. Also note the incoming wave boundary condition $-i\varepsilon$. This suggests that we define

$$\chi_\alpha^{(-)} = \phi_\alpha + \frac{1}{E - H_\alpha - T_\alpha - U_\alpha^\dagger - i\varepsilon} U_\alpha^\dagger \phi_\alpha, \tag{6.44a}$$

which satisfies

$$(E - H_\alpha - T_\alpha - U_\alpha^\dagger)\chi_\alpha^{(-)} = 0. \tag{6.44b}$$

With this definition we then obtain the equality corresponding to (6.32):

$$\mathscr{T}_{\alpha',\alpha}^{U_\alpha} = \langle \chi_{\alpha'}^{(-)} | U_\alpha | \phi_\alpha \rangle = \langle \phi_{\alpha'} | U_\alpha | \chi_\alpha^{(+)} \rangle. \tag{6.45}$$

Thus (6.32) holds for non-Hermitian Hamiltonians if the convention (6.44) is adopted.

Having defined what we mean by an incoming wave solution for a non-Hermitian Hamiltonian, we can derive, analogous to (6.39), the equation

$$\mathscr{T}_{\beta\alpha} = \langle \chi_\beta^{(-)} | V_\alpha - V_\beta + U_\beta | \phi_\alpha \rangle + \langle \chi_\beta^{(-)} | V_\beta - U_\beta | \Psi_\alpha^{(+)} \rangle. \tag{6.46}$$

Equations (6.39) and (6.46) are generalizations of the Gell-Mann–Goldberger transformation for a reaction involving two partitions, i.e., a rearrangement reaction.

The result derived above is completely general, and holds true for any choice of potential U. Of course, it is a useful result only if U is chosen so as to pose a solvable problem whose solution represents some of the essential features of the physical situation. The simplest case is the one already treated in Chapter 5, where U was assumed to be a one-body optical potential whose parameters were chosen so that the elastic cross section was well described. In Chapter 17 we shall take up the subject of inelastic contributions to particle transfer reactions. In this more general case, U will be assumed to have properties that induce elastic and inelastic scattering, and can be referred to as a generalized optical potential. In either of these cases, U does not cause rearrangement reactions. Consequently, the first term of (6.39) vanishes if β is different from the entrance channel α.‡

At this point note that there is an equation analogous to (6.38) for $\Psi_\beta^{(-)}$. It is

$$\Psi_\beta^{(-)} = \chi_\beta^{(-)} + \frac{1}{E - H - i\varepsilon}(V_\beta - U_\beta^\dagger)\chi_\beta^{(-)}, \tag{6.47}$$

where, in accord with the convention adopted for non-Hermitian potentials, $\chi_\beta^{(-)}$ satisfies

$$(E - H_\beta - T_\beta - U_\beta^\dagger)\chi_\beta^{(-)} = 0. \tag{6.48a}$$

That (6.47) is a solution of (6.1) can be proven in several ways. For example, by direct substitution into (6.1) we obtain,

$$(E - H)\Psi_\beta^{(-)} = \frac{i\varepsilon}{E - H - i\varepsilon}(V_\beta - U_\beta^\dagger)\chi_\beta^{(-)}.$$

The norm of the right side can be proven to vanish in the limit $\varepsilon \to 0$, just as was done for a similar object earlier in the chapter. Alternatively, note the integral equation and solution corresponding to (6.48a):

$$\chi_\beta^{(-)} = \phi_\beta + \frac{1}{E - H_\beta - T_\beta - i\varepsilon} U_\beta^\dagger \chi_\beta^{(-)} \tag{6.48b}$$

$$\chi_\beta^{(-)} = \phi_\beta + \frac{1}{E - H_\beta - T_\beta - U_\beta^\dagger - i\varepsilon} U_\beta^\dagger \phi_\beta. \tag{6.48c}$$

‡ This is clear for example from (6.40). If U_α is an optical potential, then $\chi_\alpha^{(+)}$ cannot contain any outgoing waves in channel β. Therefore $\chi_\alpha^{(+)}$ can have no rearrangement amplitude. Formally, the first term of (6.39) can be proven to vanish by rewriting $V_\beta - V_\alpha + U_\alpha = (H_\alpha + T_\alpha + U_\alpha) - (H_\beta + T_\beta)$. Hermiticity can be proven in the case in which U is an optical potential as in the last section of this chapter. Consequently the first term of (6.39) vanishes on the energy shell.

Inserting (6.48c) into (6.47) we reproduce, after some manipulation, the solution (6.21).

G. DISTORTED-WAVE GREEN'S FUNCTION

There is another Green's function of interest beside the free one (6.24). It is the distorted-wave or optical potential Green's function, which was explicitly constructed in the last chapter (5.13). We can deduce a result similar to (6.24b) for the distorted-wave Green's function. First, however, note a trivial difference in definition. Usually, the distorted-wave function χ concerns the relative motion only, as in Chapters 4 and 5 and throughout most of the book. However, in this chapter the auxiliary potential, for most of the development, was allowed to be more general—a sum of two-body interactions just as V, but simply constructed from different potentials. Consequently, in the Schrödinger equation, the relative and intrinsic degrees of freedom do not, in general, separate. Thus, contrast (6.36) and (4.2). In this and the last section, however, we have derived results explicitly for an optical potential. We continue to use the notation of this chapter. The relationship is obviously

$$\chi^{(\pm)}(\mathbf{r}_\alpha, \mathbf{A}, \mathbf{a}) = \chi^{(\pm)}(\mathbf{r}_\alpha)\Phi_\alpha(\mathbf{A}, \mathbf{a}), \qquad \text{(if } U = \text{optical potential)} \quad (6.49)$$

where the notation of this chapter (in coordinate space) is written on the left. The Green's function operator corresponding to the optical potential U_α is

$$G_\alpha^{(+)} = \frac{1}{E - H_\alpha - T_\alpha - U_\alpha + i\varepsilon}, \qquad G_\alpha^{(-)} = (G_\alpha^{(+)})^\dagger. \quad (6.50a)$$

In the convention of the last chapter, the nuclear Hamiltonian H_α would be absent. The Green's function operator of the last chapter is simply the matrix element of the present one:

$$(\Phi_{\alpha'}|G_\alpha^{(+)}|\Phi_\alpha) = \frac{1}{E_\alpha - T_\alpha - U_\alpha + i\varepsilon}\,\delta_{\alpha'\alpha}, \qquad (U = \text{optical potential}) \quad (6.50b)$$

This is diagonal because U_α is now supposed to be a one-body optical potential.

We can use the identity (6.20a) to extract the free Green's function from (6.50):

$$\frac{1}{E - H_\alpha - T_\alpha - U_\alpha + i\varepsilon}$$
$$= \frac{1}{E - H_\alpha - T_\alpha + i\varepsilon}\left(1 + U_\alpha \frac{1}{E - H_\alpha - T_\alpha - U_\alpha + i\varepsilon}\right).$$

Now take the representation $\mathbf{r}_\alpha\Phi_\alpha$ of $G_\alpha^{(+)}$ acting on an arbitrary vector $X\rangle$.

$$\langle \mathbf{r}_\alpha\Phi_{\alpha'} | G_\alpha^{(+)} X\rangle$$

$$= \sum_{\alpha''} \int d\mathbf{r}'_\alpha \langle \mathbf{r}_\alpha\Phi_{\alpha'} | G_\alpha^{0(+)} | r'_\alpha\Phi_{\alpha''}\rangle \left\langle \mathbf{r}'_\alpha\Phi_{\alpha''} \middle| \left(1 + U_\alpha \frac{1}{E - H_\alpha - T_\alpha - U_\alpha + i\varepsilon}\right) X \right\rangle$$

$$= \int d\mathbf{r}'_\alpha G_\alpha^{0(+)}(\mathbf{r}_\alpha, \mathbf{r}'_\alpha) \left\langle r'_\alpha\Phi_{\alpha'} \middle| \left(1 + U_\alpha \frac{1}{E - H_\alpha - T_\alpha - U_\alpha + i\varepsilon}\right) X \right\rangle$$

$$= \frac{-m_\alpha}{2\pi\hbar^2} \frac{\exp(ik'_\alpha r_\alpha)}{r} \left\langle \phi_{\alpha'} \middle| \left(1 + U_\alpha \frac{1}{E - H_\alpha - T_\alpha - U_\alpha + i\varepsilon}\right) X \right\rangle$$

$$= \frac{-m_\alpha}{2\pi\hbar^2} \frac{\exp(ik'_\alpha r_\alpha)}{r_\alpha} \langle \chi_{\alpha'}^{(-)} | X\rangle \qquad (U = \text{optical potential}). \tag{6.51}$$

Thus the appearance of an operator like (6.50a) permits us to extract the scattering amplitude in terms of distorted waves, just as the appearance of the free Green's function operator permitted its extraction according to (6.24b) for plane waves.

H. DISTORTED-WAVE BORN APPROXIMATION

We can use (6.38) in (6.46) to replace the exact wave function Ψ_α by solutions χ_α of the auxiliary problem defined by (6.36). In practice, U is chosen as an optical potential such as was introduced in Chapter 4, or as a generalized optical potential that can cause both elastic and inelastic scattering. Let us now restrict U to be an optical potential. Then the first term of (6.46) is diagonal and we have,

$$\mathscr{T}_{\beta\alpha} = \langle \chi_\alpha^{(-)} | U_\alpha | \phi_\alpha\rangle \, \delta_{\alpha\beta}$$

$$+ \langle \chi_\beta^{(-)} | (V_\beta - U_\beta) + (V_\beta - U_\beta) \frac{1}{E - H + i\varepsilon} (V_\alpha - U_\alpha) | \chi_\alpha^{(+)}\rangle. \tag{6.52a}$$

Similarly, (6.47) can be used in (6.39) with the result

$$\mathscr{T}_{\beta\alpha} = \langle \phi_\alpha | U_\alpha | \chi_\alpha^{(+)}\rangle \, \delta_{\alpha\beta}$$

$$+ \langle \chi_\beta^{(-)} | (V_\alpha - U_\alpha) + (V_\beta - U_\beta) \frac{1}{E - H + i\varepsilon} (V_\alpha - U_\alpha) | \chi_\alpha^{(+)}\rangle. \tag{6.52b}$$

We obtain the distorted-wave Born approximation by neglecting the second- and higher-order terms in Eqs. (6.52a) and (6.52b). Depending on whether we use (6.52a) or (6.52b), we obtain, for the first distorted-wave Born approximation,

$$\mathscr{T}^{\text{DWBA}}_{\beta\alpha(\text{post})} = \langle \chi_\alpha^{(-)} | U_\alpha | \phi_\alpha \rangle \, \delta_{\alpha\beta} + \langle \chi_\beta^{(-)} | V_\beta - U_\beta | \chi_\alpha^{(+)} \rangle, \tag{6.53a}$$

and

$$\mathscr{T}^{\text{DWBA}}_{\beta\alpha(\text{prior})} = \langle \phi_\alpha | U_\alpha | \chi_\alpha^{(+)} \rangle \, \delta_{\alpha\beta} + \langle \chi_\beta^{(-)} | V_\alpha - U_\alpha | \chi_\alpha^{(+)} \rangle. \tag{6.53b}$$

These two formulas are referred to as the post and prior forms of the distorted-wave Born approximation. They are equal to each other as will be shown. It is, therefore, a matter of convenience which formula is used. However, though equivalent as they stand, subsequent approximations can lead to different results.

The first terms in (6.53a) and (6.53b) were already proven equivalent in (6.45). To see that the second terms are also equal note that because

$$H = H_\alpha + T_\alpha + V_\alpha = H_\beta + T_\beta + V_\beta,$$

then

$$V_\beta - V_\alpha = H_\alpha + T_\alpha - (H_\beta + T_\beta)$$

Consequently the difference between the second terms of (6.53) can be rewritten,

$$\begin{aligned} &\chi_\beta^{(-)} | (V_\beta - U_\beta) - (V_\alpha - U_\alpha) | \chi_\alpha^{(+)} \rangle \\ &\quad = \langle \chi_\beta^{(-)} | (H_\alpha + T_\alpha + U_\alpha) - (H_\beta + T_\beta + U_\beta) | \chi_\alpha^{(+)} \rangle. \end{aligned}$$

By virtue of (6.36), the Hamiltonian in the first parenthesis on the right can be replaced by E. We shall prove in the last section of this chapter that $H_\beta + T_\beta$ is Hermitian in the above context. Therefore by virtue of (6.48a) the quantity in the second parenthesis can also be replaced by E. Therefore,

$$\langle \chi_\beta^{(-)} | V_\beta - U_\beta | \chi_\alpha^{(+)} \rangle = \langle \chi_\beta^{(-)} | V_\alpha - U_\alpha | \chi_\alpha^{(+)} \rangle. \tag{6.54}$$

Thus, as asserted (6-53a) and (6-53b) are equivalent.

Finally, we add a note on the relationship between $\chi^{(+)}$ and $\chi^{(-)}$, which was derived in Chapter 5. We can rederive it by noting the solutions (6.41b) and (6.44a). Thus, because ϕ_α has a plane wave part $\exp(i\mathbf{k}_\alpha \cdot \mathbf{r}_\alpha)$, we have

$$\chi^{(+)}(\mathbf{k}_\alpha, \mathbf{r}_\alpha) = \left(1 + \frac{1}{E - H_\alpha - T_\alpha - U_\alpha + i\varepsilon} U_\alpha\right) \exp(i\mathbf{k}_\alpha \cdot \mathbf{r}_\alpha) \Phi_\alpha,$$

and

$$\chi_\alpha^{(-)}(\mathbf{k}_\alpha, \mathbf{r}_\alpha) = \left(1 + \frac{1}{E - H_\alpha - T_\alpha - U_\alpha^\dagger - i\varepsilon} U_\alpha^\dagger\right) \exp(i\mathbf{k}_\alpha \cdot \mathbf{r}_\alpha) \Phi_\alpha,$$

from which it follows immediately that

$$\chi_\alpha^{(-)*}(\mathbf{k}_\alpha, \mathbf{r}_\alpha) = \chi_\alpha^{(+)}(-\mathbf{k}_\alpha, \mathbf{r}_\alpha). \tag{6.55}$$

I. SECOND DISTORTED-WAVE BORN APPROXIMATION

The expressions (6.52) recast the problem posed in (6.39) or (6.46). In the one case the exact wave function appears in the expression for the amplitude. In the other, the exact Green's function appears. It can be expanded in a variety of ways, by using the identity (6.20a). Clearly an expansion in terms of an optical potential Green's function is interesting, since the optical model is solvable. In (5.13) we have an explicit construction of such a Green's function. Let us denote by γ any channel, which could be α, or β, or some other one. Use the identity (6.20a) to write the exact Green's function as

$$\frac{1}{E - H + i\varepsilon} = \frac{1}{E - H_\gamma - T_\gamma - U_\gamma + i\varepsilon}\left(1 + (V_\gamma - U_\gamma)\frac{1}{E - H + i\varepsilon}\right).$$

Let us denote the exact Green's function by

$$G^{(+)} = \frac{1}{E - H + i\varepsilon} \tag{6.56}$$

Then the above identity can be written

$$G^{(+)} = G_\gamma^{(+)}(1 + (V_\gamma - U_\gamma)G^{(+)}) \tag{6.57}$$

and can be iterated in a number of ways,

$$\begin{aligned} G^{(+)} &= G_\gamma^{(+)} + G_\gamma^{(+)}(V_\gamma - U_\gamma)G_\gamma^{(+)} + \cdots \\ &= G_\gamma^{(+)} + G_\gamma^{(+)}(V_\gamma - U_\gamma)G_\delta^{(+)} + \cdots \end{aligned} \tag{6.58}$$

etc., where γ and δ are any two-body channels, which could be chosen as α or β or as any other channels. If summed to infinity, all such expansions are identical. However, this is not possible in practice. The optimum choice must therefore depend on the physics of the situation at hand.

We obtain series expansions of the transition amplitude when such expansions for $G^{(+)}$ are used in (6.52). If we terminate the series for the amplitude we obtain

$$\mathscr{T}_{\beta\alpha}^{(2)} = \mathscr{T}_{\beta\alpha}^{(1)} + \mathscr{T}_{\beta\gamma\alpha}^{(2)} \qquad (\text{any } \gamma) \tag{6.59a}$$

$$\mathscr{T}_{\beta\alpha}^{(1)} = \langle \chi_\beta^{(-)} | (V_\beta - U_\beta) | \chi_\alpha^{(+)} \rangle \tag{6.59b}$$

$$\mathscr{T}_{\beta\gamma\alpha}^{(2)} = \langle \chi_\beta^{(-)} | (V_\beta - U_\beta) G_\gamma (V_\alpha - U_\alpha) | \chi_\alpha^{(+)} \rangle \tag{6.59c}$$

$\mathscr{T}_{\beta\alpha}^{(1)}$ is just the DWBA amplitude. I call it by this new name to systematize the notation of this section. It obviously describes the direct transition from initial to final state. The second term represents two-step processes passing through the intermediate channels of partition γ. Interestingly the two terms can be combined into a single one. For $\gamma \neq \alpha$, notice

$$(V_\alpha - U_\alpha - V_\gamma + U_\gamma)\chi_\alpha = (H_\gamma + T_\gamma + U_\gamma - H_\alpha - T_\alpha - U_\alpha)\chi_\alpha$$
$$= (H_\gamma + T_\gamma + U_\gamma - E)\chi_\alpha.$$

Thereforc

$$G_\gamma^{(+)}(V_\alpha - U_\alpha)\chi_\alpha^{(+)} = G_\gamma^{(+)}(V_\gamma - U_\gamma)\chi_\alpha^{(+)} - \chi_\alpha^{(+)}.$$

Hence,

$$\mathscr{T}_{\beta\gamma\alpha}^{(2)} = -\mathscr{T}_{\beta\alpha}^{(1)} + \langle\chi_\beta^{(-)}|(V_\beta - U_\beta)G_\gamma^{(+)}(V_\gamma - U_\gamma)|\chi_\alpha^{(+)}\rangle. \tag{6.60}$$

Using this result in (6.59a) yields

$$\mathscr{T}_{\beta\alpha}^{(2)} = \langle\chi_\beta^{(-)}|(V_\beta - U_\beta)G_\gamma^{(+)}(V_\gamma - U_\gamma)|\chi_\alpha^{(+)}\rangle \tag{6.61}$$

This second order amplitude is in fact the *sum* of the first- and second-order processes described above and was obtained first by Kunz and Rost (1973). In fact, the result can be generalized to any order, as Nagarajan (pers. com. 1982) has shown. Thus, for example,

$$\mathscr{T}_{\beta\alpha}^{(3)} \equiv \mathscr{T}_{\beta\alpha}^{(1)} + \mathscr{T}_{\beta\gamma\alpha}^{(2)} + \mathscr{T}_{\beta\gamma\delta\alpha}^{(3)}$$
$$= \langle\chi_\beta^{(-)}|(V_\beta - U_\beta)G_\gamma^{(+)}(V_\gamma - U_\gamma)G_\delta^{(+)}(V_\delta - U_\delta)|\chi_\alpha^{(+)}\rangle \tag{6.62}$$

These formulas are very compact in appearance, but before they can be used in an actual calculation, the full implication of the notation must be realized. Generally a calculation is performed in coordinate space. It will be useful at this point to recall the material of Chapter 5 Section *H*, especially the discussion of relative coordinates. To give more precise meaning to (6.59c), as an example, we must use the completeness relation for the channel wave functions Φ_γ on both sides of the Green's function G_γ. Note that it is diagonal (6.50b). The coordinate space representation of (6.50b) was obtained in (5.13). Thus we can write

$$\mathscr{T}_{\beta\gamma\alpha}^{(2)} = -(2m_\gamma/\hbar^2)\sum_\gamma \int d\mathbf{r}_\alpha\, d\mathbf{r}_\beta\, d\mathbf{r}_\gamma\, d\mathbf{r}'_\gamma\, \chi_\beta^{*(-)}(\mathbf{k}_\beta, \mathbf{r}_\beta)$$
$$\times (\Phi_\beta|V_\beta - U_\beta|\Phi_\gamma)_{\mathbf{r}_\beta \mathbf{r}_\gamma} f_{\ell_\gamma}(r_{\gamma<}) h_{\ell_\gamma}^{(+)}(r_{\gamma>})/(k_\gamma r_{\gamma>} r_{\gamma<})$$
$$\times Y_{\ell_\gamma}(\hat{\mathbf{r}}_\gamma) \cdot Y_{\ell_\gamma}(\hat{\mathbf{r}}'_\gamma)(\Phi_\gamma|V_\alpha - U_\alpha|\Phi_\alpha)_{\mathbf{r}'_\gamma \mathbf{r}_\alpha} \chi_\alpha^{(+)}(\mathbf{k}_\alpha, \mathbf{r}_\alpha)$$

Here $r_{\gamma<}$ and $r_{\gamma>}$ mean the lesser and greater of the radial coordinates r_γ and r'_γ, The coordinate dependences of the nuclear matrix elements are indicated as subscripts, and we have reverted to the notation of Chapter 5 as concerns the product wave function (6.49). The sum on γ is over the nuclear states Φ_γ and relative angular momentum ℓ_γ.

J. MULTIPLE-SCATTERING SERIES OF WATSON

A very elegant and powerful formalism was developed by K. M. Watson (1953), which achieves a formal solution to the general scattering problem. In effect, what Watson succeeded in doing was rearranging the terms in the series (6.35) so that the first term in his series sums an important set of the terms in (6.35) to all orders. Such a rearrangement plays a central role in Breuckner's many-body theory of the nucleus. It will be important to us in later chapters, where the actual two-nucleon interaction is related to nucleon–nucleus scattering, to have an acquaintance with Watson's theory.

Instead of deriving Watson's solution to the integral equation for $\Psi^{(+)}$ and then deriving the multiple scattering series for the T operator from it, we derive the latter directly, so as to exhibit the most important result for our purpose first. Later we shall prove Watson's solution for the state vector.

Recall the integral equation for the T operator for the channel α is (6.34):

$$T = V + V\frac{1}{E - H_{\mathrm{A}} - T_{\alpha} + i\varepsilon} T. \tag{6.63}$$

Because we consider only a single channel, as is appropriate for elastic and inelastic scattering, we do not label V. Recall that the channel interaction V, in case the projectile is simply the nucleon i, is

$$V = \sum_{j \in \mathrm{A}} V_{ij}. \tag{6.64a}$$

We usually discuss the theory in the context of nucleon–nucleus scattering, in which case a is just the projectile nucleon. This specialization is not necessary. In either case, we seek to express T in terms of elementary quantities involving the projectile and individual target nucleons, of which one is given the label j. Call this interaction V_j, where

$$V_j = \begin{cases} V_{ij}, & \text{if a is a nucleon} \\ \sum_{i \in \mathrm{a}} V_{ij}, & \text{otherwise,} \end{cases} \tag{6.65}$$

Consequently, we write (6.64a) as

$$V = \sum_{j=1}^{\mathrm{A}} V_j. \tag{6.64b}$$

Remember that T is an operator that acts in the space of A + a nucleons. We can write (6.63) as

$$T = V(1 + g^{(+)}T) = \sum_j V_j(1 + g^{(+)}T), \tag{6.66}$$

where we define, for convenience, the Green's function

$$g^{(+)} = \frac{1}{E - H_0 + i\varepsilon}, \qquad H_0 = H_A + T_\alpha. \tag{6.67}$$

(This is the same as (6.24a). Here we simplify the notation.)

The t operator corresponding to the scattering of two free nucleons labeled i and j is given by (6.63) with H_A absent:

$$t_{ij} = V_{ij} + V_{ij}\frac{1}{E - T_\alpha + i\varepsilon}t_{ij}, \tag{6.68}$$

where T_α is the relative kinetic energy operator.‡ The corresponding operator if the nucleon j is bound in a nucleus A is

$$\tau_{ij} = V_{ij} + V_{ij}\frac{1}{E - H_A - T_\alpha + i\varepsilon}\tau_{ij}. \tag{6.69a}$$

In the notation of (6.65) and (6.67)

$$\tau_j = V_j + V_j g^{(+)}\tau_j = V_j(1 + g^{(+)}\tau_j). \tag{6.69b}$$

Now define a two-body operator ω_j by the identity

$$V_j(1 + g^{(+)}T) \equiv \tau_j\omega_j. \tag{6.70}$$

This is motivated by the fact that the left side is one term in the sum that appears in (6.66). Thus (6.66) can be rewritten as

$$T = \sum_{j=1}^{A} \tau_j\omega_j. \tag{6.71}$$

To find out what ω_j is, substitute (6.69b) into the right side of (6.70) to obtain

$$1 + g^{(+)}T = (1 + g^{(+)}\tau_j)\omega_j. \tag{6.72}$$

Now rearrange this as

$$\omega_j = 1 + g^{(+)}T - g^{(+)}\tau_j\omega_j. \tag{6.73}$$

In view of (6.71), we now have

$$\omega_j = 1 + g^{(+)}\sum_{k\neq j}\tau_k\omega_k. \tag{6.74}$$

This can be iterated to obtain

$$\omega_j = 1 + g^{(+)}\sum_{k\neq j}\tau_k + g^{(+)}\sum_{k\neq j}\tau_k g^{(+)}\sum_{l\neq k}\tau_l + \cdots. \tag{6.75}$$

‡ No confusion need arise by the use of T as a T operator and T as a kinetic energy because the latter will occur, as in (6.68), in the denominator.

When this is inserted into (6.71), the very important result for T is obtained:

$$T = \sum_{j=1}^{A} \tau_j + \sum_{j,\,k \neq j}^{A} \tau_j g^{(+)} \tau_k + \sum_{j,\,k \neq j,\,l \neq k}^{A} \tau_j g^{(+)} \tau_k g^{(+)} \tau_l + \cdots, \quad (6.76)$$

which expresses the many-body operator in terms of a sum of two-body operators. This is Watson's multiple scattering series, which is of fundamental importance in nuclear physics. We give a preview of why this is so. First, it is a series in terms of τ_j not V_j. Thus at each order in the series (6.76), the full scattering sequence between projectile and target nucleon j enters, represented by the series

$$\tau_j = V_j + V_j g^{(+)} V_j + V_j g^{(+)} V_j g^{(+)} V_j + \cdots. \quad (6.77)$$

This follows by iterating (6.69b). Thus (6.76) is a remarkable reordering of the series that one obtains by iterating (6.63), namely,

$$T = \sum_{j=1}^{A} V_j + \sum_{j,k} V_j g^{(+)} V_k + \sum_{jkl} V_j g^{(+)} V_k g^{(+)} V_l + \cdots. \quad (6.78)$$

Both (6.76) and (6.78) are exact. However, for a singular interaction such as the nucleon–nucleon interaction (6.78) is of no use because in the matrix element for the amplitude (6.27) it diverges term by term. In contrast, (6.76) is written in terms of the two-body τ operator. The operator is nonsingular, even though V_j is singular. There are several ways of understanding this. One is simply by noting that the scattering amplitude, which is perfectly well defined even for scattering by a hard-core potential, is given by the plane-wave matrix element of τ.‡ Another way of making the same point is by manipulating (6.69b) to the form

$$\tau_j = (1 - V_j g^{(+)})^{-1} V_j. \quad (6.79)$$

The ratio of numerator and denominator remains finite even in the region of the hard core! Thus (6.76) is a series development in terms of a well-behaved operator. Second, Watson's series represents the successive scattering of the projectile from different target nucleons. The first term represents the complete scattering from any single nucleon in the target, summed over all target nucleons. The second term represents the complete scattering of the incident nucleon by any two different bound nucleons: first by nucleon k, its propagation in the nucleus until it scatters from nucleon j and then emerges, and such events summed over all pairs. The third term represents the complete scattering from three different bound nucleons, and so on.

Thus Watson's series is a very powerful reordering of the T matrix series, which, at the first term, sums a subset of the infinite series of (6.78) and,

‡ The Schrödinger equation for scattering by an infinite repulsive potential is easy to solve.

moreover, does it in such a way as to achieve a nonsingular operator at each term. Such a reordering plays a central role in Breuckner's many-body theory of the nucleus.

At this point we mention a basic assumption that underlies the development of scattering theory in the preceding chapters. We know that the vacuum nucleon–nucleon interaction is singular, possessing strong (if not infinite) repulsion at short distance. It is certainly not possible to construct a meaningful perturbation theory based on such an interaction. The matrix elements, taken with respect to uncorrelated wave functions, would be infinite in the case of a hard-core potential and otherwise would lead to divergences for a strongly repulsive interaction. Therefore, we have assumed that there exists an effective interaction that is well behaved and for which the perturbative approach can converge. The DWBA is of course based on a power series in the interaction. In Chapter 8, using Watson's theory, we shall prove that the concept of an effective interaction has more than phenomenological meaning. It exists as a formal theoretical construct, and it can be developed, as in Watson's series, in terms of the nonsingular two-body τ matrix.

The preceding results are the main ones as far as we are concerned. However, for completeness, let us derive Watson's solution for the state vector, which satisfies

$$\Psi_\alpha^{(+)} = \phi_\alpha + \frac{1}{E - H_A - T_\alpha + i\varepsilon} V_\alpha \Psi_\alpha^{(+)}. \tag{6.80}$$

Earlier we derived the formal solution

$$\Psi_\alpha^{(+)} = \phi_\alpha + \frac{1}{E - H + i\varepsilon} V_\alpha \phi_\alpha. \tag{6.81}$$

Using the identity (6.20a) it is trivial to rewrite this in terms of the T operator as

$$\Psi_\alpha^{(+)} = \phi_\alpha + \frac{1}{E - H_0 - T_\alpha + i\varepsilon} T\phi_\alpha, \tag{6.82a}$$

or in the preceding abbreviated notation as

$$\Psi_\alpha = (1 + g^{(+)}T)\phi_\alpha. \tag{6.82b}$$

Use (6.72) in this to obtain

$$\Psi_\alpha = (1 + g^{(+)}\tau_j)\omega_j\phi_\alpha = (1 + g^{(+)}\tau_j)\psi_j^{(+)}, \tag{6.83}$$

where we define

$$\begin{aligned} \psi_j^{(+)} &= \omega_j\phi_\alpha = \phi_\alpha + g^{(+)} \sum_{k \neq j} \tau_k \omega_k \phi_\alpha \\ &= \phi_\alpha + g^{(+)} \sum_{k \neq j} \tau_k \psi_k^{(+)}. \end{aligned} \tag{6.84}$$

We used (6.74) to rewrite ω_j in the derivation. From (6.70), (6.83), and (6.84), note the relation

$$V_j \Psi_\alpha^{(+)} = \tau_j \psi_j^{(+)}, \tag{6.85}$$

which is analogous to (6.33). Consequently,

$$V_\alpha \Psi_\alpha^{(+)} = \sum_j V_j \psi_\alpha^{(+)} = \sum_j \tau_j \psi_j^{(+)}.$$

Insert this into the right side of (6.80) to obtain

$$\Psi_\alpha^{(+)} = \phi_\alpha + g^{(+)} \sum_{j=1}^{A} \tau_j \psi_j^{(+)}. \tag{6.86}$$

Equations (6.84) and (6.86) are Watson's solution to the integral equation of $\Psi_\alpha^{(+)}$, written in terms of the two-body τ_k. He used them to derive the multiple scattering series, which we obtained directly.

K. GREEN'S THEOREM AND THE HERMITIAN H

The proof that the Laplacian operator in H is Hermitian involves the use of Green's theorem. In scattering theory, however, the very existence of a nonvanishing transition amplitude depends on the non-Hermiticity of H with respect to its matrix element between a plane-wave state and a scattering state. This fact, is worth emphasizing. For example, consider

$$\langle \phi_\beta | V_\alpha - V_\beta | \phi_\alpha \rangle = \langle \phi_\beta | (H_\beta + T_\beta) - (H_\alpha + T_\alpha) | \phi_\alpha \rangle. \tag{6.87}$$

Here ϕ_α and ϕ_β have the same boundary condition; they are both plane waves in relative motion. The Hamiltonian $H_\beta + T_\beta$ is Hermitian, and the preceding result is zero on the energy shell. This was the proof of (6.29).

Now consider, by contrast, the general expression for the transition amplitude, and rewrite the potential in terms of the Hamiltonians:

$$\langle \phi_\beta | V_\beta | \Psi_\alpha^{(+)} \rangle = \langle \phi_\beta | H - (H_\beta + T_\beta) | \Psi_\alpha^{(+)} \rangle. \tag{6.88}$$

If we assume that $H_\beta + T_\beta$ is Hermitian here, we obtain zero for the amplitude. Either our theory is nonsense or $H_\beta + T_\beta$ is not Hermitian in this context. The latter is of course true. This Hamiltonian is non-Hermitian if

$$\langle (H_\beta + T_\beta)\phi_\beta | \Psi_\alpha^{(+)} \rangle - \langle \phi_\beta | H_\beta + T_\beta | \Psi_\alpha^{(+)} \rangle \neq 0. \tag{6.89}$$

We now prove explicitly that the difference is finite. The angle brackets of course represent integrals over the $3(A + a - 1)$ relative coordinates $\xi_1 \cdots \xi_{A+a-1}$. The potential terms drop out of the difference if they are

Hermitian, which we assume to be the case, leaving only the kinetic terms. The difference is therefore

$$D = \sum_{\gamma} \frac{\hbar^2}{2m_\gamma} \int [\phi_\beta^*(\nabla_\gamma^2 \Psi_\alpha^{(+)}) - (\nabla_\gamma^2 \phi_\beta^*)\Psi_\alpha^{(+)}]\, d\xi_1 \cdots d\xi_{\mathrm{A+a-1}}, \tag{6.90}$$

where γ is summed over all the relative coordinates and m_γ is the reduced mass for this coordinate. Let the region of integration by initially bounded by a sphere in each coordinate. Later we let the boundary go to infinity. Consider a typical term in the sum. The integral on the γ relative coordinate can be converted into a surface integral on the boundary. The surface element is $R_\gamma^2\, d\Omega_\gamma$, where R_γ is the radius of the sphere and Ω_γ the solid angle. There remains of course the volume integral on each of the other coordinates. Denote this by dV_γ. Then the γth term in the sum is

$$D_\gamma = \frac{\hbar^2}{2m_\gamma} \int \left(\phi_\beta^* \frac{d\Psi_\alpha^{(+)}}{dr_\gamma} - \frac{d\phi_\beta^*}{dr_\gamma} \Psi_\alpha^{(+)} \right)_{r_\gamma = R_\gamma} R_\gamma^2\, d\Omega_\gamma\, dV_\gamma. \tag{6.91}$$

Now recall that

$$\begin{aligned} \phi_\beta &= \Phi_\beta \exp(i\mathbf{k}_\beta \cdot \mathbf{r}_\beta), \\ \Psi_\alpha^{(+)} &\xrightarrow[r_\gamma \to \infty]{} \Phi_\alpha \exp(i\mathbf{k}_\alpha \cdot \mathbf{r}_\alpha)\, \delta_{\gamma\alpha} + \sum_{\gamma'} \Phi_{\gamma'} f_{\gamma'\alpha}(\hat{\mathbf{r}}_\gamma) \frac{\exp(ik'_\gamma r_\gamma)}{r_\gamma}. \end{aligned} \tag{6.92}$$

The second expression is the standard asymptotic form of a scattering state with a plane-wave state in channel α and an outgoing wave in open channels, which we saw in (3.34). The plane-wave part contributes nothing. This follows because it is proportional to $(\Phi_\beta | \Phi_\alpha)$ evaluated on the surface $r_\gamma = R_\gamma$. As $R_\gamma \to \infty$, $\Phi_\beta \to 0$ because Φ_β is bound in all relative coordinates except r_β. Therefore, any contribution must come from the outgoing wave parts of (6.92). In (6.91) these involve the overlap $(\Phi_\beta | \Phi_{\gamma'})$, which again vanishes in the limit $R_\gamma \to \infty$ unless $\gamma \equiv \beta$. Therefore, all terms D_γ vanish except D_β, which is finite if β is an open channel, that is, if $f_{\beta\alpha} \neq 0$. Therefore, (6.90) is

$$\begin{aligned} D = \frac{\hbar^2}{2m_\beta} \int \Bigg[& \exp(-i\mathbf{k}_\beta \cdot \mathbf{r}_\beta) f_{\beta\alpha} \frac{d}{dr_\beta} \left(\frac{\exp(ik_\beta r_\beta)}{r_\beta} \right) \\ & - \frac{d}{dr_\beta} [\exp(-i\mathbf{k}_\beta \cdot \mathbf{r}_\beta] f_{\beta\alpha} \frac{\exp(ik_\beta r_\beta)}{r_\beta} \Bigg]_{r_\beta = R_\beta} R_\beta^2\, d\Omega_\beta. \end{aligned} \tag{6.93}$$

We have already anticipated $R_\beta \to \infty$ by setting

$$(\Phi_\beta | \Phi_{\beta'}) = \int \Phi_\beta^* \Phi_{\beta'}\, dV_\beta = \delta_{\beta\beta'}.$$

To take the radial derivative of the plane wave it is useful to have the following asymptotic form:

$$\exp(i\mathbf{k}\cdot\mathbf{r}) \underset{r\to\infty}{\longrightarrow} \frac{2\pi}{ikr}[\delta(\Omega_r-\Omega_k)\exp(ikr)-\delta(\Omega_r+\Omega_k)\exp(-ikr)]. \quad (6.94)$$

The δ functions are two dimensional over the polar coordinates. This limit can be derived by examining the spherical harmonic expansion of $\exp(i\mathbf{k}\cdot\mathbf{r})$ and using the closure property of the spherical harmonics. Carrying out the derivatives now and letting $R_\beta\to\infty$, we find

$$D = -\frac{2\pi\hbar^2}{m_\beta} f_{\beta\alpha}(\hat{\mathbf{k}}_\beta). \quad (6.95)$$

Thus

$$\begin{aligned}\langle\phi_\beta|V_\beta|\Psi_\alpha^{(+)}\rangle &= \langle\phi_\beta|H-(H_\beta+T_\beta)|\Psi_\alpha^{(+)}\rangle\\ &= E\langle\phi_\beta|\Psi_\alpha^{(+)}\rangle - E\langle\phi_\beta|\Psi_\alpha^{(+)}\rangle - \frac{2\pi\hbar^2}{m_\beta} f_{\beta\alpha}(\hat{\mathbf{k}}_\beta), \end{aligned} \quad (6.96)$$

which proves that $\langle\phi_\beta|V_\beta|\Psi_\alpha^{(+)}\rangle$ does not vanish if the channel β is open and is proportional, as expected, to the scattering amplitude in this channel.

We can prove (6.54) by Green's theorem under the stated condition that U_α and U_β are optical potentials that cannot cause rearrangement scattering. The key point is that they do not possess any common channel in such a case. First, note that

$$V_\beta - V_\alpha = H_\alpha + T_\alpha - (H_\beta + T_\beta). \quad (6.97)$$

To determine if $H_\beta + T_\beta$ is Hermitian examine

$$\begin{aligned}&\langle(H_\beta+T_\beta)\chi_\beta^{(-)}|\chi_\beta^{(+)}\rangle - \langle\chi_\beta^{(-)}|H_\beta+T_\beta|\chi_\alpha^{(+)}\rangle\\ &= \lim_{R_\gamma\to\infty}\sum_\gamma \frac{\hbar^2}{2m_\gamma}\int\left(\chi_\beta^{(-)*}\frac{d\chi_\alpha^{(+)}}{dr_\gamma} - \frac{d\chi_\beta^{(-)*}}{dr_\gamma}\chi_\alpha^{(+)}\right)_{r_\gamma=R_\gamma} R_\gamma^2\, d\Omega_\gamma\, dV_\gamma. \end{aligned} \quad (6.98)$$

Because there are no common channels in χ_β and χ_α and because,

$$\int(\Phi_\beta|\Phi_\alpha)_{r_\gamma=R_\gamma}\, dV_\gamma \underset{R_\gamma\to\infty}{\longrightarrow} 0, \quad (6.99)$$

the difference (6.98) vanishes. As a consequence $H_\beta + T_\beta$ is Hermitian in this context. Therefore

$$\begin{aligned}&\langle\chi_\beta^{(-)}|(V_\beta-U_\beta)-(V_\alpha-U_\alpha)|\chi_\alpha^{(+)}\rangle\\ &= \langle\chi_\beta^{(-)}|(H_\alpha+T_\alpha+U_\alpha)-(H_\beta+T_\beta+U_\beta)|\chi_\alpha^{(+)}\rangle\\ &= E\langle\chi_\beta^{(-)}|\chi_\alpha^{(+)}\rangle - \langle(H_\beta+T_\beta+U_\beta^\dagger)\chi_\beta^{(-)}|\chi_\alpha^{(+)}\rangle\\ &= 0 \end{aligned} \quad (6.100)$$

The equality to zero holds on the energy shell. This proves (6.54).

Notes to Chapter 6

Representation theory in quantum mechanics is due to P. A. M. Dirac and a complete account of it can be found in his book "The Principles of Quantum Mechanics" (1947). For a briefer account see Rodberg and Thaler (1967). For a rigorous treatment of scattering theory using wave packets, see Chapters 3 and 4 of Goldberger and Watson (1964) The original derivation of the distorted wave Born approximation was performed in the famous book "The Theory of Atomic Collisions" by Mott and Massey (1933).

Chapter 7

Calculation of the DWBA Amplitude

A. INTRODUCTION

In the preceding chapters, we derived an expression for the transition amplitude for a general rearrangement reaction involving two channels. This involves the complete (and therefore unattainable) exact wave function. The foundations for a very useful approximate scheme were then laid, which involved first the optical potential that describes the relative motion in the elastic channel. The direct reactions are regarded as a perturbation on this. The resulting approximate transition amplitude, involving as it does the distortion (from plane waves) of the wave function of relative motion in incident and outgoing channels, is known as the distorted-wave Born approximation amplitude (DWBA). It has proved enormously useful in the interpretation of experimental data. Whereas closed-form expressions for the cross sections can be obtained in the plane-wave approximation, this is not the case for the DWBA amplitude. Its calculation requires a large-scale computational effort that is practicable only on fast electronic computers. Such computations have been possible for many years. The general strategy of the calculation is common to all direct reactions. The elastic scattering data in the initial and final partitions at or near the energy of the direct reaction are needed for the purpose of determining the optical-potential parameters. The distorted waves are the solution to these two optical-model problems. This describes the wave functions of relative motion. The other ingredient of the transition amplitude is a nuclear overlap function whose form depends on the particular reaction under consideration. It was denoted by $(\Phi_\beta|V_\beta - U_\beta|\Phi_\alpha)$ and involves an integration over all intrinsic nuclear coordinates and spins. It remains a function of the two channel coordinates $\mathbf{r}_\alpha$ and $\mathbf{r}_\beta$. The DWBA amplitude is then the integral of this nuclear overlap function with the distorted waves.

Various techniques are involved in the evaluation of the nuclear overlap function, depending on the nature of the reaction under consideration and on the nuclear description. The object of this chapter is to make explicit the ingredients of the distorted-wave amplitude for a few reactions, which at this stage are chosen to be fairly simple.

B. THE (d, p) STRIPPING REACTION

The stripping reaction and its inverse, the pickup reaction, are among the most important, especially for nuclear spectroscopy. The spins and parities of an enormous number of nuclear states have been determined through the analysis of these reactions. They were instrumental in confirming the shell-model description of nuclei, and they provide a direct measure of the single-particle character of nuclear states, as will become evident.

The existence of a direct reaction mechanism, which is distinct from compound nucleus formation and decay, was first recognized by Butler (1950, 1951), based on his interpretation of data on the (d, p) reaction. The early estimates of the distortion effects made by Horowitz and Messiah (1953) and by Butler and Austern (1954) were soon followed by Tobocman's (1954) complete formulation and extensive numerical calculations.

The general form of the DWBA amplitude to which we must now give explicit meaning is

$$\mathscr{T}_{\alpha\beta} = W\langle\chi_\beta^{(-)}\Phi_\beta|V_\beta - U_\beta|\Phi_\alpha\chi_\alpha^{(+)}\rangle, \tag{7.1}$$

where W is the statistical weight of the direct terms (see Chapter 5). The reaction is denoted by

$$\mathrm{d} + \mathrm{A} \rightarrow \mathrm{p} + (\mathrm{A} + 1) \equiv \mathrm{p} + \mathrm{B}. \tag{7.2}$$

From the definition (3.6) we have

$$V_\beta = V_{\mathrm{pn}} + \sum_{i=1}^{\mathrm{A}} V_{\mathrm{p}i} \equiv V_{\mathrm{pn}} + V_{\mathrm{pA}}. \tag{7.3}$$

This full interaction is seldom if ever evaluated. One may argue that V_{pA} should have the general effect that is described by the one-body optical potential U_{p} and that the difference

$$\langle V_{\mathrm{pA}} - U_{\mathrm{p}} \rangle \simeq 0 \tag{7.4}$$

may therefore be neglected. From the meaning ascribed to Φ_α and Φ_β, they take the particular forms

$$\Phi_\alpha = \Phi_{J_{\mathrm{A}}}^{M_{\mathrm{A}}}(\mathbf{A})\phi_{\mathrm{d}}(r)X_1^{\mu_{\mathrm{d}}}(\sigma_{\mathrm{n}}, \sigma_{\mathrm{p}}), \tag{7.5}$$

$$\Phi_\beta = \Phi_{J_{\mathrm{B}}}^{M_{\mathrm{B}}}(\mathbf{A}, \mathbf{r}_{\mathrm{n}})X_{1/2}^{\mu_{\mathrm{p}}}(\sigma_{\mathrm{p}}), \tag{7.6}$$

where $\mathbf{A}$ refers to the nucleon coordinates and spins, X denotes spin wave functions, σ_{n} and σ_{p} denote neutron and proton spin coordinates, and ϕ_{d} denotes the deuteron radial function (assumed to be a triplet S-state). Finally, $\mathbf{r} = \mathbf{r}_{\mathrm{n}} - \mathbf{r}_{\mathrm{p}}$ is the relative coordinate on which the deuteron wave function depends.

With approximation (7.4), we recognize that the remaining interaction does not depend on the $\mathbf{A}$ coordinates, so that we can consider separately the integral

$$\int \Phi_{J_{\mathrm{B}}}^{M_{\mathrm{B}}^*}(\mathbf{A}, \mathbf{r}_{\mathrm{n}})\Phi_{J_{\mathrm{A}}}^{M_{\mathrm{A}}}(\mathbf{A})\, d\mathbf{A} \tag{7.7}$$

It is customary to introduce a fractional parentage expansion of the nuclear wave function, which for the present purpose would be a single particle parentage, namely, (see Macfarlane and French, 1960; Glendenning, 1963).

$$\Phi_{J_{\mathrm{B}}}^{M_{\mathrm{B}}}(\mathbf{A}, \mathbf{r}_{\mathrm{n}}) = \sum_{\mathrm{A}'jl} \beta_{j\ell}(\mathrm{B}, \mathrm{A}')\mathscr{A}[\Phi_{J_{\mathrm{A}'}}(\mathbf{A})\phi_{n\ell j}(\mathbf{r}_{\mathrm{n}})]_{J_{\mathrm{B}}}^{M_{\mathrm{B}}}, \tag{7.8}$$

where $\mathscr{A}$ is an antisymmetrization operator [see (5.28) for its form]. The β are generalized coefficients of fractional parentage, and their value depends on the detailed structure of the nuclear wave function. In principle there would be a sum on the radial quantum number n in (7.8) and (7.13). However, the shell model tells us which radial state is in question. As a particularly trivial example, suppose B is a closed-shell nucleus plus one neutron. In that case,

$$\Phi_{J_{\mathrm{B}}}^{M_{\mathrm{B}}}(\mathbf{A}, \mathbf{r}_{\mathrm{n}}) = \mathscr{A}\{\Phi_0^0(\mathbf{A})\phi_{n\ell j}^m(\mathbf{r}_{\mathrm{n}})\}\delta_{jJ_{\mathrm{B}}}\delta_{mM_{\mathrm{B}}}. \tag{7.9}$$

In expression (7.8) the square bracket denotes vector coupling:

$$[\Phi_{J'}\phi_j]_J^M \equiv \sum_{M'm} C_{M'mM}^{J'\,j\,J}\Phi_{J'}^{M'}\phi_j^m, \tag{7.10}$$

and $\phi_{n\ell j}$ denotes a spin-orbit function for a shell-model state:

$$\phi_{n\ell j}^m(\mathbf{r}_{\mathrm{n}}, \boldsymbol{\sigma}_{\mathrm{n}}) = [\phi_{n\ell}(\mathbf{r}_{\mathrm{n}})X_{1/2}(\boldsymbol{\sigma}_{\mathrm{n}})]_j^m, \tag{7.11}$$

$$\phi_{n\ell}^{m_\ell}(\mathbf{r}_{\mathrm{n}}) = u_{n\ell}(r_{\mathrm{n}})Y_\ell^{m_\ell}(\hat{\mathbf{r}}_{\mathrm{n}}) \tag{7.12}$$

Here $u_{n\ell}$ is the radial wave function. Now we have from (7.8)

$$\binom{A+1}{1}^{1/2} \int \Phi_{J_B}^{M_B^*}(\mathbf{A}, \mathbf{r}_n)\Phi_{J_A}^{M_A}(\mathbf{A})\, d\mathbf{A}$$
$$= \sum_{j\ell} \beta_{\ell j}(B, A) C_{M_A m_j M_B}^{J_A\ j\ J_B} \phi_{n\ell j}^{m_j^*}(\mathbf{r}_n \sigma_n). \tag{7.13}$$

Note the very important property that the parentage amplitudes accessible to study by the direct single-step transition are those for which the core nucleons A are unchanged by the reaction. In general there are of course many other amplitudes in the wave function (7.8).

The square of the amplitude β is often called the spectroscopic factor (MacFarlane and French, 1960; Glendenning, 1963)

$$\mathscr{S}_{\ell j}^{1/2} = \beta_{\ell j} = \int d\mathbf{A}\, d\mathbf{r}_n\, \Phi_{J_B}^{M_B*}(\mathbf{A}+\mathbf{1}) \mathscr{A}[\Phi_{J_A}(A)\phi_{n\ell j}(\mathbf{r}_n)]_{J_B}^{M_B}$$
$$= \binom{A+1}{1}^{1/2} \int d\mathbf{A}\, dr_n\, \Phi_{J_B}^{M_B*}(\mathbf{A}+\mathbf{1})[\Phi_{J_A}(A)\phi_{n\ell j}(\mathbf{r}_n)]_{J_B}^{M_B}. \tag{7.14}$$

We shall discuss this important quantity later in the chapter.

Recall that the statistical weight of the direct term is to be understood in the sense as described in Chapter 5. Thus, if the neutrons and protons are distinguished, as in this example, then

$$\binom{A+x}{x} \to \binom{N+\nu}{\nu} \binom{Z+\pi}{\pi}, \tag{7.15}$$

where ν and π are the number of neutrons and protons transferred.

Combining these results in (7.1) leads to

$$\mathscr{T}_{pd}(J_A M_A \mu_d \to J_B M_B \mu_p) = \sum_{\ell j} C_{M_A m_j M_B}^{J_A\ j\ J_B} C_{\mu_p\ \mu_n\ \mu_d}^{(1/2)(1/2)(1)} C_{m_\ell\ \mu_n\ m_j}^{\ell\ (1/2)\ j}$$
$$\times\, i^\ell (2\ell+1)^{1/2} B_\ell^{m_\ell} \mathscr{S}_{\ell j}^{1/2} \tag{7.16}$$

where

$$B_\ell^{m_\ell}(\mathbf{k}_p, \mathbf{k}_d)$$
$$= i^{-\ell}(2\ell+1)^{-(1/2)} \int \chi_p^{(-)*}(\mathbf{k}_p, \mathbf{r}_p)\phi_{n\ell}^{m_\ell^*}(\mathbf{r}_n) V_{np}(r) \chi_d^{(+)}(\mathbf{k}_d, \mathbf{r}_d)\phi_d(r)\, d\mathbf{r}_n\, d\mathbf{r}_p. \tag{7.17}$$

The cross section is (3.39)

$$\frac{d\sigma}{d\Omega}(J_A M_A \mu_d \to J_B M_B \mu_p) = \frac{m_d m_p}{(2\pi\hbar^2)^2} \frac{k_p}{k_d} |\mathscr{T}|^2, \tag{7.18}$$

where m_d and m_p are the reduced masses. Usually one does not measure the spin directions $\mu_p M_B$, and additionally the target and projectile are unoriented. Therefore the cross section that is usually measured is a sum over all final orientations produced by an average initial orientation, (see 5.50)

$$\frac{d\sigma}{d\Omega} = \frac{1}{3(2J_A + 1)} \sum_{M_B \mu_p} \sum_{M_A \mu_d} \frac{d\sigma}{d\Omega} (J_A M_A \mu_d \to J_B M_B \mu_p)$$

$$= \frac{1}{2} \frac{m_d m_p}{(2\pi\hbar^2)^2} \frac{k_p}{k_d} \frac{2J_B + 1}{2J_A + 1} \sum_{\ell j m_\ell} \mathcal{S}_{\ell j} |B_\ell^{m_\ell}|^2. \tag{7.19}$$

This expression exhibits a division into two factors: $\mathcal{S}_{\ell j}$, the spectroscopic factor, which is determined solely by the properties of the nuclei; and a factor $|B|^2$, which contains all of the kinematic dependence through the wave functions of the relative motion and their overlap with each other. In addition, it contains the nuclear wave function of the transferred neutron. Together these factors, the distorted waves, the bound-state wave function, and the interaction, control where the reaction occurs, or more aptly stated, its radial distribution.

As a condition that the Clebsch–Gordan coefficients in (7.16) should not vanish, the following selection rules on ℓ and j must hold:

$$\mathbf{J}_B = \mathbf{J}_A + \mathbf{j} = \mathbf{J}_A + \ell + \tfrac{1}{2}. \tag{7.20}$$

In addition, the parities are connected by

$$\pi_B = \pi_A(-)^\ell. \tag{7.21}$$

From the discussion in Chapter 2, we anticipate that $|B_\ell^{m_\ell}|^2$ will have an angular dependence that depends on ℓ and that by a comparison with experiment the value of ℓ can be determined. In this event, the selection rules give explicit relations between initial and final angular momenta and parity of the nuclear states, such that, if those of the initial state are known, then those of the final state can be inferred. This is a most important application of reactions.

To calculate the cross section, we must still calculate B_ℓ^m. This was done for plane waves in Chapter 2. The task is now much more difficult because a six-fold integration is involved. Several mathematical methods for dealing with this six-fold integration have been derived,‡ and this subject will be taken up in Chapter 16. However, many of the calculations and analysis of experiments reported in the literature employed a range approximation. This is what we now consider.

‡ See Austern *et* al., 1964; Sawaguri and Tobocman, 1967; Charlton, 1973; Robson and Koshel, 1972; and Glendenning and Nagarajan, 1974.

C. ZERO-RANGE APPROXIMATION

To render the integral defining the amplitude B_ℓ^m more tractable, we now consider the approximation

$$D(\mathbf{r}) = V_{pn}(\mathbf{r})\phi_d(r) = D_0\delta(\mathbf{r}). \tag{7.22}$$

There are several possible ways of estimating the magnitude of D_0. If a Hulthen wave function is employed as the deuteron wave function, one finds

$$D_0 = \int r^2 V_{np}(\mathbf{r})\phi_0(r)\,dr \simeq 1.5 \times 10^4 \quad (\mathrm{MeV^2 fm^3})^{1/2}. \tag{7.23}$$

Alternatively, it can be evaluated using effective range theory; the result is the same (Austern, 1970).

We are assuming that the dueteron exists only in the S state. There is a small D-state component, and its effects have been investigated by several authors (Johnson and Santos, 1967 and Delic and Robson, 1970). They are small but may be more important for transfers involving high angular momentum ℓ than for low values.

With reference to the coordinates in (2.5) one can see that the effect of the zero-range approximation is

$$r_d \to r_n, \qquad r_p \to \frac{A}{A+1} r_n, \tag{7.24}$$

so that B becomes

$$B_\ell^m = i^{-\ell}(2\ell+1)^{-(1/2)} D_0 \int \chi_p^{(-)*}\left(\mathbf{k}_p, \frac{A}{A+1}\mathbf{r}\right)\phi_\ell^{m*}(\mathbf{r})\chi_d^{(+)}(\mathbf{k}_d, \mathbf{r})\,d\mathbf{r}, \tag{7.25}$$

which is now a three-dimensional integral. The two angle integrals can be done very easily be introducing the partial-wave expansions for the distorted waves (4.7) and (5.14) and making use of well-known results of angular momentum algebra (see A11), with the result

$$\begin{aligned} B_\ell^m = {} & \frac{(4\pi)^{3/2}}{k_p k_d} D_0 \sum_{\ell_d \ell_p m_d m_p} i^{\ell_d - \ell_p - \ell}\left(\frac{2\ell_p + 1}{2\ell_d + 1}\right)^{1/2} Y_{\ell_d}^{m_d *}(\hat{\mathbf{k}}_d) Y_{\ell_p}^{m_p}(\hat{\mathbf{k}}_p) \\ & \times C_{m_p m m_d}^{\ell_p \ell \ell_d} C_{0\,0\,0}^{\ell_p \ell \ell_d} R_{\ell_p \ell \ell_d} \exp[i(\sigma_{\ell_p} + \sigma_{\ell_d})], \end{aligned} \tag{7.26}$$

where

$$R_{\ell_p \ell \ell_d} = \frac{A+1}{A}\int_0^\infty f_{\ell_p}\left(k_p, \frac{A}{A+1} r\right) u_{n\ell}(r) f_{\ell_d}(k_d, r)\,dr. \tag{7.27}$$

To summarize, the zero-range approximation has allowed the reduction to a single radial integral. That integral involves the bound state radial function and the radial functions for the distorted waves that are solutions to the optical-model wave equation (4.6) and (4.7). That involves, as discussed, the complex optical potentials U_p and U_d. Except for square-well potentials, the solutions can be obtained only by numerical integration of the complex Schrödinger equations. This is done quite routinely on an electronic computer.

D. EXAMPLES

Before the theory can be applied to an analysis of data, it is necessary to obtain satisfactory optical-model parameters for both initial and final channels of the reaction. They are chosen, as discussed, to reproduce the elastic scattering at the same energies as the reaction under consideration or, if such elastic data is not available, then under the approximate conditions of the reaction. Any deviation at this point introduces an inaccuracy into the subsequent analysis.

The analysis of the

$$^{56}\mathrm{Fe(d,\,p)}^{57}\mathrm{Fe}, \qquad E_d = 15 \quad \mathrm{MeV}$$

reaction affords a good example. In a case such as this, where the target is an even nucleus, then only one value of ℓ and j is consistent with conservation of angular momentum and parity. In a shell-model interpretation of the nucleus, these define the shell-model state into which the particle is stripped. There may be other components in the wave function. For example,

$$\Phi(^{57}\mathrm{Fe}) = \beta_1 \Phi_0(^{56}\mathrm{Fe})\phi_{n\ell j}(r_n) + \beta_2[\Phi_2(^{56}\mathrm{Fe})\phi_{n'\ell' j'}(r_n)]_j$$

is a possible wave function. The reaction can, to the order we consider, populate only the first component and thus provide a measure of β_1^2 or the spectroscopic factor as it is frequently called. Figure 7.1 shows four transitions of two different shell-model states in ^{57}Fe. Both the data and the theory illustrate the dependence of the angular distribution on ℓ. In particular, the qualitative rule obtained in Chapter 2 from the consideration of angular momentum conservation and also the expectations based on the plane wave theory are born out.

To achieve the agreement between theory and data, one calculates the cross section using shell-model wave functions $\phi_{n\ell}$ with various values of ℓ until the shape of the cross section is found to agree. The factor by which theory must be multiplied to achieve agreement in magnitude provides the value of the spectroscopic factor $\mathscr{S}$.

The agreement of the shapes of the angular distributions of the zero-range theory with the data is quite convincing. Thus one feels that the identification

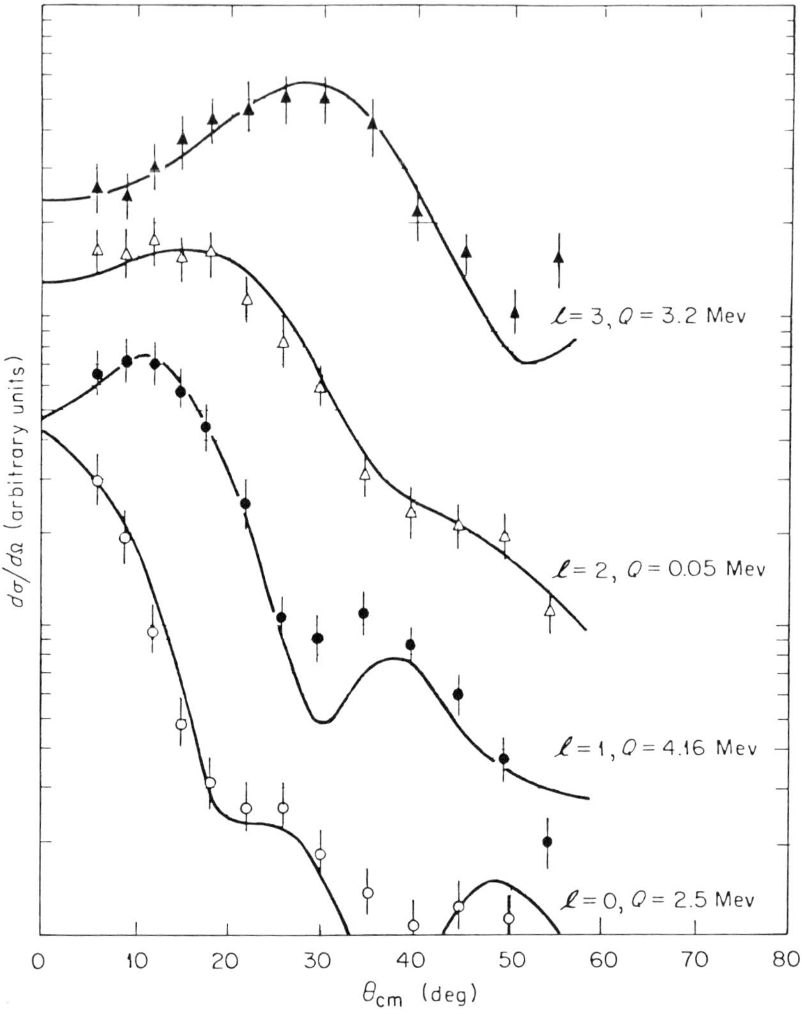

Fig. 7.1. The DWBA stripping theory corresponding to neutron transfer into states with $\ell = 0, 1, 2, 3$ are compared with data for ^{56}Fe(d, p)^{57}Fe, $E_d = 15$ MeV (Cohen *et al.*, 1962.)

of the spins and parities $\ell, j, (-)^\ell$ are quite sound. However, the normalization by which the spectroscopic factor $\mathscr{S}_{\ell j}$ is determined may be incorrectly estimated. A comparison of the zero-range theory with a full finite-range calculation is usually reassuring (Drisco and Satchler, 1964) for the (d, p) reaction.

E. IMPROVEMENTS WITHIN THE FRAMEWORK OF THE DWBA

The basic expression for the T matrix with which a DWBA calculation begins is given by Eq. (7.1). Even this expression is not trivial to compute. Various assumptions, depending on the particular reaction, are often made in evaluating the DWBA amplitude. These are in addition to the basic assumptions discussed in Chapter 5, which underly this amplitude. For example, in the application discussed in this chapter, it was assumed that

(1) the optical potential is local,
(2) there is an average cancellation of the optical potential against the proton–nucleus interaction,
(3) $V_{\rm pn}$ is the interaction responsible for the reaction,
(4) $V_{\rm pn}$ is of zero range, and
(5) the neutron is removed from a shell model state of the nucleus.

The last point may not appear to be an assumption because, in principle, the parentage expansion (7.8) exists. In practice it is not known (see Philpott *et al.*, 1968). To know it would be tantamount to having the solution of the (A + 1)-body problem. However, in regions of closed shells one feels confident that it is dominated by a few terms that can be approximated by shell-model wave functions, and that the single-particle wave functions are approximately eigenstates of a potential well.

A review of how some shortcomings have been handled can be found in the work of Glendenning (1975b) and can be used as a guide to the original literature on the subject. The finite range of the interaction is the subject of Chapter 16.

F. SPECTROSCOPIC FACTORS FOR ONE-NUCLEON TRANSFER

A quantity of great importance in one-particle transfer reactions is the spectroscopic factor, which is the square of the parentage amplitude β defined in Eqs. (7.8), (7.13), and (7.14). The reason for its importance is evidently that it measures the degree to which a state populated in a transfer reaction is a single-particle state. In the ideal shell model, the low-lying states of an odd nucleus would be pure single-particle shell model states. This situation is realized to a high degree near doubly magic nuclei such as ^{208}Pb. The addition of one neutron or proton would form ^{209}Pb and ^{209}Bi, respectively, whose lowest states are good examples of nearly pure single-particle states. The removal of a neutron or proton likewise produces nearly pure hole states.

However, in nuclei removed from such a special situation, the single-particle nature of the ideal shell-model state is spread over several, or even many, nuclear states. This is already implied by our parentage expansion (7.8), in which many states of the nucleus (A + 1) can share a fraction of the single-particle orbital labeled by $n\ell j$.

Expression (7.14) for the spectroscopic factor, written in terms of wave functions, can also be written as a reduced matrix element of a particle creation operator. This provides a concise statement and is particularly useful in the quasi-particle picture for treating the pairing interaction in nuclei. Thus, from its definition (7.14), we may write

$$\begin{aligned}\beta_{\ell j}(\mathrm{B}, \mathrm{A}) &= \langle \Phi_{J_\mathrm{B}}^{M_\mathrm{B}} | \mathscr{A}[\Phi_{J_\mathrm{A}} \phi_j]_{J_\mathrm{B}}^{M_\mathrm{B}} \rangle \\ &= \sum_{M_\mathrm{A} m} C_{M_\mathrm{A} m M_\mathrm{B}}^{J_\mathrm{A}\, j\, J_\mathrm{B}} \langle J_\mathrm{B} M_\mathrm{B} | a_{jm}^\dagger | J_\mathrm{A} M_\mathrm{A} \rangle \\ &= \sum_{M_\mathrm{A} m} (-)^{J_\mathrm{A} - j + M_\mathrm{B}} \sqrt{\hat{J}_\mathrm{B}} \begin{pmatrix} J_\mathrm{A} & j & J_\mathrm{B} \\ M_\mathrm{A} & m & -M_\mathrm{B} \end{pmatrix} \langle J_\mathrm{B} M_\mathrm{B} | a_{jm}^\dagger | J_\mathrm{A} M_\mathrm{A} \rangle,\end{aligned} \tag{7.28}$$

where $a_{jm}^\dagger$ denotes a creation operator for a particle in the state $n\ell jm$. The Wigner–Eckart theorem (Appendix) for its matrix elements reads

$$\langle J_\mathrm{B} M_\mathrm{B} | a_{jm}^\dagger | J_\mathrm{A} M_\mathrm{A} \rangle = (-)^{J_\mathrm{B} - M_\mathrm{B}} \begin{pmatrix} J_\mathrm{B} & j & J_\mathrm{A} \\ -M_\mathrm{B} & m & M_\mathrm{A} \end{pmatrix} \langle J_\mathrm{B} \| a_j^\dagger \| J_\mathrm{A} \rangle. \tag{7.29}$$

Using the symmetry and summation properties of the 3-j coefficients, we find

$$\beta_{\ell j}(\mathrm{B}, \mathrm{A}) = (-)^{2j} \langle J_\mathrm{B} \| a_j^\dagger \| J_\mathrm{A} \rangle \Big/ \sqrt{\hat{J}_\mathrm{B}}\,. \tag{7.30}$$

Thus we see that $\beta_{\ell j}$ is independent of any projections of angular momentum. The phase is of no consequence because the sum on j is always incoherent in the cross section. Note that $a_{jm}^\dagger$ is correctly treated as a tensor operator of rank j and projection m because it creates a particle state having these quantum numbers. The destruction operator a_{jm} is not a tensor operator with the indicated projection. However, $(-)^m a_{j,-m}$ and $(-)^{j \pm m} a_{j,-m}$ are tensor operators that satisfy the rules, mentioned in the Appendix, by which a tensor operator T_j^m is defined. To avoid a complex phase we can adopt the definition

$$\tilde{a}_{jm}^\dagger \equiv (-)^{j-m} a_{j,-m}. \tag{7.31}$$

This operator can be called the hole-creation operator. The other choice of phase, $(-)^{j+m}$, is equally possible. The operators so defined simply differ by a sign.

Useful sum rules can be obtained from the Wigner–Eckart theorem and the anticommutation relations satisfied by the particle operators $a^\dagger$ and a.

From (7.29) and (7.30),

$$\langle J_B M_B | a_{jm}^\dagger | J_A M_A \rangle = (-)^{J_B - M_B} \begin{pmatrix} J_B & j & J_A \\ -M_B & m & M_A \end{pmatrix} (-)^{2j} \sqrt{\hat{J}_B} \beta_{\ell j}(\mathrm{B}, \mathrm{A}). \tag{7.32}$$

Multiply this equation by its adjoint. Thus we may write

$$\begin{aligned} \hat{J}_B \begin{pmatrix} J_B & j & J_A \\ -M_B & m & M_A \end{pmatrix}^2 \beta_{\ell j}^2(\mathrm{B}, \mathrm{A}) &= \langle J_A M_M | a_{jm} | J_B M_B \rangle \langle J_B M_B | a_{jm}^\dagger | J_A M_A \rangle \\ &= \langle J_B M_B | a_{jm}^\dagger | J_A M_A \rangle \langle J_A M_A | a_{jm} | J_B M_B \rangle. \end{aligned} \tag{7.33}$$

In stripping reactions (7.2), J_A is the target ground state and J_B any state in the final nucleus that can be reached by the reaction. We now obtain a rule for the sum of spectroscopic factors of these final states. Sum first over M_B and m. The 3-j symbol sums to $\hat{J}_A^{-1}$. Then sum over J_B. This gives

$$\sum_{J_B} (\hat{J}_B / \hat{J}_A) \beta_{\ell j}^2(\mathrm{B}, \mathrm{A}) = \langle J_A M_A | \sum_m a_{jm} a_{jm}^\dagger | J_A M_A \rangle,$$

where the completeness relation

$$\sum_{J_B M_B} |J_B M_B\rangle \langle J_B M_B| = 1 \tag{7.34}$$

was used in the first line in the right side of (7.33). Now use the anticommutator

$$a_{jm} a_{jm}^\dagger = 1 - a_{jm}^\dagger a_{jm}. \tag{7.35}$$

Then we find

$$\sum_{J_B} (\hat{J}_B / \hat{J}_A) \mathscr{S}_{\ell j}(\mathrm{B}, \mathrm{A}) = 2j + 1 - n_j(\mathrm{A}), \tag{7.36}$$

where

$$n_j(\mathrm{A}) = \langle J_A M_A | \sum_m a_{jm}^\dagger a_{jm} | J_A M_A \rangle \tag{7.37}$$

is the target ground state expectation value of the counting operator for particles in the shell-model state j. That is, n_j is the number of neutrons occupying the shell j in the target ground state.

On the other hand, the pickup reaction

$$\mathrm{p} + \mathrm{B} \equiv \mathrm{p} + (\mathrm{A} + 1) \to \mathrm{A} + \mathrm{d} \tag{7.38}$$

populates states in the nucleus A that are built on the ground state of B. Again from (7.33) we obtain a sum rule. Sum first over m and M_A and then sum over J_A. The second line on the right of (7.33) then yields

$$\sum_{J_A} \mathscr{S}_{\ell j}(\mathrm{B}, \mathrm{A}) = n_j(\mathrm{B}), \tag{7.39}$$

where now

$$n_j(\mathrm{B}) = \langle J_B M_B | \sum_m a_{jm}^\dagger a_{jm} | J_B M_B \rangle \tag{7.40}$$

is the number of neutrons occupying the shell j in the ground state of nucleus B.

It might appear at first sight that the number operator matrix elements depend on the M projection. Physically, the expectation for occupying the shell-model orbital j cannot depend on the orientation of the nucleus. To prove that this is so, note that we can write

$$\sum_m a_{jm}^\dagger a_{jm} = \sum (-)^{j+m} a_{jm}^\dagger \tilde{a}_{j-m}^\dagger = -(2j+1)^{1/2} [a_j^\dagger \tilde{a}_j^\dagger]_0^0, \tag{7.41}$$

where the special value of a Clebsch–Gordan coefficient (A19) was used. Therefore, the counting operator is a scalar, as it must be. By the Wigner–Eckart theorem,

$$\begin{aligned} &\langle J_B M_B | \sum_m a_{jm}^\dagger a_{jm} | J_B M_B \rangle \\ &= -(-)^{J_B - M_B} \begin{pmatrix} J & 0 & J_B \\ -M_B & 0 & M_B \end{pmatrix} \langle J_B \| [a_j^\dagger \tilde{a}_j^\dagger]_0 \| J_B \rangle \sqrt{\hat{j}} \\ &= -(\hat{j}/\hat{J}_B)^{1/2} \langle J_B \| [a_j^\dagger \tilde{a}_j^\dagger]_0 \| J_B \rangle \end{aligned} \tag{7.42}$$

is independent of projections.

A comprehensive study of sum rules such as (7.36) and (7.39) has been provided by Macfarlane and French (1960). Here we discuss only a few simple examples. In certain cases, such as in the neighborhood of closed shells, the single-particle character may be concentrated in one nuclear state. However, in general it will be shared by a number of states, and then the sum rules can be employed to determine when all states sharing a certain single-particle character have been found. The reason that individual spectroscopic factors $\mathscr{S}_{\ell j}(\mathrm{B}, \mathrm{A})$ can be determined in many cases, and their sum, (7.36) or (7.39), formed to compare against the sum rule, is the following. In (7.19) we have an expression for the cross section. In case the target nucleus has spin zero, only one value of j is allowed according to the selection rules (7.20). As has been emphasized many times, the angular distribution of the reaction is very sensitive to the angular momentum transfer ℓ. One can guess the angular momentum and then test whether the guess is correct.

Corresponding to the orbital angular momentum ℓ, the other two quantum numbers of the orbit, n and j, can usually be inferred from the level ordering of the shell model. In any case, the ambiguity $j = \ell \pm \frac{1}{2}$ is not serious because the radial wave function of the orbital $u_{n\ell j}(r)$ depends only weakly on j through the presence of the spin-orbit interaction. The radial function satisfies a Schrödinger equation like (4.43), with U the shell-model central potential. The behavior of $u_{n\ell j}(r)$ for large r is most important for direct reactions because the absorption of the optical potential tends to concentrate the reactions on the surface. The appropriate asymptotic behavior can be ensured by adjusting the depth of the shell-model potential to yield the separation energy as determined by the measured Q-value of the reaction. Thus, the radial function entering the transfer amplitude $\mathrm{B}_\ell^{m_\ell}$, given by (7.17), is known with little uncertainty. The distorted waves can be determined through an analysis of the relevant elastic scattering, as described in Chapter 4. thus completing the determination of the ingredients of $\mathrm{B}_\ell^{m_\ell}$. Once this has been calculated and the assumed value of ℓ is confirmed by comparison with the experimental angular distribution, then a comparison of (7.19) with the measured cross section yields the spectroscopic factor $\mathscr{S}_{\ell j}(\mathrm{B}, \mathrm{A})$.

In certain cases, theory can predict the spectroscopic factors. Then comparison with experiment can be made on the basis of individual spectroscopic factors, not only their sum. Let us first consider several simple shell-model examples corresponding to closed shells plus several additional nucleons in the same orbital $n\ell j$. For a stripping reaction on an even target, the nuclear wave functions would therefore be

$$\begin{aligned} \Phi_{\mathrm{A}}(\mathrm{A}) &= \mathscr{A}_{\mathrm{A}}[\Phi_0(\mathrm{A}-n)\Phi_{(j^n)_0}(1,\ldots,n)], \\ \Phi_{\mathrm{B}}(\mathrm{A}+1) &= \mathscr{A}_{\mathrm{B}}[\Phi_0(\mathrm{A}-n)\Phi_{(j^{n+1})J_{\mathrm{B}}}(1,\ldots,n+1)], \end{aligned} \tag{7.43}$$

where $\mathscr{A}_{\mathrm{A}}$ and $\mathscr{A}_{\mathrm{B}}$ are the antisymmetrization operators of (5.28) and $\Phi_0(\mathrm{A}-n)$ denotes a wave function of $(\mathrm{A}-n)$ nucleons occupying closed shells and therefore having zero angular momentum, whereas $\Phi_0(j^n)$ and $\Phi_{J_{\mathrm{B}}}(j^{n+1})$ describe the valence nucleons in the shell j. The parentage amplitude from (7.8) is

$$\begin{aligned} \beta_{\ell j}(\mathrm{B}, \mathrm{A}) &= \langle \Phi_{J_{\mathrm{B}}}^{M_{\mathrm{B}}} | \mathscr{A}[\Phi_{J_{\mathrm{A}}}\Phi_j]_{J_{\mathrm{B}}}^{M_{\mathrm{B}}}\rangle \\ &= \binom{\mathrm{A}+1}{1}^{1/2} \langle \Phi_{J_{\mathrm{B}}}^{M_{\mathrm{B}}} | [\Phi_{J_{\mathrm{A}}}\phi_j(\mathrm{A}+1)]_{J_{\mathrm{B}}}^{M_{\mathrm{B}}}\rangle, \end{aligned} \tag{7.44}$$

where in the second line the $(\mathrm{A}+1)$st particle occupies the ϕ_j wave function. Each of the terms in the sum on permutations implied by $\mathscr{A}$ is equal to the typical one written. Remember that $\Phi_{J_{\mathrm{B}}}(\mathrm{A}+1)$ and $\Phi_{J_{\mathrm{A}}}(\mathrm{A})$ are always assumed to be antisymmetrized wave functions. For our special case of a common core we have

$$\beta_{\ell j}(\mathrm{B}, \mathrm{A})$$
$$= \binom{n+1}{1}^{1/2} \langle \Phi_{(j^{n+1})J_\mathrm{B}}(1, 2, \ldots, n+1) \,|\, \Phi_{(j^n)0}(1, 2, \ldots, n)\phi_j(n+1)\rangle. \tag{7.45}$$

The multiplying factor is obtained by correctly counting the number of nonvanishing equal overlaps when (7.43) is used in (7.44). We can work with either the first or second line of (7.44). Writing out the second line, it is

$$\binom{\mathrm{A}+1}{1}^{1/2} \binom{\mathrm{A}+1}{n+1}^{-1/2} \binom{\mathrm{A}}{n}^{-1/2} \sum (-)^{P_\mathrm{A}+P_\mathrm{B}}$$
$$\times \langle P_\mathrm{B}[\Phi_0(\mathrm{A}-n)\Phi_{(j^{n+1})J_\mathrm{B}}] \,|\, P_\mathrm{A}[\Phi_0(\mathrm{A}-n)\Phi_{(j^n)0}]\phi_j(n+1)\rangle.$$

This is a sum over the indicated permutations. Each of the $\binom{\mathrm{A}}{n}$ terms in the P_A sum has nonzero overlap with only one term in the sum over the P_B permutations on the left, namely the one in which the same group occupies the core. All of these nonvanishing overlaps are equal because each of the wave functions Φ is itself antisymmetric. Thus the factor multiplying one of the typical elements is

$$\binom{\mathrm{A}+1}{1}^{1/2} \binom{\mathrm{A}+1}{n+1}^{-1/2} \binom{\mathrm{A}}{n}^{-1/2} \binom{\mathrm{A}}{n} = \binom{n+1}{1}^{1/2},$$

as was to be proved. Because in applications of the shell model one usually employs an antisymmetrized wave function for only an active subset of the nucleons, (7.45) is an important result.

To proceed with the evaluation of (7.45), we now introduce the coefficients of fractional parentage (cfp) first employed by Racah (1942) and reviewed by deShalit and Talmi (1963). These are coefficients that permit the partitioning of a function such as $\Phi_J(j^{n+1})$, which is antisymmetric, into two subsets, each of which is internally antisymmetrized. Thus

$$\Phi_{(j^{n+1})\alpha J} = \sum_{\alpha' J'} [(j^n)\alpha' J', j; J| \}(j^{n+1})\alpha J][\Phi_{(j^n)\alpha' J'}(1, \ldots, n)\phi_j(n+1)]_J. \tag{7.46a}$$

A label α, in addition to J, has been appended because J is not always sufficient to specify a j^n configuration completely. The main point of (7.46) is that a specific particle coordinate occupies the wave function ϕ_j, but nevertheless, the coefficients of fractional parentage are constructed so that the sum of such unsymmetrized products is totally antisymmetric. Having written the coordinate dependences so they can be understood, we can now use a more concise notation for the preceding:

$$|(j^{n+1})\alpha J\rangle = \sum_{\alpha'J'} [(j^n)\alpha'J', j; J|\}(j^{n+1})\alpha J]|(j^n)\alpha'J', j; J\rangle. \tag{7.46b}$$

The comma and semicolon on the right signify that J' and j are coupled to J.

Equation (7.45) can now be written immediately in terms of the cfp as

$$\beta_{\ell j}(\alpha J, 0) = \sqrt{n+1}[(j^n)0, j; J|\}(j^{n+1})\alpha J], \tag{7.47}$$

or more generally, if the j^n configuration is coupled to J',

$$\beta_{\ell j}(\alpha J, \alpha'J') = \sqrt{n+1}[(j^n)\alpha'J', j; J|\}(j^{n+1})\alpha J]. \tag{7.48}$$

In a number of cases there are simple explicit formulas for the cfp. For example,

$$[(j^n)0, j; j|\}(j^{n+1})j] = \left(\frac{2j+1-n}{(n+1)(2j+1)}\right)^{1/2}, \qquad n = \text{even}, \tag{7.49a}$$

$$[(j^n)j, j; 0|\}(j^{n+1})0] = 1, \qquad n = \text{odd}, \tag{7.50a}$$

for states of the lowest seniority. Thus

$$\mathscr{S}_{\ell j}(n+1, n) = \begin{cases} 1 - n/(2j+1), & n = \text{even}, \quad (7.49\text{b}) \\ n+1, & n = \text{odd}. \quad (7.50\text{b}) \end{cases}$$

In these special cases there is only one spin state possible for B and one for A, respectively, corresponding to (7.49) and (7.50). The rules (7.36) and (7.39) agree with the preceding results, as they should.

For most nuclei, removed as they are by more than a few nucleons from closed shells, the shell model does not provide a simple means of calculating the spectroscopic factors. In such cases, experiments have exhibited a high degree of fragmentation of the single-particle strength. The ^{60}Ni(d, p)^{61}Ni reaction (Fulmer *et al.*, 1964; Cohen and Price, 1961) is a good example. There are approximately 40 levels in ^{61}Ni that share the $d_{3/2}$ and $d_{5/2}$ strength (which were not distinguished from each other). The spreading of the single-particle strength is illustrated in Fig. 7.2. The nickel nuclei have a magic number, $Z = 28$ of protons, and the valence neutrons occupy the shell-model states beyond the $N = 28$ magic core. The nickel isotopes such as ^{59}Ni and ^{61}Ni are therefore not expected to exhibit a simple single-particle structure.

A convenient theory for describing the structure of single–closed-shell nuclei with a few valence nucleons will now be briefly touched on, for it pertains to such nuclei as Ni. The sum rules yield values for the sums of spectroscopic amplitudes. It is, of course, desirable to have theoretical estimates

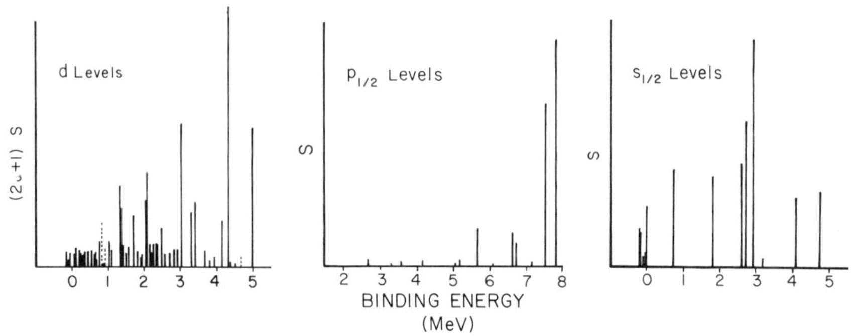

Fig. 7.2. Relative strengths of transitions in the (d, p) reaction showing the fragmentation of the shell-model levels and their spreading over many states. (From Fulmer *et al.*, 1964.)

for individual spectroscopic amplitudes, which our simple consideration of the shell model does not provide in regions removed from closed shells.

G. PAIRING THEORY IN NUCLEAR STRUCTURE

In regions of the periodic table that are removed by more than several nucleons from closed shells, the ordinary shell model becomes very cumbersome. Nuclear wave functions would have to be described in terms of a number of valence particles occupying a number of shell-model states. The number of configurations having the same spin becomes very large for low spins, such as are found near the ground state of even nuclei. The conventional shell model would therefore involve diagonalizations of large matrices. Although this can be done, for many purposes, considerable physical insight is gained, as well as a great simplification, by considering an alternative solution to the interacting shell-model problem. This involves the recognition of the prominent part played by the large pair interaction of an angular momentum zero pair $(jj)0$. Such a state is a superposition of pairwise occupation of states having opposite projections of angular momentum

$$
\begin{aligned}
|(j^2)0\rangle &= \sum_m C^{j\ \ j\ \ 0}_{m\,-m\,0} a^\dagger_{jm} a^\dagger_{j-m}|0\rangle \\
&= \hat{j}^{-1/2} \sum_m (-)^{j-m} a^\dagger_{jm} a^\dagger_{j-m}|0\rangle.
\end{aligned} \tag{7.51}
$$

The mathematical methods for solving a many-body problem in which the pair interactions between particles in conjugate states are especially important were developed by Bardeen *et al.* (1957), Bogolyubov (1958), and Valatin (1958) in connection with the theory of superconductivity. Perhaps the main point to emphasize is the possibility of a phase transition in which the ground state is considerably lowered in energy relative to the normal state by a structural change in the wave function. The normal state in this context is understood to be a quasi-gaseous state or shell-model state, with the typical correlation or configuration mixing that can be treated in the shell model. A phase transition cannot be treated as a perturbation on such a state. Bohr *et al.* (1958) observed that, in nuclei, the energy of the first excited state (exclusive of collective states) lies at considerably more than twice the excitation energy in odd nuclei, suggesting that the ground state possesses special correlations that lower the energy by more than the interaction of a pair in a given shell-model state. At a fundamental level, theory ought to demonstrate that, for the nuclear forces, such a phase transition takes place. A criterion for superfluidity has been established by Cooper *et al.* (1959) and Mills *et al.* (1959), who show that, in nuclear matter, the (Gammel–Thaler) nucleon–nucleon potential leads to a superfluid ground state at the normal density of nuclei, but not much above. The pairing theory has been formulated by several authors for nuclei (Belyaev, 1959; Migdal, 1959; Kisslinger and Sorensen, 1960). In many applications it has been assumed from the beginning that the ground state is superfluid. In such cases the parameters of the effective pairing Hamiltonian are fitted to observed data.

The theory is capable of describing not only particle excitations but also collective vibrational states (Baranger, 1960). Yoshida (1961) employed the pairing theory to obtain the spectroscopic amplitudes for individual states, as well as for vibrational collective states. His results for single quasi-particle states are

$$\begin{aligned} \mathscr{S}_j(j, 0) &= U_j^2 \\ \mathscr{S}_j(0, j) &= (2j + 1)V_j^2, \end{aligned} \tag{7.52}$$

where V_j^2 is the probability that the orbital j in the even nucleus is occupied and U_j the probability that it is empty. These probabilities can be calculated from the theory extracted from the experiments through (7.19), in which $\mathscr{S}$ is treated as an unknown, while the transfer probability $|B_\ell^{m_\ell}|^2$ is calculated. In the special case of a pure configuration, they reduce to our previous results. Note, however, that in the pairing theory the even nucleus is not assumed to be a degenerate pure state. Instead, a number of valence orbitals are occupied with varying probabilities V_j^2.

The interpretation of (7.52) is straightforward and sensible. For the reactions

$$\left.\begin{matrix} \text{even (d, p) odd} \\ \text{odd (p, d) even} \end{matrix}\right\}, \qquad \mathscr{S} = U_j^2,$$

the cross section for depositing a nucleon onto the even target in the orbital j is proportional to the probability that the orbital is empty. The time-reversed reaction must involve the same quantity. For the reactions

$$\left\{\begin{array}{l}\text{even (d, p) odd} \\ \text{odd (p, d) cvcn}\end{array}\right\}, \qquad \mathscr{S} = (2j + 1)V_j^2,$$

the cross section for picking a nucleon in the even nucleus out of the orbital j is proportional to the number of nucleons in that orbital.

H. INELASTIC EXCITATION OF SURFACE VIBRATIONS

It was discovered in early direct reaction experiments that the so-called vibrational states in spherical nuclei had larger cross sections than were otherwise observed in the low-energy spectra of either even or odd nuclei. (see Cohen, 1957; and Cohen and Rubin, 1958.) The existence of vibrational-like states was inferred by Bohr and Mottelson (1975) because certain energy levels, connected by fast electric quadrupole transitions, had equal energy spacings suggestive of a quantum-mechanical oscillator. Such states are associated with the coherent motion of many nucleons and are therefore not particularly simple to describe by a wave function of the nucleons. The collective picture provides a simple description that we can use for the inelastic excitation of such levels if, for the time being, we forgo a description in terms of the nucleonic motion. We suppose, therefore, that the nuclear density can undergo shape oscillations corresponding simply to variations in the position of the nuclear surface about a sphere, with the density as measured from this moving surface retaining a constant profile. The equation for the surface can be written as

$$R = R_0\left[1 + \sum_{\mu} \alpha_{\lambda\mu} Y^*_{\lambda\mu}(\theta, \phi)\right] \equiv R_0 + \delta R(\theta, \phi). \tag{7.53}$$

For R to be real, $\alpha_{\lambda\mu}$ must have the same transformation properties as $Y_{\lambda\mu}$, namely,

$$\alpha^\dagger_{\lambda\mu} = (-)^\mu \alpha_{\lambda-\mu}, \tag{7.54}$$

In the quantized theory of vibrations, the amplitudes α of the surface vibration are associated with phonon creation and destruction operators $c^\dagger_{\lambda\mu}$ and $c_{\lambda\mu}$ (Bohr and Mottelson, 1975).

The scattered particle interacts with the nucleus through the optical potential U, which follows the nuclear density as discussed in Chapter 4. In the case of the vibrational modes, the scattered particle can excite a phonon through its interaction with the nucleus, and the distance of the scattered particle from the deformable surface is the crucial quantity. We can expand this interaction in a Taylor series about the equilibrium position of the surface, thus

$$U(r-R) = U(r) - \delta R \frac{dU}{dr} + \cdots$$
$$= U(r) - R_0 \frac{dU}{dr} \alpha_{\lambda\mu} Y^*_{\lambda\mu}(\hat{\mathbf{r}}) + \cdots. \tag{7.55}$$

The first term has only diagonal matrix elements between nuclear states. However, the second term, being linear in the amplitude, can connect vibrational modes differing by one unit in the number of phonons. Thus the transition amplitude in the distorted-wave Born approximation in (5.25) becomes, in this case,

$$\mathscr{T} = -\langle N_f J_f M_f | \alpha^\dagger_{\lambda\mu} | N_i J_i M_i \rangle \int \chi_f^{(-)*}(\mathbf{k}_f, \mathbf{r}) R_0 \frac{dU}{dr} Y_{\lambda\mu}(\hat{\mathbf{r}}) \chi_i^{(+)}(\mathbf{k}_i, \mathbf{r})\, d\mathbf{r}, \tag{7.56}$$

corresponding to the nuclear excitation

$$N_i J_i M_i \to N_f J_f M_f, \tag{7.57}$$

where N is the phonon number of the nuclear state. For the ground state of an even nucleus, $N_i = 0$. Note that the spherical part $U(r)$ of (7.55) is cancelled in the matrix element of (5.25) but appears in the distorted waves χ.

It is convenient to express the nuclear matrix element in terms of the zero-point amplitude β_λ, defined as the root mean square of α:‡

$$\beta_\lambda^2 = \langle 0 | \alpha_\lambda \cdot \alpha_\lambda | 0 \rangle = \sum_\mu \langle 0 | \alpha_{\lambda\mu} \alpha^\dagger_{\lambda\mu} | 0 \rangle. \tag{7.58}$$

This is easily evaluated by inserting a complete set of oscillator states between the operators α. Because they have nonvanishing matrix elements only between states differing by one phonon, we obtain

$$\beta_\lambda^2 = \sum_\mu \langle 0 | \alpha_{\lambda\mu} | 1_\mu \rangle \langle 1_\mu | \alpha^\dagger_{\lambda\mu} | 0 \rangle, \tag{7.59}$$

where $|1\mu\rangle$ denotes a one-phonon state with angular momentum and projection λ and μ. Using the Wigner–Eckhart theorem (Appendix A39),

$$\langle 1_\mu | \alpha^\dagger_{\lambda\mu} | 0 \rangle = (-)^{\lambda-\mu} \begin{pmatrix} \lambda & \lambda & 0 \\ -\mu & \mu & 0 \end{pmatrix} \langle 1 \| \alpha^\dagger_\lambda \| 0 \rangle$$
$$= \hat{\lambda}^{-1/2} \langle 1 \| \alpha^\dagger_\lambda \| 0 \rangle. \tag{7.60}$$

‡ The scalar product of two tensors denoted by the dot, is defined in the appendix, A.38.

Thus

$$\beta_\lambda^2 = |\langle 1\|\alpha_\lambda^\dagger\|0\rangle|^2, \tag{7.61}$$

so the transition amplitude for exciting a vibrational state λ_μ in an even nucleus is

$$\mathscr{T} = -\hat{\lambda}^{-1/2} \int \chi_{\rm f}^{(-)*}(\mathbf{k}_{\rm f}, \mathbf{r})\, \beta_\lambda R_0 \frac{dU}{dr} Y_{\lambda\mu}(\hat{r}) \chi_{\rm i}^{(+)}(\mathbf{k}_{\rm i}, \mathbf{r})\, d\mathbf{r}, \tag{7.62}$$

and the cross section is given by

$$\left(\frac{d\sigma}{d\Omega}\right)_{0\to\lambda} = \left(\frac{m}{2\pi\hbar^2}\right)^2 \frac{k_{\rm f}}{k_{\rm i}} \frac{(\beta_\lambda R_0)^2}{2\lambda+1} \sum_\mu \left| \int \chi_{\rm f}^{(-)*}(\mathbf{k}_{\rm f}, \mathbf{r}) \frac{dU}{dr} Y_\lambda^\mu(\hat{\mathbf{r}}) \chi_{\rm i}^{(+)}(\mathbf{k}_{\rm i}, \mathbf{r})\, d\mathbf{r} \right|^2 \tag{7.63}$$

The potential U, being the optical potential, is complex valued. The real part of dU/dr is concentrated on the nuclear surface because U is of the Woods–Saxon form. The imaginary part of U is either of Woods–Saxon form, or the derivative thereof, so that the imaginary part of dU/dr is also concentrated on the surface. These are illustrated schematically in Fig. 7.3 for the nuclear part of the interaction. The derivative "form factor" dU/dr is often referred to in this context as the collective form factor.

The multipolarity of the lowest-lying vibration is $\lambda = 2$, that is, a 2^+ nuclear excitation. Higher-order terms in (7.55) can cause a direct transition from the ground state to a two-phonon state. However, this would be of the same order in β_λ as the two-step transition going through the one phonon

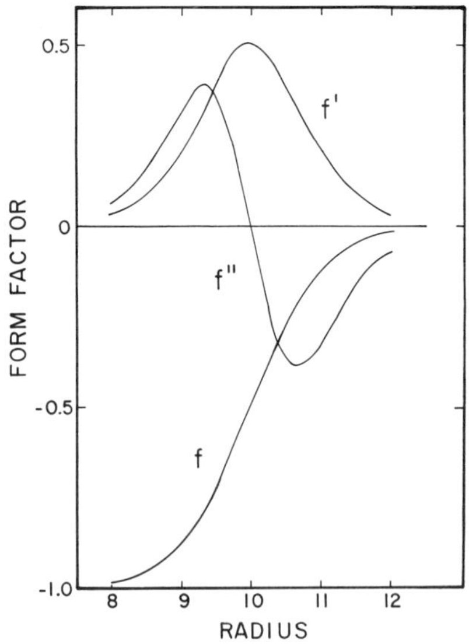

Fig. 7.3. Woods–Saxon form factor for optical potential and its derivative form factors for a nominal 10 fm radius and 0.5 fm diffuseness.

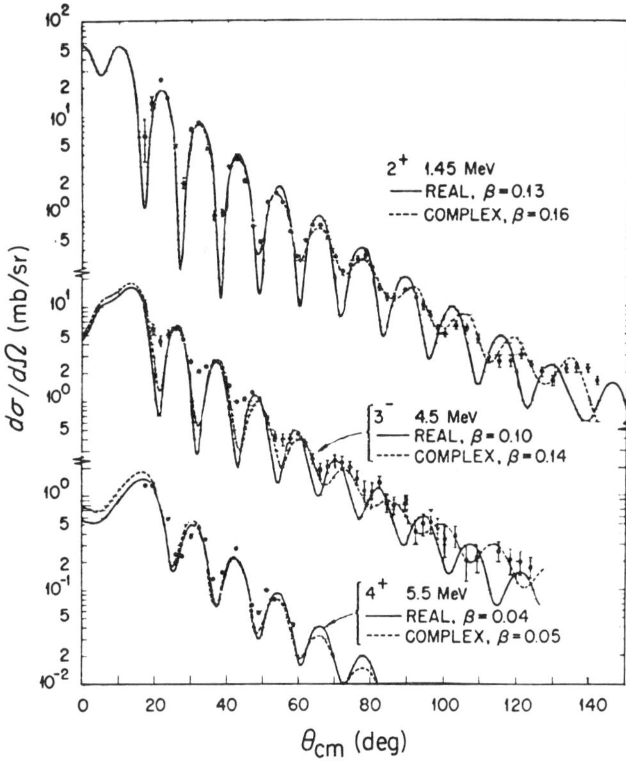

Fig. 7.4. Cross sections for collective states calculated as vibrational levels for the reaction $^{58}Ni(\alpha, \alpha')$ with $E_\alpha = 43$ MeV. Two calculations for each level correspond to the use of a purely real and a complex form factor. (From Broek *et al.*, 1965.)

state $0 \to 1\lambda \to 2\lambda$. Thus both such processes would need to be considered together in second or higher order in β_λ. In later chapters we shall study the coupled-channels formalism, which is suitable for calculating such higher-order processes.

In Fig. 7.4 is shown a typical calculation for the excitation of a collective 2^+ state in a spherical nucleus (see Broek *et al.*, 1965). It is free of adjustable parameters save for the amplitude β_λ, corresponding to the zero-point motion, the parameters of the optical potential having been determined from an analysis of the elastic cross section. By comparison with experiment, this characteristic quantity of a vibrational nucleus can be determined. It can also be determined from other types of experiments, such as the electric quadrupole transition rate, either by decay of the state or by Coulomb excitation. The phase relations between the oscillations in cross section to even and odd states is particularly striking in Fig. 7.4 (Blair phase rule, Blair 1959).

We have not explicitly mentioned the role of the Coulomb force. It also is a part of the interaction between the target nucleus and a charged projectile. It is included as part of the optical potential and should be included in (7.63). In Fig. 7.3 we have shown only the nuclear part. The Coulomb part can be obtained by expanding the Coulomb potential about the equilibrium position of the surface. We shall take up a similar problem in greater detail in connection with deformed nuclei in Chapter 13. For the present, we state the result in lowest order:

$$\frac{dU}{dr} = \frac{dU_{\mathrm{N}}}{dr} - \frac{3ZZ'e^2}{2\lambda + 1}\begin{cases} r^{\lambda}/R_{\mathrm{c}}^{\lambda+1}, & r < R_{\mathrm{c}}, \\ R_{\mathrm{c}}^{\lambda}/r^{\lambda+1}, & r > R_{\mathrm{c}}, \end{cases} \tag{7.64}$$

where R_{c} is the radius of a uniform nuclear charge distribution, ZZ' the product of the projectile and target charges, and U_N the nuclear part of the optical potential.

I. INELASTIC EXCITATION OF NUCLEAR ROTATIONAL LEVELS

The excitation of rotational levels of permanently deformed nuclei can be described very similarly to the excitation of vibrations because they both involve the interaction of the projectile with the target through its deviation from sphericity. The shape of a permanently deformed nucleus having a quadrupole distortion and axial symmetry is given by (7.53), with $\alpha_{\lambda\mu} = \beta_{\lambda}\delta_{\mu 0}$ and with θ, ϕ replaced by θ' and ϕ', which refer to a coordinate system fixed in the nucleus with z' axis along the symmetry axis of the nucleus. The DWBA amplitude, which is of first order in the deformation, is identical to (7.56) with the nuclear matrix element replaced by β_{λ}. However, for permanently deformed nuclei, β_{λ} is of such a magnitude that expansion (7.55) must be taken to higher order. At the same time, to treat the problem consistently, second-order transitions passing through intermediate states must be taken into account. This requires that we go beyond the DWBA. In Chapter 6 we gave an expansion of the transition amplitude that could be used to compute second- or higher-order contributions term by term. However, the problem can be solved to all orders in a given subspace of nuclear states by solving the coupled equations that can be derived from the Schrödinger equation. This method of solving the scattering problem will be taken up in the next chapter, and the explicit treatment of the very important problem of scattering by deformed nuclei will be given a detailed account in Chapter 13.

J. INELASTIC EXCITATION OF SINGLE-NUCLEON STATES

It is natural to expect that an incident projectile can excite a nucleon in the target from one single-particle state to another through their mutual interaction. The cross section for such a reaction was calculated by Austern *et al.* (1953) at a very early date in the development of the theory, employing the plane-wave approximation for spinless nucleons interacting through a zero-range surface interaction. The reader might find it profitable to repeat their calculation because, as in Chapter 2, it can be carried through in analytic form and will illustrate the characteristic strong dependence on the angular-momentum transfer. The first distorted-wave Born calculations were carried out soon afterward (see Levinson and Banerjee, 1958; Glendenning, 1959).

For simplicity, consider an odd nucleus with one nucleon outside of closed shells. We shall defer the general case to Chapter 9. Denote the scattered nucleon by the label 1 and the nucleons in the target by 2, 3, . . . , (A + 1). The antisymmetrized nuclear wave function has the form

$$\begin{aligned}\Phi_{\mathrm{i}}(\mathrm{A}) &= \mathscr{A}[\phi_{\mathrm{i}}(2)\Phi_0(3, 4, \ldots, \mathrm{A}+1)] \\ &= \mathrm{A}^{-1/2} \sum_P (-)^P P[\phi_{\mathrm{i}}(2)\Phi_0(3, 4, \ldots, \mathrm{A}+1)], \qquad (7.65)\end{aligned}$$

where $\mathscr{A}$ is the antisymmetrization operator defined by (5.28) and (5.29), ϕ_i the wave function of the odd particle state and Φ_0 that of the closed shell core having, therefore, zero angular momentum. We shall need matrix elements of the interaction between the scattered particle and those of the nucleus. The nuclear matrix element is

$$\begin{aligned}\langle\Phi_{\mathrm{f}}(\mathrm{A})|\sum_{j=2}^{\mathrm{A}+1} V(1,j)|\Phi_{\mathrm{i}}(\mathrm{A})\rangle &= \mathrm{A}^{-1}\sum_{j=2}^{\mathrm{A}+1}\langle\phi_{\mathrm{f}}(j)|V(1,j)|\phi_{\mathrm{i}}(j)\rangle \\ &= \langle\phi_{\mathrm{f}}(2)|V(1, 2)|\phi_{\mathrm{i}}(2)\rangle. \qquad (7.66)\end{aligned}$$

The orthonormality of the wave functions has required that the coordinates occupying ϕ_{f} and ϕ_{i} be the same as the nuclear coordinate in V. The normalization factor has been cancelled by the A equal terms in the sum over the target nucleons. The optical potential term of (5.25) vanishes identically because of the orthogonality of ϕ_{f} and ϕ_{i}.

In the distorted-wave Born approximation, the free nucleon is represented by a distorted wave in the relative coordinate between scattered nucleon and target and by its spin and isospin function

$$\psi_{\mathrm{i}}^{(+)}(1) = \chi_{\mathrm{i}}^{(+)}(\mathbf{k}_{\mathrm{i}}, \mathbf{r}_1)X_{1/2}^{m_{\mathrm{i}}}(\boldsymbol{\sigma}_1)\, X_{1/2}^{m_{t_{\mathrm{i}}}}(\tau_1) \qquad (7.67)$$

It is trivial to show that the antisymmetrization between projectile and target nucleons can be achieved by inserting the factor $1 - P_{12}$ in the matrix element, where P_{12} interchanges all coordinates of particle 1 and 2. Thus the transition amplitude is

$$\begin{aligned}\mathscr{T} &= \langle \psi_f^{(-)}(1)\phi_f(2)|V_{12}(1 - P_{12})|\phi_i(2)\psi_i^{(+)}(1)\rangle \\ &= \langle \psi_f^{(-)}(1)\phi_f(2)|V_{12}|\phi_i(2)\psi_i^{(+)}(1)\rangle - \langle \psi_f^{(-)}(1)\phi_f(2)|V_{12}|\phi_i(1)\psi_i^{(+)}(2)\rangle.\end{aligned} \tag{7.68}$$

The first matrix element is referred to as the direct and the second as the exchange amplitude. The latter is sometimes said to be nonlocal with respect to the distorted waves because (for the moment dropping reference to spin and isospin) $\mathscr{T}$ could be rewritten as

$$\begin{aligned}\mathscr{T} &= \int d\mathbf{r}_1\, d\mathbf{r}_1' \chi_i^{(-)*}(\mathbf{r}_1)\Big[\delta(\mathbf{r}_1 - \mathbf{r}_1')\int d\mathbf{r}_2\, \phi_f^*(\mathbf{r}_2)V(\mathbf{r}_1, \mathbf{r}_2)\phi_i(\mathbf{r}_2) \\ &\quad - \phi_f^*(\mathbf{r}_1')V(\mathbf{r}_1, \mathbf{r}_1')\phi_i(\mathbf{r}_1)\Big]\chi_f^{(+)}(\mathbf{r}_1') \\ &\equiv \int d\mathbf{r}_1\, d\mathbf{r}_1'\, \chi_i^{(-)*}(\mathbf{r}_1)F(\mathbf{r}_1, \mathbf{r}_1')\chi_f^{(+)}(\mathbf{r}_1').\end{aligned} \tag{7.69}$$

We will soon see that it is possible to reduce the direct term to a two-dimensional radial integration, whereas the exchange integral cannot be reduced (without approximation) from a six-fold coordinate-space integration. For this reason the early calculations neglected the exchange integral, on the expectation that it would be smaller than the direct one. This expectation was based on the orthogonality that should ideally hold between the bound and scattered states. This would not make it vanish identically because of the presence of the interaction but might make it small because of poor overlap. The neglect of exchange turns out often to be unjustified, but for the present we calculate only the direct amplitude. In Chapter 12 we shall discuss an approximation that can be used to reduce the exchange amplitude to the form of a direct one, with a suitably redefined interaction. Then, we can apply our knowledge about calculating the direct amplitude to the exchange amplitude.

The interaction that we use in (7.68) should ideally be related to the free interaction, but in general it should not be the free interaction itself. There are several reasons for this, including a very practical one: The free interaction is believed to have a strong repulsive core, and thus any perturbative treatment based on uncorrelated wave functions that do not tend to zero in the core region will fail. Second, and also very important, the scattering of two nucleons in the nuclear medium cannot take place as in free space, if for no other reason than that some of the momentum states into which the particles might scatter are already occupied. The medium is therefore said to polarize the interaction. These are subjects that we shall take up in greater detail in later chapters. For the present we take the interaction to be a phenomenological effective interaction, having a general central form.

The central interaction can have different strengths in the four different states of relative motion of two nucleons. These four states could be characterized as singlet even and odd and triplet even and odd, referring to the total spin and the even or odd character of the wave function of relative motion under interchange of the spatial coordinates. Alternatively it can be written as

$$V(r) = V_{00}(r) + V_{10}(r)\boldsymbol{\sigma}_1 \cdot \boldsymbol{\sigma}_2 + V_{01}(r)\boldsymbol{\tau}_1 \cdot \boldsymbol{\tau}_2 + V_{11}(r)\boldsymbol{\sigma}_1 \cdot \boldsymbol{\sigma}_2 \boldsymbol{\tau}_1 \cdot \boldsymbol{\tau}_2, \quad (7.70)$$

where σ and τ are the Pauli matrices for spin and isospin. Some authors use a different notation for the potentials in the four terms, namely, V_0, V_σ, V_τ, $V_{\sigma\tau}$. There are other ways of writing (7.70). For example, the spin and isospin products could be represented in terms of the exchange operators,

$$P^\sigma = (1 + \boldsymbol{\sigma}_1 \cdot \boldsymbol{\sigma}_2)/2, \qquad P^\tau = (1 + \boldsymbol{\tau}_1 \cdot \boldsymbol{\tau}_2)/2, \quad (7.71)$$

which have the eigenvalues

$$P^\sigma = \begin{cases} -1, & S = 0 \\ +1, & S = 1, \end{cases} \quad (7.72)$$

when acting on a two-particle state with spins coupled to $S = 0$ or 1. The various ways of writing the four-component character of the general central interaction are most convenient depending on the context. For our purpose (7.70) is a convenient characterization because we now show how to write it in terms of spherical tensors in orbital and spin angular momentum which are useful in characterizing the exchange of angular momentum between the relative motion and the nuclear states. To do this we need to express (7.70) as products of two operators, one acting on the nuclear coordinates and one on the coordinates of the scattered particle. We now seek such a representation.

Because the potentials $V(r)$ are functions of the distance between the two particles,

$$r = |\mathbf{r}_1 - \mathbf{r}_2| = (r_1^2 + r_2^2 - 2r_1 r_2 \cos\theta)^{1/2}, \quad (7.73)$$

the complete set of Legendre polynomials $P_L(\cos\theta)$ can be used to represent it:

$$V(r) = \sum_L v_L(r_1, r_2) P_L(\cos\theta). \quad (7.74)$$

The scalar functions v_L can be obtained by inversion of this equation. For certain functional forms of $V(r)$ they have closed forms (Levinson and

Banerjee, 1958; Glendenning, 1959) or can be obtained through recursion relations.

The Legendre polynomial can be expressed in terms of spherical harmonics depending on the directions $\hat{\mathbf{r}}_1$ and $\hat{\mathbf{r}}_2$ (see A9): Thus

$$V(r) = \sum_{LM} \frac{4\pi}{2L+1} v_L(r_1, r_2) Y_L^M(\hat{\mathbf{r}}_1) Y_L^{M^*}(\hat{\mathbf{r}}_2). \tag{7.75}$$

By virtue of the phase relation (A8), the complex conjugate sign can be placed on either of the spherical harmonics. It is also sometimes convenient to use the scalar product notation (A38)

$$Y_L(\hat{\mathbf{r}}_1) \cdot Y_L(\hat{\mathbf{r}}_2) \equiv \sum_M (-)^M Y_L^M(\hat{\mathbf{r}}_1) Y_L^{-M}(\hat{\mathbf{r}}_2). \tag{7.76}$$

One can also easily verify, using the special value for the Clebsch–Gordan coefficient

$$C_{M\,-M\,0}^{L\;L\;0} = (-)^{L-M}(2L+1)^{-1/2}, \tag{7.77}$$

that

$$\begin{aligned} Y_L(\hat{\mathbf{r}}_1) \cdot Y_L(\hat{\mathbf{r}}_2) &= (-)^L \sqrt{2L+1} \sum_M C_{M\,-M\,0}^{L\;L\;0} Y_L^M(\hat{\mathbf{r}}_1) Y_L^{-M}(\hat{\mathbf{r}}_2) \\ &\equiv (-)^L \sqrt{2L+1} [Y_L(\hat{\mathbf{r}}_1) Y_L(\hat{\mathbf{r}}_2)]_0^0. \end{aligned} \tag{7.78}$$

This expresses the scalar product in terms of a tensor product with coupling to angular momentum zero. Definition (7.76) and relation (7.78) hold quite generally for spherical tensors, not just for the spherical harmonics. In particular

$$\boldsymbol{\sigma}_1 \cdot \boldsymbol{\sigma}_2 = -\sqrt{3}[\boldsymbol{\sigma}_1 \boldsymbol{\sigma}_2]_0^0. \tag{7.79}$$

Consequently, when the expansion in (7.74) is employed in (7.70), the products of spherical harmonics and spin matrices will appear. Using the tensor product notation we can write

$$\begin{aligned} &Y_L(\hat{\mathbf{r}}_1) \cdot Y_L(\hat{\mathbf{r}}_2) \boldsymbol{\sigma}_1 \cdot \boldsymbol{\sigma}_2 \\ &\quad = (-)^{L+1} \sqrt{3(2L+1)} [Y_L(\hat{\mathbf{r}}_1) Y_L(\hat{\mathbf{r}}_2)]_0^0 [\boldsymbol{\sigma}_1 \boldsymbol{\sigma}_2]_0^0 \\ &\quad = (-)^{L+1} \sqrt{3(2L+1)} [[Y_L(\hat{\mathbf{r}}_1) Y_L(\hat{\mathbf{r}}_2)]_0^0 [\boldsymbol{\sigma}_1 \boldsymbol{\sigma}_2]_0^0]_0^0 \\ &\quad = (-)^{L+1} \sqrt{3(2L+1)} \sum_J \begin{bmatrix} L & L & 0 \\ 1 & 1 & 0 \\ J & J & 0 \end{bmatrix} [[Y_L(\hat{\mathbf{r}}_1) \boldsymbol{\sigma}_1]_J [Y_L(\hat{\mathbf{r}}_2) \boldsymbol{\sigma}_2]_J]_0^0. \end{aligned} \tag{7.80}$$

In the second line on the right we have expressed in terms of a sum over coupled states (A14′), of which, in this case of zero angular momentum, there is only one term. In the last line we then used the recoupling coefficients (A33) to obtain an expression involving the product of two operators, one acting on particle 1 and the other on particle 2. Define, for convenience, the tensor

$$T_{L1J}^{\;\;M}(1) \equiv [Y_L(\hat{\mathbf{r}}_1)\boldsymbol{\sigma}_1]_J^M, \tag{7.81}$$

so that the spin-independent terms in (7.70) can be written in a unified notation. Define also

$$T_{L0L}^{\;\;M}(1) \equiv Y_L^M(\hat{\mathbf{r}}_1) \tag{7.82}$$

or, alternatively, define

$$\Sigma_S = \begin{cases} \boldsymbol{\sigma}_1 & S = 1 \\ 1, & S = 0, \end{cases} \tag{7.83}$$

with

$$T_{LSJ}^{\;\;M} \equiv [Y_L \Sigma_S]_J^M = (-)^{L+S+J+M} T_{LSJ}^{\;\;-M\dagger}. \tag{7.84}$$

Inserting the value of the particular transformation coefficient in (7.66), we can write

$$Y_L(\hat{\mathbf{r}}_1) \cdot Y_L(\hat{\mathbf{r}}_2)\boldsymbol{\sigma}_1 \cdot \boldsymbol{\sigma}_2 = \sum_J (-)^{L+1+J} T_{L1J}(1) \cdot T_{L1J}(2). \tag{7.85}$$

Now, if we employ a common radial form for each term in (7.70), we can write

$$V_{00} + V_{10}\boldsymbol{\sigma}_1 \cdot \boldsymbol{\sigma}_2 = \sum_{LSJ} (-)^{L+S+J} \frac{4\pi}{2L+1} V_{S0} v_L(r_1, r_2) T_{LSJ}(1) \cdot T_{LSJ}(2), \tag{7.86}$$

and similarly for the $\tau_1 \cdot \tau_2$ terms.

We have achieved in (7.86) what we sought, the expression for the interaction that has the transformation properties under rotations expressed as a scalar product of two tensor operators acting separately on the spaces of particle 1 and 2. This can be referred to as the multipole expansion of the interaction.

Return now to the evaluation of the direct term of the transition matrix (7.68)

$$\begin{aligned} \mathscr{T}_D &= \langle \psi_f^{(-)}(1)\phi_f(2)|V_{12}|\phi_i(2)\psi_i^{(+)}(1)\rangle \\ &= \sum_{LSJ} \frac{4\pi}{2L+1} (-)^{L+S+J} \langle m_{t_f} M_{t_f}| V_{S0} + V_{S1}\boldsymbol{\tau}_1 \cdot \boldsymbol{\tau}_2 | m_{t_i} M_{t_i}\rangle \\ &\quad \times \int \chi_f^{(-)*}(\mathbf{k}_f, \mathbf{r}_1) \langle n_f \ell_f j_f M_f | v_L(r_1, r_2) T_{LSJ}^{\;\;M}(2) | n_i \ell_i j_i M_i\rangle \\ &\quad \times (-)^M \langle m_f | T_{LSJ}^{\;\;-M}(1) | m_i \rangle \chi_i^{(+)}(\mathbf{k}_i, \mathbf{r}_1)\, d\mathbf{r}_1, \end{aligned} \tag{7.87}$$

where

$$|m_t M_t\rangle \equiv x^{m_t}_{1/2}(\tau_1) X^{M_T}_{1/2}(\tau_2) \tag{7.88}$$

$$\begin{aligned}|n\ell jM\rangle &= \phi_{nl\,j}^{\ M} \\ &= u_{nl}(r_2)[Y_\ell(\hat{\mathbf{r}}_2) X_{1/2}(\sigma_2)]_j^M\end{aligned} \tag{7.89}$$

$$|m\rangle = X^m_{1/2}(\sigma_1). \tag{7.90}$$

Equation (7.89) represents a shell-model spin-orbit wave function with radial wave function $u_{n\ell}(r)$. The transition amplitude corresponds to the reaction

$$(m_i m_{t_i})_1 (M_i M_{t_i})_2 \rightarrow (m_f m_{t_f})_1 (M_f M_{t_f})_2, \tag{7.91}$$

where the doublets (mm_t), etc., denote the spin and isospin projection of scattered particle 1 and nuclear particle 2.

Equation (7.87) is now factorized in a very convenient way. The isospin matrix element is trivial. If both scattered and bound nucleons are alike, then it yields an effective interaction,

$$V = (V_{00} + V_{10}) + (V_{10} + V_{11})(\sigma_1 \cdot \sigma_2), \qquad \text{(n–n or p–p)}, \tag{7.92}$$

that is,

$$V_S \equiv V_{S0} + V_{S1},$$

whereas if they are unalike,

$$V = (V_{00} - V_{10}) + (V_{10} - V_{11})(\sigma_1 \cdot \sigma_2), \qquad \text{(n–p)}, \tag{7.93}$$

that is,

$$V_S \equiv V_{S0} - V_{S1}.$$

The nuclear matrix element can be expressed, through the Wigner–Eckart theorem (A39), in terms of a reduced matrix element that is independent of magnetic quantum numbers as

$$\begin{aligned}&\langle n_f \ell_f j_f M_f | v_L T_{LS\,J}^{\ M}(2) | n_i \ell_i j_i M_i \rangle \\ &= (-)^{j_f - M_f} \begin{pmatrix} j_f & J & j_i \\ -M_f & M & M_i \end{pmatrix} \langle n_f \ell_f j_f \| v_L(r_1, r_2) T_{LSJ}(2) \| n_i \ell_i j_i \rangle,\end{aligned} \tag{7.94}$$

where, in view of (7.89),

$$\begin{aligned}&\langle n_f \ell_f j_f \| v_L(r_1, r_2) T_{LS\,J}^{\ M}(2) \| n_i \ell_i j_i \rangle \\ &= \int u_{n_f \ell_f}(r_2) v_L(r_1, r_2) u_{n_i \ell_i}(r_2) r_2^2 \, dr_2 \langle (\ell_f \tfrac{1}{2}) j_f \| T_{LSJ} \| (\ell_i \tfrac{1}{2}) j_i \rangle.\end{aligned} \tag{7.95}$$

We can for convenience define

$$R^L_{n_f\ell_f n_i\ell_i}(r_1) = \frac{1}{2L+1}\int u_{n_f\ell_f}(r_2)v_L(r_1, r_2)u_{n_i\ell_i}(r_2)r_2^2\,dr_2. \tag{7.96}$$

The reduced matrix element of the spherical tensor can be evaluated using the results of the appendix (A45, A46, A53) to obtain

$$\langle(\ell_f\tfrac{1}{2})j_f\|[Y_L\Sigma_S]_J\|(\ell_i\tfrac{1}{2})j_i\rangle$$
$$= (-)^{\ell_f}\left[\frac{\hat{\ell}_f\hat{j}_f\hat{\ell}_i\hat{j}_i\hat{L}\hat{S}\hat{J}}{2\pi}\right]^{1/2}\begin{pmatrix}\ell_f & L & \ell_i\\ 0 & 0 & 0\end{pmatrix}\begin{Bmatrix}\ell_f & \ell_i & L\\ \frac{1}{2} & \frac{1}{2} & S\\ j_f & j_i & J\end{Bmatrix}, \tag{7.97}$$

where the large bracket is a $9-j$ symbol. (This particular one is related to the transformation coefficients from jj to LS coupling.) It is convenient to define the quantity

$$\mathscr{F}^{fi}_{LSJ}(r_1) \equiv \frac{4\pi}{2L+1}\langle n_f\ell_f j_f\|v_L(r_1, r_2)T_{LSJ}(2)\|n_i\ell_i j_i\rangle$$
$$= 4\pi R^L_{fi}(r_1)\langle j_f\|T_{LSJ}\|j_i\rangle \tag{7.98}$$

as the single-particle form factor for the indicated transition. We have used an obvious abbreviation to denote (7.97). For our simple example of single-nucleon excitation it corresponds to promotion of a nuclear particle from a single-particle state $n_i\ell_i j_i$ to $n_f\ell_f j_f$. We shall find that the corresponding form factor for a general nuclear transition induced by a one-body operator (such as is the interaction with the external particle) is simply a linear combination of such quantities as (7.98). How could it be otherwise? Thus our special example is much more instructive than might have been thought.

The last matrix element in (7.87) is the spin matrix element for the scattered particle. We evaluate it as follows:

$$(-)^M\langle m_f|T^{-M}_{LSJ}(1)|m_i\rangle$$
$$= (-)^M\langle m_f|[Y_L(\hat{\mathbf{r}})\Sigma_S(1)]_J^{-M}|m_i\rangle$$
$$= (-)^{L-S}\sqrt{2J+1}\sum_{M_L M_S}\begin{pmatrix}L & S & J\\ M_L & M_S & M\end{pmatrix}Y_L^{M_L}(\hat{\mathbf{r}}_1)\langle m_f|\Sigma_S^{M_S}|m_i\rangle$$
$$= (-)^{L-S+\frac{1}{2}-m_f}\sqrt{2J+1}\sum_{M_L M_S}\begin{pmatrix}L & S & J\\ M_L & M_S & M\end{pmatrix}\begin{pmatrix}\frac{1}{2} & S & \frac{1}{2}\\ -m_f & M_s & m_i\end{pmatrix}$$
$$\times\ \langle\tfrac{1}{2}\|\Sigma_S(1)\|\tfrac{1}{2}\rangle\, Y_L^{M_L}(\hat{\mathbf{r}}). \tag{7.99}$$

More generally, if the projectile is a composite particle with spin s_i and s_f, then

$$\langle s_f \| \Sigma_s \| s_i \rangle = \delta_{s_i s_f} \sqrt{s_i(s_i + 1)(2s_i + 1)}, \tag{7.100}$$

as was derived in the appendix (A54). The phase $(-)^{1/2 - m_f}$ should in this case be replaced by $(-)^{s_f - m_f}$.

Assembling these results,

$$\mathscr{T}_D = \sum_{LSJM_LM_SM} V_S(-)^{J + j_f - M_f + s_f - m_f} \sqrt{2J + 1}$$
$$\times \begin{pmatrix} j_f & J & j_i \\ -M_f & M & M_i \end{pmatrix} \begin{pmatrix} L & S & J \\ M_L & M_S & M \end{pmatrix} \begin{pmatrix} s_f & S & s_i \\ -m_f & M_S & m_i \end{pmatrix} \langle s_f \| \Sigma_S(1) \| s_i \rangle$$
$$\times \int \chi_f^{(-)*}(\mathbf{k}_f, \mathbf{r}_1) \mathscr{F}_{LSJ}^{fi}(r_1) Y_L^{M_L}(\hat{\mathbf{r}}_1) \chi_i^{(+)}(\mathbf{k}_i, \mathbf{r}_1)\, d\mathbf{r}_1. \tag{7.101}$$

Note that this has the structure discussed in Chapter 5 for the multipole expansion of the nuclear matrix element, as it must.

In the calculation of the cross section the magnetic sums eliminate the cross terms, just as in Chapter 5, and we find

$$\frac{d\sigma}{d\Omega} = \left(\frac{m}{2\pi h^2}\right)^2 \frac{k_f}{k_i} \frac{1}{(2J_i + 1)(2s_i + 1)}$$
$$\times \sum_{LSJM_L} \frac{\langle s_f \| \Sigma_s \| s_i \rangle^2 V_S^2}{(2S + 1)} |B_{LSJ}^{M_L}(\mathbf{k}_f, \mathbf{k}_i)|^2, \tag{7.102}$$

where we have defined

$$B_{LSJ}^{M_L}(\mathbf{k}_f, \mathbf{k}_i) = \frac{i^L}{(2L + 1)^{1/2}} \int \chi_f^{(-)*}(\mathbf{k}_f, \mathbf{r}) \mathscr{F}_{LSJ}^{f,i}(r) Y_L^{M_L}(\hat{\mathbf{r}}) \chi_i^{(+)}(\mathbf{k}_i, \mathbf{r})\, d\mathbf{r}. \tag{7.103}$$

This three-dimensional integral can be reduced to a sum of one-dimensional radial integrals by introducing the partial wave expansions for the distorted waves and then doing the angular integrals over the spherical harmonics analytically as in (7.26), which is precisely in the same form as (7.101). This then is our final result, because further evaluation would require numerical integration over the radial functions of the distorted waves, as in (7.27). We shall defer a detailed discussion and application of these results until Chapter 9.

For the present, let us merely observe that the angular distribution of the scattered particle is determined, as always in direct reactions, by integrals of the form (7.103). We further observe that this integral over distorted waves of the scattered particle is weighted by the form factor $\mathscr{F}(r)$ (a property of the nuclear states and the interaction), that its multipolarity is defined by the orbital angular momentum L, and that the orbital, spin, and total angular momenta L, S, J correspond to the transfer of these quantities between the target nucleus and the relative motion between target and projectile. Through the properties of the angular momentum factors appearing in the nuclear form factors (7.97) and (7.98) we can infer

$$
\begin{aligned}
(-)^{\ell_i+\ell_f} &= (-)^{L}, \qquad \mathbf{L}+\mathbf{S}=\mathbf{J}, \\
\boldsymbol{\ell}_f+\boldsymbol{\ell}_i &= \mathbf{L}, \qquad \tfrac{1}{2}+\tfrac{1}{2}=\mathbf{S}, \qquad \mathbf{j}_f+\mathbf{j}_i=\mathbf{J}.
\end{aligned}
\tag{7.104}
$$

The parity of the initial and final state are

$$\pi_i = (-)^{\ell_i}\pi_0, \qquad \pi_f = (-)^{\ell_f}\pi_0, \tag{7.105}$$

where π_0 is the parity of Φ_0, the unaltered state of the closed shells. The first equation (7.104) connects L to the change in parity induced in the nucleus by the reaction, because the left side is the change in parity of the nuclear states. Therefore we expect to be able to infer, through observations made on the scattered particle the parity and angular momentum transferred in the reaction.

K. CHARGE-EXCHANGE REACTIONS

The interaction between the projectile and scattered particle was written in (7.70). Alternatively, it can be written in terms of the spin and isospin exchange operator (7.71) as

$$
\begin{aligned}
V(r) = {} & (V_{00} - V_{10} - V_{01} + V_{11}) + 2(V_{10} - V_{11})P^{\sigma} \\
& + 2(V_{01} - V_{11})P^{\tau} + 4V_{11}P^{\sigma}P^{\tau}
\end{aligned}
\tag{7.106}
$$

Because this is written in isospin variables, the wave function on which it acts is assumed to be antisymmetrized. Therefore, the product of space exchange P^r and spin and isospin exchange must yield -1, so we could substitute

$$P^r = -P^{\sigma}P^{\tau}. \tag{7.107}$$

When written as in (7.106), the four terms are often referred to as Wigner (V_W), Bartlett (V_B), Heisenberg ($-V_H$), and Majorana ($-V_M$), respectively. This is most useful for the discussion of charge-exchange reactions such as (p, n) or (He^3, t). It was pointed out by Bloom *et al.* (1959) that the charge-exchange reaction is analogous to inelastic scattering in the sense that the charge exchange parts of the interaction will yield a transition amplitude of precisely the same structure as the direct amplitude of inelastic scattering. The only difference, an important one, is that a different combination of the components of the force are projected by the isospin matrix element in (7.87).

In the charge exchange case it is

$$V = 2(V_{01} - V_{11}) + 4V_{11}P^{\sigma} = 2V_{01} + 2V_{11}\boldsymbol{\sigma}_1 \cdot \boldsymbol{\sigma}_2, \qquad \text{(p, n)}, \quad (7.108)$$

which enters the direct integral, compared with (7.92) or (7.93) for inelastic scattering of like or unalike nucleons.[‡] Thus the different reactions can in principle be exploited to disentangle the components of the effective interaction. The cross section corresponding to the direct (i.e., charge exchange) amplitude of the (p, n) or other charge exchange reactions, can be calculated precisely as in the last section, with V_S given in this case by

$$V_0 = 2V_{01}, \qquad V_1 = 2V_{11}, \qquad \text{(p, n)}. \qquad (7.109)$$

The direct amplitude for (p, n), calculated as described, corresponds to the incident proton exchanging its charge through the interaction with a neutron in the nucleus. The exchange amplitude corresponds to the neutron being knocked out and the proton captured. Direct charge exchange reactions were expected to be especially strong for ground-state transitions between mirror nuclei because the other quantum numbers of the nuclear state remain unchanged.

L. DISTORTED-WAVE IMPULSE APPROXIMATION[§]

As the bombarding energy increases, the mean free path of a nucleon in nuclear matter becomes longer and its de Broglie wave length becomes shorter. Under ideal circumstances, in which the mean free path is long compared with the nuclear size and the wavelength is short compared with the average distance between nucleons in the nucleus (~ 2 fm), the scattering of a free and bound nucleon would take place as though both were free. For example, at 1000 MeV the mean free path (Chapter 4) is ~ 7 fm and the wavelength is $\lambda = 2\pi/k \sim 1$ fm.

The free nucleon–nucleon scattering can be described by the matrix element between plane-wave states of the T matrix, which was discussed in Chapter 6. For the scattering of two nucleons, this will be called t. Recall that it sums the scattering caused by the two-body potential to all orders. Thus, even if the nuclear potential were one that could be treated in perturbation theory, we would want to use t, not v, under conditions where we believe that the scattering takes place for free nucleons.

In practice we do not know the vacuum interaction, but we have considerable knowledge of the t matrix. This is provided by the phase shift

[‡] The same result can be obtained from (7.70) by writing $\boldsymbol{\tau}_1 \cdot \boldsymbol{\tau}_2$ in terms of raising and lowering operators $t^{\pm} = t^x \pm it^y$ and noting $\tau_1 \cdot \tau_2 = 2(t_1^+ t_2^- + t_1^- t_2^+) + 4t_1^z t_2^z$.

[§] See Kawai *et al.*, 1964, and Haybron and McManus, 1964.

analysis of nucleon–nucleon scattering data (Stapp *et al.*, 1957). We suppose therefore that such a matrix $\langle \mathbf{k}_1', \mathbf{k}_2'|t|\mathbf{k}_1, \mathbf{k}_2\rangle$ is available in the desired energy and momentum-transfer region. As far as possible, we want to use this as the interaction for our transition amplitude.

We follow closely the formulation of Kawai *et al.*, (1964) in what follows. Being a physical quantity, momentum is conserved by t, so we can write

$$\langle \mathbf{k}_1', \mathbf{k}_2'|t|\mathbf{k}_1, \mathbf{k}_2\rangle = \langle \mathbf{k}_1' - \mathbf{k}_2'|t^0|\mathbf{k}_1 - \mathbf{k}_2\rangle\, \delta(\mathbf{k}_1' + \mathbf{k}_2' - \mathbf{k}_1 - \mathbf{k}_2). \quad (7.110)$$

Moreover, at high energy the momentum of the nucleon (say, 2) in the nucleus is small compared with the other. Therefore

$$\langle \mathbf{k}_1', \mathbf{k}_2'|t|\mathbf{k}_1, \mathbf{k}_2\rangle \simeq \langle \mathbf{k}_1'|t^0|\mathbf{k}_1\rangle\, \delta(\mathbf{k}_1' + \mathbf{k}_2' - \mathbf{k}_1 - \mathbf{k}_2). \quad (7.111)$$

The scattering matrix $\langle k'|t^0|k\rangle$ is an operator in the spin space of the two nucleons and will be written explicitly later. For the present this need not concern us. We seek instead a coordinate representation to use in our theory. In its unapproximated form it is nonlocal in free and bound particle coordinates 1 and 2:

$$\begin{aligned}
\langle \mathbf{r}_1', \mathbf{r}_2'|t|\mathbf{r}_1, \mathbf{r}_2\rangle &= (2\pi)^{-12} \int d\mathbf{k}_1'\, d\mathbf{k}_2'\, d\mathbf{k}_1\, d\mathbf{k}_2 \\
&\quad \times \exp(i\mathbf{k}_1' \cdot \mathbf{r}_1') \exp(i\mathbf{k}_2' \cdot \mathbf{r}_2') \langle \mathbf{k}_1'\mathbf{k}_2'|t|\mathbf{k}_1\mathbf{k}_2\rangle \\
&\quad \times \exp(-i\mathbf{k}_1 \cdot \mathbf{r}_1) \exp(-i\mathbf{k}_2 \cdot \mathbf{r}_2) \qquad (7.112) \\
&\simeq (2\pi)^{-12} \int d\mathbf{k}_1'\, d\mathbf{k}_1\, d\mathbf{k}_2 \exp(i\mathbf{k}_1' \cdot \mathbf{r}_1') \\
&\quad \times \exp[i(\mathbf{k}_1 + \mathbf{k}_2 - \mathbf{k}_1') \cdot \mathbf{r}_2'] \langle \mathbf{k}_1'|t^0|\mathbf{k}_1\rangle \\
&\quad \times \exp(-i\mathbf{k}_1 \cdot \mathbf{r}_1) \exp(-i\mathbf{k}_2 \cdot \mathbf{r}_2) \\
&= (2\pi)^{-9} \delta(\mathbf{r}_2 - \mathbf{r}_2') \int d\mathbf{k}_1'\, d\mathbf{k}_1 \exp(i\mathbf{k}_1' \cdot \mathbf{r}_1') \\
&\quad \times \exp[i(\mathbf{k}_1 - \mathbf{k}_1') \cdot \mathbf{r}_2'] \langle \mathbf{k}_1'|t^0|\mathbf{k}_1\rangle \exp(i\mathbf{k}_1 \cdot \mathbf{r}_1). \qquad (7.113)
\end{aligned}$$

Now, in accordance with our high-energy assumption, the momenta $\mathbf{k}_1'$ and $\mathbf{k}_1$ should never differ much from their asymptotic values $\mathbf{k}_\mathrm{f}$ and $\mathbf{k}_\mathrm{i}$, respectively. For example, when the preceding result is substituted for the interaction in $\mathscr{T}$, we shall obtain factors

$$\left(\int d\mathbf{r}_1' \exp(i\mathbf{k}_1' \cdot \mathbf{r}_1')\, \chi_\mathrm{f}^{(-)}(\mathbf{k}_\mathrm{f}, \mathbf{r}_1')\right)^* \left(\int d\mathbf{r}_1 \exp(i\mathbf{k}_1 \cdot \mathbf{r}_1)\, \chi_\mathrm{i}^{(+)}(\mathbf{k}_\mathrm{i}, \mathbf{r}_1)\right).$$

If χ were a plane wave, which is an approximation that improves at high energy, these integrals would yield $\delta(\mathbf{k}_1' - \mathbf{k}_\mathrm{f})\, \delta(\mathbf{k}_1 - \mathbf{k}_i)$. Therefore, instead of using the local values of the momenta, we may use the asymptotic values

$k_1' \to k_f$ and $k_1 \to k_i$ in the preceding t matrix to obtain, upon setting

$$\mathbf{K} = \tfrac{1}{2}(\mathbf{k}_1' + \mathbf{k}_1), \qquad \mathbf{q} = \mathbf{k}_1 - \mathbf{k}_1', \tag{7.114}$$

the result

$$\begin{aligned} \langle \mathbf{r}_1', \mathbf{r}_2' | t | \mathbf{r}_1, \mathbf{r}_2 \rangle &= (2\pi)^{-9}\, \delta(\mathbf{r}_2 - \mathbf{r}_2')\, \langle \mathbf{k}_f | t^0 | \mathbf{k}_i \rangle \int d\mathbf{k}\, d\mathbf{q} \exp[-i\mathbf{K}\cdot(\mathbf{r}_1 - \mathbf{r}_1')] \\ &\quad \times \exp[-i\mathbf{q}\cdot[\tfrac{1}{2}(\mathbf{r}_1 + \mathbf{r}_1') - \mathbf{r}_2]] \\ &= (2\pi)^{-3}\, \delta(\mathbf{r}_2 - \mathbf{r}_2')\, \delta(\mathbf{r}_1 - \mathbf{r}_1')\, \delta(\mathbf{r}_1 - \mathbf{r}_2) \langle \mathbf{k}_f | t^0 | \mathbf{k}_i \rangle. \end{aligned} \tag{7.115}$$

Thus our approximations have yielded a local interaction of zero range, with a strength $\langle \mathbf{k}_f | t^0 | \mathbf{k}_i \rangle$, which depends on energy and the momentum transfer

$$q = |\mathbf{k}_f - \mathbf{k}_i| = (k_f^2 + k_i^2 - 2k_f k_i \cos\theta)^{1/2} \simeq 2k_i \sin\theta/2. \tag{7.116}$$

The most general form that $\langle k_f | t^0 | k_i \rangle$ can take, consistent with charge independence and invariance under space and time reflection (Wolfenstein, 1956; Dalitz, 1952; Clegg and Satchler, 1961; Kerman *et al.* 1959), is

$$\langle \mathbf{k}_f | t^0 | \mathbf{k}_i \rangle = a + b\boldsymbol{\sigma}_1 \cdot \hat{\mathbf{n}} \boldsymbol{\sigma}_2 \cdot \hat{\mathbf{n}} + c(\boldsymbol{\sigma}_1 + \boldsymbol{\sigma}_2) \cdot \hat{\mathbf{n}} + e\boldsymbol{\sigma}_1 \cdot \hat{\mathbf{q}} \boldsymbol{\sigma}_2 \cdot \hat{q} + f\boldsymbol{\sigma}_1 \cdot \hat{\mathbf{Q}} \boldsymbol{\sigma}_2 \cdot \hat{\mathbf{Q}}, \tag{7.117}$$

where

$$\mathbf{q} = \mathbf{k}_i - \mathbf{k}_f, \qquad \mathbf{Q} = \mathbf{k}_i + \mathbf{k}_f, \qquad \mathbf{n} = \mathbf{q} \times \mathbf{Q} \tag{7.118}$$

and $\hat{\mathbf{q}}$, $\hat{\mathbf{Q}}$, and $\hat{\mathbf{n}}$ are unit vectors. Each of the coefficients $a, b, \ldots, f$ are functions of energy and scattering angle and also of isospin, thus

$$a = a_0 + a_1\, \boldsymbol{\tau}_1 \cdot \boldsymbol{\tau}_2. \tag{7.119}$$

This spin dependence is more general than we have treated earlier, containing spin–orbit and tensor components in addition to the central part. These are fairly straightforward to handle in the preceding approximation (See Kawai *et al.*, 1964; Haybron and McManus, 1964). A less severe approximation, which retains a finite range for the interaction, will be taken up in Chapters 11 and 12.

Approximation (7.115) has been employed both in the plane-wave approximation (Kerman *et al.*, 1959) and in the distorted-wave approximation (Kawai *et al.*, 1964; Haybron and McManus, 1964) of the transition amplitude. In this latter case the approximate transition amplitude is referred to as the distorted-wave impulse approximation. We avoid use of the designation "Born" because this term refers to a calculation of first order in the potential. Use of the free t matrix implies that the free potential has been summed to all orders. This must be so because there is nothing in the observation of the scattering that prevents successive interactions, and t is taken directly from observation. In the last chapter we saw this in a formal expansion of t in terms of V(6.35).

M. COULOMB EXCITATION

Coulomb excitation refers to inelastic scattering in which the Coulomb interaction, rather than the nuclear interaction, is principally responsible for the excitation. In the sense that the Coulomb interaction is better known than the nuclear, this can have some advantage as far as the measurement of the electric multipole moments of nuclear states is concerned. It is at low energy that the Coulomb interaction dominates. To understand why, note that the optical potential, which describes the average interaction between two scattering partners, has a nuclear and Coulomb part which are given typically by (4.34) and (4.35). The nuclear part is attractive and of short range, following as it does the nuclear density, whereas the Coulomb part is repulsive and of long range. As a consequence, the optical potential is repulsive at large distance and attractive at shorter distance. The maximum in the potential between these regions is called the barrier energy. A classical particle with energy below this value cannot enter the nucleus. Of course, there is a finite but small quantum-mechanical probability for penetration. However, it is true that at a bombarding energy below the barrier, the main interaction between the scattering partners must be the Coulomb interaction. If the product of charges is small, then the cross sections for all inelastic and reaction processes will simply be small. However, if the product of charges is large, there may be appreciable inelastic cross sections. The excitation of vibrational or rotational states, which involves the coherent participation of many nucleons and therefore appreciable charge, is a good example in which the cross section can be large and nearly purely Coulombic. Such processes are often referred to as Coulomb excitation. The quantum-mechanical theory for excitation of surface vibrations, discussed earlier in this chapter, is quite general, applying at energies above or below the barrier. Therefore, nothing need be added to the theory developed so far, if the first-order treatment is adequate.

There are, however, practical differences in the numerical calculation of the cross section. Because of the repulsion of the optical potential under circumstances in which the Coulomb interaction dominates, large partial waves, which are only slightly scattered by the repulsive field, scatter to forward angles. Small partial waves, which are strongly repelled, scatter to large angles. Therefore a large number of partial waves may be needed to obtain convergence for small scattering angles. To describe accurately the cross sections for angles θ greater than θ_0, one would need all partial waves up to the value corresponding to classical scattering at θ_0. This is $l_0 = \eta \cot(\theta_0/2) \simeq 2\eta/\theta_0$, where η is the Coulomb parameter (4.10). For low energy and large charges, η can be of the order of 100. Therefore, l_0 can be very large. Although fast electronic computers can be used to perform such calculations, semiclassical methods were first developed for handling Coulomb excitation (Alder *et al.*, 1956; Alder and Winther, 1966). The classical description of elastic scattering should be valid if the wavelength λ associated with the relative motion is small compared with the distance of closest approach of the classical orbit. For the Coulomb potential this can be translated into the condition $\eta \gg 1$. Under such circumstances the Coulomb process can be approximated by calculating the classical trajectory of the scattering nuclei. Along such a trajectory the electromagnetic field between them is changing. The time variation of the field at the location of one of the nuclei can excite it. This is calculated in time-dependent perturbation theory. Nörenberg and Weidenmüller (1976) give a nice account of classical scattering and the Coulomb excitation process and references to the more detailed literature.

As the bombarding energy is raised to the vicinity of the barrier, usually called the Coulomb barrier, the nuclear interaction becomes effective. Its opposite sign can partially cancel the effect of the Coulomb field for a certain group of orbits, giving rise to what is referred to as nuclear–Coulomb interference or a reduction in cross section near the classical scattering angle corresponding to these orbits. When the energy is at or above the barrier, the classical deflection function ceases to be a monotonic function. It becomes multivalued. In such cases the interference between trajectories that have the same scattering angle becomes an important feature. These matters are discussed in greater detail in connection with heavy-ion scattering (Chapter 18).

Notes to Chapter 7

A review of transfer reactions can be found in Austern's (1970) book and in the review of Glendenning (1963, 1975b).

Chapter 8

Coupled Equations and the Effective Interaction

A. COUPLED EQUATIONS FOR INELASTIC SCATTERING

In Chapters 3, 5, and 6 we derived exact expressions for the $\mathcal{T}$ matrix, and we also discussed two approximations, one based on the Born series and one on the distorted-wave Born series. The DWBA is a very valuable approximation which is widely employed in the interpretation of data from direct reactions. It is valid for transitions that are not too strong, although it is difficult to give a quantitative measure to this statement. Generally we know that transitions involving collective states are too strong to be treated in first order. The series expansions of the transition amplitude, whether the Born (6.35) or distorted Born series (6.52, 6.58), are not convenient as a means of calculating higher-order corrections, generally speaking. Each order is increasingly more cumbersome. There is an alternative to summing the series term by term that consists of developing the Schrödinger equation as a system of coupled equations. Then, subject to the approximations with which this system is obtained, its solution accounts for the interaction to all orders.

As in Chapter 3, denote by H_α the intrinsic nuclear Hamiltonian(s)

$$H_\alpha = H_{\mathrm{A}} + H_{\mathrm{a}}. \tag{8.1}$$

Denote the kinetic energy operator of relative motion by T, and the interaction by V. (Because we discuss only elastic and inelastic scattering, the channel subscript α is not needed on T and V.)

Suppose that we have model Hamiltonian(s) for the target H_A (and projectile H_a) and we wish to use scattering experiments to test the appropriateness of the model. In other words, the solution to

$$(H_A - \varepsilon_A)\Phi_A = 0 \tag{8.2}$$

(and the corresponding equation for a) is known from a nuclear model. The Hamiltonian for the entire system, including relative motion and interaction between target and projectile, is

$$H = H_\alpha + T + V. \tag{8.3}$$

The potential V is a sum of two-body interactions between particles in the projectile and those in the target, as in (3.6). At this point, V_{ij} is still regarded as a well-behaved potential of a phenomenological character and is not the vacuum nucleon–nucleon interaction. The connection with the vacuum interaction will be alluded to later in this chapter and taken up in greater detail in Chapters 11 and 12.

To embed our model solutions into the problem it is natural to expand the solution to the Schrödinger equation

$$(E - H)\Psi = 0 \tag{8.4}$$

in terms of the model wave functions

$$\Psi = \sum_{\alpha'} \psi_{\alpha'}(\mathbf{r})\Phi_{\alpha'}(\mathbf{A}, \mathbf{a}), \tag{8.5}$$

where

$$\Phi_\alpha = \Phi_A \Phi_a \tag{8.6}$$

denotes our model nuclear wave functions. Insert the expansion into (8.4), multiply by Φ_α and integrate on internal coordinates, to obtain

$$(E_\alpha - T - V)\psi_\alpha(\mathbf{r}) = \sum_{\alpha' \neq \alpha} V_{\alpha\alpha'}(\mathbf{r})\psi_{\alpha'}(\mathbf{r}), \tag{8.7}$$

where

$$V_{\alpha\alpha'}(\mathbf{r}) = (\Phi_\alpha | V | \Phi_{\alpha'}) \tag{8.8}$$

is a matrix element over the internal coordinates. There is clearly one such equation for each state $\alpha, \alpha', \ldots,$ and they are coupled by the terms on the right-hand side. It is important to notice that all of the transitions represented on the right of (8.7) contribute coherently.

The solution of these equations for the ψ's would provide us with the solution to H in terms of the model wave functions. Although in principle the model wave functions are a complete set, in practice we cannot solve the infinite set of equations implied by (8.7). Therefore we must recognize that, at best, a practical calculation is a truncated and therefore approximate one.

B. TRUNCATION AND EFFECTIVE INTERACTIONS

We recognized previously that a straightforward derivation of coupled equations leads to an infinite set. This is just a manifestation of the fact that we are dealing with a many-body scattering problem. A generalization (Glendenning, 1969a) of Feshbach's (1958, 1962) discussion of the optical model permits a reformulation so that the problem of solving an infinite set of coupled equations is replaced by that of solving a finite set with a modified or effective interaction.

We recognize that the cross sections to only a few of the many channels that are open in a typical cyclotron or tandem accelerator experiment can be resolved from each other. Therefore, the amount of precise information available is always limited. We shall now build a theory that places an emphasis on the limited data that can be obtained. Accordingly, let P refer to the elastic and a few inelastic channels that can be resolved and that are open, and let Q refer to all the others, which could be tens of thousands in a medium to heavy nucleus at tens of mega electron volts bombarding energy (see Fig. 8.1). Also, let P and Q denote operators that project onto

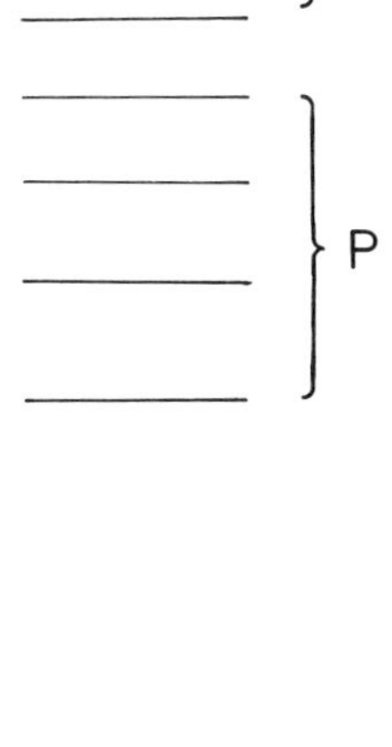

Fig. 8.1 Schematic level diagram of nucleus and a division into included and excluded levels P and Q.

the included and excluded parts of the wave functions, so that $P\Psi$ asymptotically contains only the P channels. Clearly

$$P + Q = 1. \tag{8.9}$$

Therefore (8.4) can also be written as

$$(E - H)(P + Q)\Psi = 0. \tag{8.10}$$

Project from the left with Q to obtain

$$(E - H_{QQ})(Q\Psi) = H_{QP}(P\Psi), \tag{8.11}$$

where

$$H_{QP} = QHP, \quad \text{etc.} \tag{8.12}$$

This equation, regarded as an operator equation, can be solved formally as in Chapter 6:

$$Q\Psi = \frac{1}{E - H_{QQ} + i\varepsilon} H_{QP}(P\Psi), \tag{8.13}$$

where, as usual, $i\varepsilon$ yields the asymptotic boundary condition on the Green's function that there are outgoing waves. The solution to the homogeneous equation $(E - H_{QQ})\chi = 0$ has not been added because any solution of the homogeneous equation has zero net flux through any closed surface about the origin and, hence, violates the conditions we wish to describe, which correspond to outgoing waves only in all but the target channel.

Similarly projecting from the left with P yields

$$(E - H_{PP})(P\Psi) = H_{PQ}(Q\Psi). \tag{8.14}$$

Inserting the preceding expression for $Q\Psi$ yields, upon rearrangement,

$$\left[E - H_{PP} - H_{PQ}\frac{1}{E - H_{QQ} + i\varepsilon} H_{QP}\right](P\Psi) = 0, \tag{8.15}$$

which is the equation that is satisfied by that part of Ψ in which we are really only interested, namely, $P\Psi$. It contains all the information necessary for calculating transitions *among* states belonging to P. We denote it more conveniently by $\Psi(P) \equiv P\Psi$.

We then rewrite (8.15) as

$$(E - H_\alpha - T - \mathscr{V})\Psi(P) = 0, \tag{8.16}$$

where

$$\mathscr{V} = V_{PP} + V_{PQ}\frac{1}{E - H_{QQ} + i\varepsilon} V_{QP}. \tag{8.17}$$

Here we have used the obvious fact that $PQ = QP = 0$. Because T commutes with P and Q (it acts only on the relative coordinate), then $H_{PQ} = V_{PQ}$.

Moreover, if we agree henceforth to solve the Schrödinger equation only in the P space, then we can drop the outer projection operators defining $\mathscr{V}$,

$$\mathscr{V} = V + V \frac{Q}{E - H + i\varepsilon} V. \tag{8.18}$$

From the identity $\Psi = P\Psi + Q\Psi$ and Eq. (8.13),

$$\Psi = \left(1 + \frac{Q}{E - H + i\varepsilon} V\right) \Psi(P). \tag{8.19}$$

Also from the general result (3.33),

$$\mathscr{T}_{\alpha'\alpha} \equiv \langle \phi_{\alpha'} | V_\alpha | \Psi_\alpha^{(+)} \rangle = \langle \phi_{\alpha'} | \mathscr{V} | \Psi_\alpha^{(+)}(P) \rangle. \tag{8.20}$$

This is an important result. It tells us that the *formally* exact expression for the parts of the T matrix referring to transitions between states α and α', both belonging to the included channels P, can be calculated from the *truncated* Schrödinger equation of Eq. (8.16) (Glendenning, 1969a). However, this truncated problem contains not the original interaction V, but an effective one $\mathscr{V}$, which is to be used in the space P. Although the problem is still no easier to solve exactly, it has been cast into a form that is more suitable for making the approximations that lead to a tractable problem, that is, a finite number of coupled channels. It also suggests, as will be discussed, some of the expected properties of the effective interaction $\mathscr{V}$ that is to be used in the truncated space.

We have referred to the transformed Schrödinger equation (8.16) as truncated because it leads to a truncated set of coupled equations. If the steps leading to (8.7) are repeated with $\mathscr{V}$ replacing V in (8.3), then the resulting coupled equations are the finite set referring explicitly to only the subset P of all states.

This clearly informs us that it would be wrong to truncate Eq. (8.7) without at the same time using the effective interaction $\mathscr{V}$ instead of V. We now discuss this very important effective interaction.

C. THE EFFECTIVE INTERACTION

Whereas the original scattering problem, governed by $H = H_\alpha + T + V$ led to an infinite set of coupled equations corresponding to all channels, a modified problem governed by $\mathscr{H} = H_\alpha + T + \mathscr{V}$, which is to be solved only within a finite subspace of channels, can be defined so that it leads to the same $\mathscr{T}$ matrix elements between channels of the subspace (Glendenning, 1969a). Of course this is a statement that is only formally correct because we have not discussed how to calculate $\mathscr{V}$. We would need the complete spectrum of H_{QQ} to do so. At the energies of interest to us, there are millions of open channels in medium to heavy nuclei so that an exact calculation of $\mathscr{V}$ is not feasible and approximations would have to be made. Important progress is being made in this direction.[‡] Nevertheless it is a very important conceptual result because we can make certain statements concerning the nature and behavior of $\mathscr{V}$. For example, the usual optical potential for elastic scattering corresponds just to the case in which P projects on the ground state alone. Only the elastic channel is treated explicitly, all other entering channels are treated implicitly through $\mathscr{V}$. Although we cannot calculate the optical potential from first principles, it can be parametrized very successfully. From its definition [Eq. (8.18)], $\mathscr{V}$ is explicitly dependent on the energy, is complex valued, and is nonlocal. The second term makes a large contribution to diagonal matrix elements but is probably small for off-diagonal ones. This can be seen by introducing the eigenstates of H_{QQ} in a typical diagonal matrix element

$$\mathscr{V}_{\alpha\alpha} \equiv (\Phi_{\alpha J}|\mathscr{V}|\Phi_{\alpha J}) \equiv V_{\alpha\alpha}(\mathbf{r}) + \sum_C \frac{(\Phi_{\alpha J}|V|\Psi_C)(\Psi_C|V|\Phi_{\alpha J})}{E - E_C + i\varepsilon}, \tag{8.21}$$

where Ψ_C is intended to denote the wave function for the A + a system and Φ_α the intrinsic wave function of the nuclei. The parentheses denote integrations over the intrinsic coordinates and remain functions of the remaining coordinate r. The sum on C runs over the space of Q and includes continuum and discrete states. The numerator is positive definite and the denominator changes sign only once. There is therefore little chance that the sum is small. However, the individual terms in the second part of $\mathscr{V}$ generally will be very small because they involve matrix elements connecting the two parts of the space P and Q, whose states will usually be of quite a different structure because the latter lie at higher excitation in the nucleus A.

This means that the sum, which is over an enormous number of terms [a continuum built on each state C of the A + a particle system], is not dominated by any particular one of them as long as no states of Q are strongly coupled to any of P. With the proviso that the effect of such strong coupling be removed from $\mathscr{V}$ by treating them explicitly in P, then the sum C in Eq. (8.21) will be dominated by the high-excitation region because of the high-level density. From this fact we see that $\mathscr{V}_{\alpha\alpha}$ will be slowly varying from

[‡] See von Geramb (ed., 1979), especially the work reported by and references to the Liege Group (Jeukenne, Lejuene, and Mahaux), the Orsay Group (Vinh Mau and Bouyssy), and the Oxford Group (Brieva, Kidwai, and Rook).

nucleus to nucleus and that the diagonal matrix elements in the space P will be similar. We do know that the optical-model treatment of elastic scattering is generally satisfactory. This suggests that we treat the *diagonal* matrix elements of (8.18) by a similar phenomenological parametrization.

For the off-diagonal matrix elements of (8.18), on the other hand, the numerator of the second term is not positive but will fluctuate in sign from term to term so that in general we may expect it to give a small contribution in comparison with the first.

D. NONLOCALITY

To demonstrate the nonlocal nature of the second term of $\mathscr{V}$, which for convenience we call $\mathscr{W}$, we examine the coordinate representation of $\mathscr{W}$,

$$\begin{aligned}\langle \mathbf{r}, \mathbf{A}, \mathbf{a}|\mathscr{W}|\mathbf{r}', \mathbf{A}', \mathbf{a}'\rangle &= \sum_C \frac{\langle \mathbf{r}, \mathbf{A}, \mathbf{a}|V|C\rangle\langle C|V|\mathbf{r}', \mathbf{A}', \mathbf{a}'\rangle}{E - E_C + i\varepsilon} \\ &= \sum_C \frac{V(\mathbf{r}, \mathbf{A}, \mathbf{a})\Psi_C(\mathbf{r}, \mathbf{A}, \mathbf{a})\Psi_C^*(\mathbf{r}', \mathbf{A}', \mathbf{a}')V(\mathbf{r}', \mathbf{A}', \mathbf{a}')}{E - E_C + i\varepsilon},\end{aligned} \tag{8.22}$$

where we have used

$$\Psi_C(\mathbf{r}, \mathbf{A}, \mathbf{a}) \equiv \langle \mathbf{r}, \mathbf{A}, \mathbf{a} \,|\, C\rangle \tag{8.23}$$

to denote a wave function of H_{QQ} for the A + a particle composite, and the fact that V is local (i.e., diagonal in the coordinate representation),

$$V(\mathbf{r}, \mathbf{A}, \mathbf{a}) \equiv \langle \mathbf{r}, \mathbf{A}, \mathbf{a}|V|\mathbf{r}, \mathbf{A}, \mathbf{a}\rangle. \tag{8.24}$$

We can now write the effect of $\mathscr{W}$ acting on the wave functions we have been using. In a coordinate representation, we have

$$\begin{aligned}\langle \mathbf{r}, \mathbf{A}, \mathbf{a}|\mathscr{W}|\psi, \Phi\rangle = \sum_C V(\mathbf{r}, \mathbf{A}, \mathbf{a})\Psi_C(\mathbf{r}, \mathbf{A}, \mathbf{a}) \frac{1}{E - E_C + i\varepsilon} \int d\mathbf{r}'\, d\mathbf{A}'\, d\mathbf{a}' \\ \times \Psi_C^*(\mathbf{r}', \mathbf{A}', \mathbf{a}')V(\mathbf{r}', \mathbf{A}', \mathbf{a}')\psi(\mathbf{r}')\Phi(\mathbf{A}', \mathbf{a}').\end{aligned} \tag{8.25}$$

E. MULTIPLE-SCATTERING SERIES FOR THE EFFECTIVE INTERACTION

In Chapter 6 we derived Watson's multiple-scattering series for the T matrix. We emphasized that it is of fundamental importance in nuclear physics. Now we shall see why this is so, using the results of Chapter 6 to derive an analogous series for the effective interaction, in terms of a nonsingular two-body t matrix. This two-body t matrix has the same significance for scattering theory that the Breuckner G matrix has for nuclear structure theory. In fact, aside from the boundary condition, they obey the same kind of equation.

Use the operator identity (6.20) to rewrite (8.18) as an integral equation for $\mathscr{V}$:

$$\mathscr{V} = V + V \frac{Q}{E - H_\alpha - T_\alpha + i\varepsilon} \mathscr{V}. \tag{8.26}$$

(We have replaced the channel label α on the kinetic energy T.) This integral equation is precisely the same as (6.63) except for the occurrence of the projection operator Q. Therefore, the results of Chapter 6 can be directly transcribed simply by multiplying the propagator (6.67) by Q. Equation (6.76) therefore tells us at once that

$$\mathscr{V} = \sum_{j=1}^{A} \tau_j + \sum_{j,\,k \neq j} \tau_j g^{(+)} \tau_k + \sum_{\substack{j,\,k \neq j \\ \ell \neq k}} \tau_j g^{(+)} \tau_k g^{(+)} \tau_\ell + \cdots, \tag{8.27}$$

where

$$g^{(+)} = \frac{Q}{(E - H_0 + i\varepsilon)}, \qquad H_0 = H_\alpha + T_\alpha, \tag{8.28}$$

and τ_j is a two-body operator acting between the projectile and the jth target nucleon. Similarly, V_j is the interaction between these two particles, and was written in (6.65), and τ_j is given by

$$\tau_j = V_j + V_j g^{(+)} \tau_j. \tag{8.29}$$

This is a nonsingular operator, as discussed in Chapter 6, even though V_j is singular. However, the τ defined here is different from Watson's (6.69) because of the projection operator Q. This is a significant difference. In the formulation of this chapter, certain low-lying levels, especially collective ones, are treated explicitly through the coupled equations. All other levels belong to the Q space. This suggests that there are simplifications possible in the calculation of (8.29), which otherwise would not be viable. One could consider, for example, introducing plane-wave states of appropriately high momentum as an approximation to the noncollective highly excited states of the nucleus, in a nuclear-matter approximation of τ.

We discussed the virtues of series like (8.27) in Chapter 6 but will briefly mention them again. The successive terms in (8.27) give rise to events in which the projectile scatters from one, two, three, etc., different target nucleons. Each such scattering is represented by τ, not V, so that the full scattering sequence in each collision is accounted for [see (6.77)].

The actual nucleon–nucleon interaction V_j is highly singular, rendering invalid any perturbative theory based on it. For this reason we have assumed from the beginning of this book that we can introduce a well-behaved effective interaction and develop a perturbative theory. Now we have the justification for this approach. We have a formal operator $\mathscr{V}$ that allows us to obtain a formally exact solution to the scattering problem in a subspace chosen at our convenience. The full effective interaction in the subspace can be developed in a multiple-scattering series (8.27) in terms of the nonsingular operator. This is the operator for scattering theory that corresponds closely to the Breuckner G matrix. Indeed, Breuckner was inspired by Watson's ingenious reordering of the series to define the G matrix.

In (8.27) we see how the full effective interaction is to be constructed from τ. In practice it may suffice to retain only the first term. This would correspond to the assumption that the nucleus is a dilute system, and therefore that the likelihood that the projectile will strike more than one different target nucleon is small. Direct reactions are of such a nature as to emphasize just such events. The replacement of $\mathscr{V}$ in (8.16) by the first term in (8.27) retrieves just such a theory as was laid out in Chapter 3.

In Chapter 11 we shall discuss the operators τ and G again. They are not easy to calculate, of course, but considerable progress on these problems, especially on the bound-state problem, has been made.

F. PARTIAL-WAVE EXPANSION

For clarity in introducing the concept of truncation and the corresponding effective interaction, we did not separate the radial from the angular coordinates in the coupled equations (8.7). In any practical calculation this is done because the angular integrations can be performed in closed form through the well-known properties of the spherical harmonics. Therefore we now develop the truncated coupled equations corresponding to the problem of Eq. (8.16) in partial-wave components.

Several different coupling schemes could be employed. For example, if the projectile a is a nucleon or light nucleus (say, up to the alpha particle), then the spin–orbit coupling scheme is a convenient one. For this case we define spin–orbit functions

$$\mathscr{Y}^m_{\ell J_a j} = [Y_\ell(\hat{\mathbf{r}})\Phi_{J_a}(\mathrm{a})]^m_j, \tag{8.30}$$

where $\hat{\mathbf{r}}$ is the angle direction of $\mathbf{r}$ (the relative coordinate between a and A) defined in Chapter 3 and Φ_{J_a} the wave function of the projectile. Here we have explicitly written its spin quantum number, which has only been implicitly implied until now. In the case is which a is simply a nucleon, then $\mathscr{Y}$ is obviously the usual spin–orbit function

$$\mathscr{Y}^m_{\ell sj} = [Y_\ell(\hat{\mathbf{r}})X_{1/2}(\boldsymbol{\sigma})]^m_j. \tag{8.31}$$

The description of the angular momentum coupling is completed by constructing the channel functions

$$\phi^M_{\alpha\pi I}(\hat{\mathbf{r}}, \mathbf{a}, \mathbf{A}) = [\mathscr{Y}_{\ell J_a j}\Phi_{J_A}(\mathrm{A})]^M_I. \tag{8.32}$$

The parity π of this partial-wave channel is clearly

$$\pi = (-)^\ell \pi_a \pi_A, \tag{8.33}$$

and I denotes the total angular momentum

$$\mathbf{I} = \mathbf{j} + \mathbf{J}_A = \boldsymbol{\ell} + \mathbf{J}_a + \mathbf{J}_A. \tag{8.34}$$

Alternatively, one could employ what is referred to as the channel–spin coupling scheme. In this case the two nuclear wave functions are coupled together first and then they are coupled to the angular momentum of relative motion ℓ:

$$[Y_\ell(\hat{\mathbf{r}})[\Phi_{J_a}\Phi_{J_A}]_S]^M_I. \tag{8.35}$$

The two schemes can be transformed, one into the other, by the techniques of Racah algebra

$$[Y_\ell[\Phi_{J_a}\Phi_{J_A}]_S]^M_I = \sum_j ((\ell J_a)j, J_A; I|\ell, (J_a J_A)S; I)[[Y_\ell \Phi_{J_a}]_j \Phi_{J_A}]^M_I, \tag{8.36}$$

where the transformation bracket is related to the Racah coefficient, or alternatively, the 6-j symbol.

$$((j_1 j_2)j_{12}, j_3; J|j_1, (j_2 j_3)j_{23}; J) = (\hat{j}_{12}\hat{j}_{23})^{1/2}(-)^{j_1+j_2+j_3+J}\begin{Bmatrix} j_1 & j_2 & j_{12} \\ j_3 & J & j_{23} \end{Bmatrix}, \tag{8.37}$$

where $\hat{j} = 2j + 1$. Any calculation performed in one can, by these relations, be expressed in the other.

The spin–orbit scheme is most convenient for nucleons because the matrix elements of the spin–orbit term in the optical potential (i.e., diagonal elements of $\mathscr{V}$) can be written immediately as

$$2\boldsymbol{\ell}\cdot\mathbf{s}\mathscr{Y}_{\ell sj} = (j^2 - \ell^2 - s^2)\mathscr{Y}_{\ell sj} = [j(j+1) - \ell(\ell+1) - \tfrac{3}{4}]\mathscr{Y}_{\ell sj}; \tag{8.38}$$

this is the scheme we shall use.

In terms of the channel functions ϕ defined by (8.32), a state Ψ of the total system with parity and angular momentum πI can be expanded as

$$\Psi_{\alpha\pi I}^{M} = 1/r \sum_{\alpha'} u_{\alpha'}^{\alpha\pi I}(r)\phi_{\alpha'\pi I}, \tag{8.39}$$

which is our partial-wave expansion corresponding to (8.5). However, we are economizing on subscripts by understanding, as suggested by the right side of Eq. (8.32), that when α is used as a channel subscript or superscript in these partial-wave expansions it stands for the collection of quantum numbers

$$\alpha = J_{\mathrm{a}}J_{\mathrm{A}}\ell j.$$

Inserting this expansion into the Schrödinger equation, we find

$$(E_{\alpha'} - \mathrm{T}_{\alpha'\ell'} - \mathscr{V}_{\alpha'\alpha'}^{I})u_{\alpha'}^{\alpha\pi I}(r) = \sum_{\alpha'' \neq \alpha'} \mathscr{V}_{\alpha'\alpha''}^{I} u_{\alpha''}^{\alpha\pi I}(r), \tag{8.40}$$

where

$$T_{\alpha\ell} = \frac{\hbar^2}{2m_\alpha}\left(-\frac{d^2}{dr^2} + \frac{\ell(\ell+1)}{r^2}\right), \tag{8.41}$$

$$E_\alpha = E - \varepsilon_\alpha = E - \varepsilon_{\mathrm{a}} - \varepsilon_{\mathrm{A}} \tag{8.42}$$

$$k_\alpha^2 = (2m_\alpha/\hbar^2)E_\alpha, \qquad v_\alpha = \hbar k_\alpha/m_\alpha \tag{8.43}$$

$$\mathscr{V}_{\alpha\alpha'}^{I}(r) = (\phi_{\alpha\pi I}|\mathscr{V}|\phi_{\alpha'\pi I}), \tag{8.44}$$

where the parentheses denote integration over all coordinates except r. Thus $\mathscr{V}_{\alpha\alpha'}^{I}$ is diagonal in π and I and independent of M if $\mathscr{V}$ is scalar. As already discussed, the diagonal elements of $\mathscr{V}$ will be parametrized as a phenomenological optical model in practice.

The superscripts $\alpha\pi J$ on u and these same subscripts on Ψ serve to remind one that the incident partial-wave channel is described by these quantum numbers.

Because of the importance of interference phenomena we stress again that the right side of (8.40) represents a sum of coherent contributions to the channel α' from all other, α'', that have finite matrix elements $\mathscr{V}_{\alpha'\alpha''}$ and that are included in the calculation. The *direct* transition from the ground state $\alpha' \to \alpha$ is represented by the $\mathscr{V}_{\alpha'\alpha}u_\alpha$ term. *Indirect* transitions through intermediate states $\alpha'' \neq \alpha$ are represented by the sum of other terms $\mathscr{V}_{\alpha'\alpha''}u_{\alpha''}$. It is because the channel $\alpha'' \neq \alpha$ is not initially populated that such terms are called *indirect*. Sometimes the transitions they represent are referred to as *two-step* transitions, because the lowest-order process contributing to them is $\alpha \to \alpha'' \to \alpha'$, which is of second order in non-diagonal matrix elements of $\mathscr{V}$.

G. BOUNDARY CONDITIONS

The solutions to second-order differential equations require the specification of two boundary conditions in order that they be uniquely defined. One of these is imposed at the origin. The function must be everywhere finite which, taking into account Eq. (8.39), implies that $u(0) = 0$. In a small neighborhood of the origin the solution can be thought of as related to the spherical Bessel function $rj_\ell(Kr)$, with K suitably defined by the relation

$$K^2 = (2m_\alpha/h^2)[E_\alpha - \mathscr{V}_{\alpha\alpha}(0)]. \tag{8.45}$$

This fact can be employed as a starting value for numerical integration.

The second boundary condition is applied at some large value of r, where the interaction has vanished or become purely Coulombic (in the case of charged particles). Hence

$$u_\alpha \to \mathrm{A}F_\ell + \mathrm{B}G_\ell,$$

where F and G are the Coulomb functions, defined in Chapter 4, that are, respectively, regular and irregular at the origin.

As already discussed in Chapter 4, it is convenient to introduce the combinations I and O:

$$I_\ell^* = O_\ell = G_\ell + iF_\ell \to \exp\{i[kr - \eta\ell n(2kr) - \ell\pi/2 + \sigma_\ell]\}. \tag{8.46}$$

Now the boundary condition for large r can be precisely stated. In the entrance channel α there are both incoming and outgoing spherical waves at infinity corresponding to the physical fact that there is a plane wave representing the incident beam and outgoing scattered waves. In all other channels, however, there are only outgoing spherical waves. Hence

$$\begin{aligned} u_{\alpha'}^{\alpha\pi I}(r) &\to -\frac{1}{2i}\left[\delta_{\alpha'\alpha}I_\ell(k_\alpha r) - \left(\frac{v_\alpha}{v_{\alpha'}}\right)^{1/2} S^I_{\alpha'\alpha}O_{\alpha'}(k_{\alpha'}r)\right] \\ &= \delta_{\alpha'\alpha}F_\ell(k_\alpha r) + \left(\frac{v_\alpha}{v_{\alpha'}}\right)^{1/2}\frac{(S^I - 1)_{\alpha'\alpha}}{2i}O_{\alpha'}(k_{\alpha'}r). \end{aligned} \tag{8.47}$$

The particular way we chose to write the coefficient of O_α allows us to refer to S as the S matrix, being the ratio of outgoing flux in channel α' to incoming flux in channel α. The requirement that the solutions to the coupled equations satisfy the preceding boundary conditions provides the numerical values of $S_{\alpha'\alpha}$ from which the cross sections can be calculated.

To see explicitly how S is specified, we note that, in general, the integrated solutions, found by starting the integration at the origin in such a way as to obey the initial values, will have both incoming and outgoing waves in all channels at infinity. Therefore a linearly independent set of solutions must be generated. Let the channels $\alpha, \alpha', \ldots$ be denoted by $1, 2, \ldots, N$ with

1 denoting the entrance channel. We now place two subscripts on each function u, say, u_{nm}, where n denotes the nth channel and m denotes one of the various N boundary conditions for which solutions will be found. Suitable initial conditions are (see Chapter 4, L)

$$u_{nm}(r) \underset{r\to 0}{\longrightarrow} \begin{cases} (Kr)^{\ell_n+1}/(2\ell_n+1)!! & \text{for} \quad n = m \\ 0 & \text{for} \quad n \neq m, \end{cases} \tag{8.48}$$

with the corresponding derivatives. By solving the coupled equations N times with $m = 1, \ldots, N$, we generate N independent solutions, some linear combinations of which can be found to satisfy the condition of (8.48) at large r. This is expressed by the linear algebraic equations at $r = R$:

$$\begin{aligned} a_1 u_{11} + a_2 u_{12} + \cdots + a_N u_{1N} + S_1 O_{\ell_1} &= I_{\ell_1} \\ a_1 u_{21} + a_2 u_{22} + \cdots + a_N u_{2N} + S_2 O_{\ell_2} &= 0 \\ &\vdots \\ a_1 u_{N1} + a_2 u_{N2} + \cdots + a_N u_{NN} + S_N O_{\ell_N} &= 0. \end{aligned} \tag{8.49}$$

These, together with the derivatives of these equations, provide $2N$ equations that can be solved for the constants $a_1, \ldots, a_N$, $S_1, \ldots, S_N$.

H. DISTORTED-WAVE BORN APPROXIMATION

It should not be surprising that the distorted-wave Born approximation is the first-order solution to the coupled equations. To first order, the coupled equations (8.40) become

$$\begin{aligned} (E_\alpha - T_{\alpha\ell} - U_\alpha)u_\alpha^0 &= 0 && \text{(target channel)} \\ (E_{\alpha'} - T_{\alpha\ell} - U_\alpha)u_{\alpha'} &= \mathscr{V}_{\alpha'\alpha} u_\alpha^0 && \text{(excited channels),} \end{aligned} \tag{8.50}$$

where $\mathscr{V}_{\alpha\alpha}$ is represented by an optical potential U_α which is taken to be the same (generally) in all inelastic channels.

The second of these can be solved in terms of the Green's functions constructed from the regular and irregular solutions $u_{\alpha'}^0$, $O_{\alpha'}$, of the corresponding homogeneous equation, just as in Chapter 5. One finds that

$$u_{\alpha'}(r) \to -\frac{2m_\alpha}{\hbar^2}\frac{O_{\alpha'}(r)}{k_{\alpha'}}\int_0^\infty u_{\alpha'}^0 \mathscr{V}_{\alpha'\alpha} u_\alpha^0 \, dr. \tag{8.51}$$

Taking account of definition (8.47), one can read the DWBA expression for the inelastic S matrix elements:

$$S_{\alpha'\alpha}^I = -\frac{4i}{\hbar^2}\frac{1}{\sqrt{v_\alpha v_{\alpha'}}}\int_0^\infty u_{\alpha'}^0 \mathscr{V}_{\alpha'\alpha} u_\alpha^0 \, dr. \tag{8.52}$$

I. CROSS SECTIONS

The preceding development in partial waves was made for a particular state of total angular momentum I and parity π of the system. The incident beam in the entrance channel, with momentum $\hbar k$ directed along the z axis, is represented by a plane wave (if there is no Coulomb force):

$$\Psi_0^{m,M} = \Phi_{J_a}^m \Phi_{J_A}^M e^{ikz}. \tag{8.53}$$

It is a superposition of all angular momenta (2.8)

$$\Psi_0^{m,M} = \Phi_{J_a}^m \Phi_{J_A}^M \frac{1}{kr} \sum_{\ell=0}^{\infty} i^\ell \sqrt{4\pi(2\ell+1)} F_\ell(kr) Y_\ell^0(\hat{\mathbf{r}}). \tag{8.54}$$

Therefore we need the particular linear combination

$$\Psi^M = \sum_{\ell jI} A_{\alpha\pi I}^M \Psi_{\alpha\pi I}^M, \tag{8.55}$$

which has the plane-wave part (8.53) when examined as $r \to \infty$. In view of the boundary condition (8.48) that the radial parts of $\Psi_{\alpha\pi I}$ (8.39) satisfy, it is easy to verify that if the amplitudes are chosen to be

$$A_{\ell j J_a J_A I}^{M+m} = \frac{1}{k} C_{0\,m\,m}^{\ell J_a j} C_{m\,M\,m+M}^{j J_A\, I} \sqrt{4\pi(2\ell+1)}\, i^\ell, \tag{8.56}$$

then (8.54) becomes precisely (8.53) if the interaction is turned off ($S_{\alpha'\alpha} \equiv \delta_{\alpha'\alpha}$). The proof involves merely noting the coupling scheme implied by $\phi_{\alpha\pi I}$, out of which $\Psi_{\alpha\pi I}$ (8.39) is constructed. In the case of a Coulomb force, the plane wave in (8.53) must be replaced by a Coulomb wave function. The only modification that needs to be made is the insertion of the factor $e^{i\sigma_\ell}$ into (8.55) and the replacement

$$F_\ell(kr) \to e^{i\sigma_\ell} F_\ell(\eta, kr)$$

in (8.54). Then, in the absence of any additional interaction, $S_{\alpha'\alpha} = \delta_{\alpha'\alpha}$, and (8.55) becomes the Coulomb wave function

$$\Psi_c^{mM} = \Phi_{J_a}^m \Phi_{J_A}^m \frac{1}{kr} \sum \sqrt{4\pi(2\ell+1)}\, i^\ell e^{i\sigma_\ell} F_\ell(\eta, kr) Y_\ell^0(\hat{\mathbf{r}}). \tag{8.57}$$

From the discussion in Chapter 4 on Coulomb scattering, we know that

$$\Psi_c^{m,M} \to \Phi_{J_a}^m \Phi_{J_A}^M \left(\exp\{i[kz - \eta \ell nk(r-z)]\} + \frac{1}{r} f_c(\theta) \exp\{i[kr - \eta(\ell n 2kr)]\} \right), \tag{8.58}$$

where f_c is the Coulomb scattering amplitude given in Chapter 4. Therefore the boundary condition (8.47) leads to the following result for (8.55):

$$\Psi^{m,M} \to \Psi_c^{m,M} + \frac{1}{r} \sum_{\alpha'} \exp\{i[k'r - \eta' \ell n(2k'r)]\} (v_\alpha/v_{\alpha'})^{1/2} F^N_{\alpha'\alpha}(\hat{\mathbf{k}}') \Phi^{m'}_{J_a} \Phi^{M'}_{J_A}, \tag{8.59}$$

where the scattering amplitude is given by

$$\begin{aligned} F_{\alpha'\alpha}(\hat{\mathbf{k}}) &= f_c(\theta) \delta_{\alpha'\alpha} \delta_{J_a J_{a'}} \delta_{J_A J_{A'}} \delta_{mm'} \delta_{MM'} + F^N_{\alpha'\alpha}(\hat{\mathbf{k}}') \\ F^N_{\alpha'\alpha}(\hat{\mathbf{k}}') &= \sum_{\ell j I} A^{m+M}_{\ell j J_a J_A I} \sum_{\ell' j'} (S^I - 1)_{\alpha'\alpha} i^{-\ell'} e^{i(\sigma_\ell + \sigma_{\ell'})} \\ &\quad \times \sum_{m_\ell} C^{\ell'\ J_{a'}\ \ j'}_{m_\ell m'\ m_\ell + m'} C^{j'\ \ \ \ J_{A'}\ \ I}_{m_\ell + m'\ M'\ M+m} Y^{m_\ell}_{\ell'}(\hat{\mathbf{k}}). \end{aligned} \tag{8.60}$$

Here recall that subscripts α mean

$$\alpha \equiv J_a J_A \ell j.$$

We replaced $\hat{\mathbf{r}}$ by $\hat{\mathbf{k}}'$ in the asymptotic region. The current associated with $(v_\alpha/v_{\alpha'})^{1/2} F_{\alpha'\alpha} e^{ik'r}/r$ is $v_\alpha |F|^2/r^2$ so that the flux in direction $\hat{\mathbf{k}}'$ through solid angle $d\Omega$ is $r^2\, d\Omega$ times the current. Thus

$$d\sigma(\alpha, m, M \to \alpha', m', M') = |F_{\alpha'\alpha}(\hat{\mathbf{k}}')|^2\, d\Omega. \tag{8.61}$$

As in (5.50), the usually measured cross section involves an unpolarized beam and unoriented target, and the final spin orientations are not measured. This cross section is the average over the initial orientations m and M and sum over the final m' and M',

$$\frac{d\sigma}{d\Omega} = \frac{1}{(2J_a + 1)(2J_A + 1)} \sum_{mMm'M'} |F_{\alpha'\alpha}(\hat{\mathbf{k}}')|^2. \tag{8.62}$$

In this section we have gone through the tedious details of showing how the scattering amplitude (8.60) is constructed from the S matrix elements under very general circumstances.

Chapter 9

Microscopic Theory of Inelastic Nucleon Scattering from Nuclei

A. THE RICHNESS OF NUCLEAR STRUCTURE

The richness and variety in nuclear structure and in the reactions between nuclei arise because the nucleus is a self-bounded, saturated system of nucleons that is held together, not collapsed or crystallized, by a complicated interaction between the nucleons. The interaction acts differently in the various states of relative motion available to two nucleons and possesses central, spin–orbit, and tensor components.

The wave functions of nuclear states are the eigenfunctions of a Hamiltonian, which are solutions to the Schrödinger equation (in the nonrelativistic approximation) corresponding to N particles. They must be exceedingly complex, more so than we will ever know, for this problem can only be solved in an approximate way. Nevertheless, however complex nuclear states may be, they could have been even more so; for through a delicate balance of several factors, nuclei are fairly dilute systems that can be described in lowest order as an almost degenerate gas.

The next corrections to this approximation involve the pairwise interactions of nucleons. The factors that lead to this relative simplicity are the repulsion in the nucleon–nucleon interaction at short distance, the relative weakness and long range of the attraction, and the Pauli principle (Gomes *et al.*, 1958). These factors permit an approximate description of nuclear states in terms of a shell model with residual pairwise interactions (Haxel *et al.*, 1949; Mayer, 1949). The interactions are usually taken into account only with respect to the nucleons in the outer shells. Even so, the appoximate wave functions can be exceedingly complicated.

The richness and complexity of nuclear structure are emphasized, in part, as a cautionary note against interpreting the success of nuclear models in accounting for nuclear spectra and other properties as evidence for their too literal truth. Any interpretation in nuclear physics must be tempered by the permanent veil that exists between us and the exact solutions of the many-body problem.

Nuclear reactions are an extremely valuable means of probing the structure of nuclei; and, having introduced some of the basic ideas of the theory in the earlier chapters, in this and many of the subsequent chapters, special attention will be given to the manner in which the structure of nuclei can be exposed to analysis by reactions. The theory has been developed to facilitate this goal.

The microscopic theory of inelastic scattering was formulated almost simultaneously and independently, with the goal stated in the preceding paragraph, by several researchers, both in the distorted-wave Born approximation and in the coupled-channel Born approximation.[‡] In this chapter, to provide a general framework and to illustrate more concretely the development of the previous chapter, the theory will be developed in the CCBA, of which the DWBA is the first-order solution. The goal of the development is to cast the theory into a form that makes it convenient to employ nuclear wave functions in the description of the scattering process, however complicated they may be, and thus to expose them to the test of consistency with scattering data. Along the way it will be pointed out how various scattering probes involve various combinations of the effective nucleon–nucleon interaction and, as a consequence, tend to populate various types of states selectively. This is a subject that has turned out to be particularly fruitful and will be the subject of Chapters 11 and 12.

B. DISCUSSION OF THE INTERACTIONS IN THE COUPLED EQUATIONS

As discussed in the preceding chapter, the scattering process is described by an infinite set of coupled equations because of the need to use a complete set in the expansion of the wave function of the entire system. On the other

[‡] For distorted-wave Born approximation, see Glendenning and Veneroni, 1965, 1966; Madsen and Tobocman, 1965; Haybron and McManus, 1965; Johnson *et al.*, 1966; Satchler, 1966; and Madsen, 1966. For coupled-channel Born approximation, see Glendenning, 1968a, 1969.

hand, information on only a few low-lying resolvable levels can be obtained. Thus a description in terms of all channels, even if it were possible, would be superfluous. Truncating these equations as they stand would be even less justified than the truncation employed in shell-model calculations. There, at least, one can invoke the argument that levels that are distant in energy from the region of interest may not have a strong effect. However, at the bombarding energies of interest in scattering at cyclotron energies, the nucleus may be excited into any one of millions of levels, and each of these states of the system corresponds to the same total energy. Therefore, there is no reasonable truncation of the *original* system. However, we showed how to transform the problem to a finite set of coupled equations in which the original interaction is replaced by an effective interaction. The usefulness of this transformation of the problem lies in the fact that the effect of the enormous number of open channels on those few of real interest is now carried by a potential operator rather than an infinite set of equations. Although we cannot exactly calculate the effective interaction because that would be tantamount to solving the many-body scattering problem which, even were it possible, we wish to defer or avoid, we can infer certain of its properties from its defining equation and seek a parametrization of it.

We observed from the structure of $\mathscr{V}$ that its *diagonal* elements are expected to vary very slowly from one nuclear state to another and, in general, from one nucleus to another. For the elastic channel we are familiar with the success of the phenomenological optical model. Therefore, for the diagonal elements, we shall adopt a common phenomenological parametrization:

$$\mathscr{V}_{\alpha'\alpha'}(r) \equiv (\Phi_{\alpha'}|\mathscr{V}|\Phi_{\alpha'}) = U_{\alpha}(r), \qquad \alpha' \in P, \tag{9.1}$$

where U_α is an optical potential as discussed in Chapter 4. It is energy dependent as is known from phenomenological analysis of elastic data and as it should be from the structure of $\mathscr{V}$.

The nondiagonal elements are a different story. We argued, on the basis of the fluctuating signs in the sum over states Q in the second term of $\mathscr{V}$, that that term would be small and that

$$\mathscr{V}_{\alpha'\alpha} \simeq V_{\alpha'\alpha}, \qquad \alpha' \neq \alpha. \tag{9.2}$$

A word now about the nucleon–nucleon interaction V_{ij} from which V is constructed,

$$V = \sum_{i \in \mathrm{a},\, j \in \mathrm{A}} V_{ij}. \tag{9.3}$$

In connection with the discussion of the shell model in Chapter 3, we alluded to the problems arising from the singular behavior of the *free* nucleon–nucleon interaction. The Brueckner Hartree–Fock theory provides the means of transforming the problem to a large shell-model basis in which the short-range correlations and the singular interaction causing them are replaced by a smooth well-behaved interaction called the *bare* G matrix.

A second possible step involves a further truncation of the large basis of the shell model to a small-model basis in which a renormalized G matrix is the shell-model effective interaction (Kuo and Brown, 1966, 1968). This second step is accomplished in a parallel manner to the development in the last chapter and is often referred to as core polarization because what is often involved is the effect on valence nucleons of the core nucleons (Love and Satchler, 1967; Petrovich *et al.*, 1977).

For the scattering problem, we saw in the last chapter that the first approximation to the effective interaction is the τ matrix. The τ and G matrices have formally similar integral equations and are therefore related. Chapter 11 will be concerned with these matters. For the present, we adopt a phenomenological view as far as applications are concerned, but the theoretical development is general.

Taking into account the discussion of the diagonal and nondiagonal elements of $\mathscr{V}$, the coupled equations that describe the elastic and inelastic scattering can now be written as

$$\left[\frac{\hbar^2}{2m}\left(\frac{d^2}{dr^2} - \frac{\ell'(\ell'+1)}{r^2}\right) - U(r) + E_{\alpha'}\right] u_{\alpha'}^{\alpha\pi I}(r) = \sum_{\alpha''\neq\alpha'} (\phi_{\alpha'\pi I}^{M}|V|\phi_{\alpha''\pi I}^{M}) u_{\alpha''}^{\alpha\pi I}(r). \tag{9.4}$$

Because we deal here only with inelastic, not particle-transfer channels, it is not necessary to carry a subscript on the optical potential U. We reserve the unprimed symbol α to refer to the ground state, whereas α', α'', ... refer to excited states. The results obtained for scattering of nucleons by a nucleus can be very easily extended to the scattering of ^{2}H, ^{3}He, and ^{4}He, and with obvious extensions to nucleus–nucleus scattering and the possible mutual excitations. The specialization leads to a simpler notation.

C. MATRIX ELEMENTS OF THE EFFECTIVE INTERACTION

We specify more precisely the interaction V in the preceding coupled equations as

$$V = \sum_{i=1}^{A} V(\mathbf{r}, \mathbf{r}_i), \tag{9.5}$$

where $\mathbf{r}$ is the coordinate of the free nucleon and $\mathbf{r}_i$ belongs to the nucleus A. Suppose that V is central,

$$V(\mathbf{r}, \mathbf{r}_i) = (V_0 + V_1 \boldsymbol{\sigma} \cdot \boldsymbol{\sigma}_i) g(|\mathbf{r} - \mathbf{r}_i|), \tag{9.6}$$

where V_0 and V_1 may in turn depend on the isospin $\boldsymbol{\tau} \cdot \boldsymbol{\tau}_i$ and

$$V_0 = V_{00} + V_{01}\boldsymbol{\tau} \cdot \boldsymbol{\tau}_i. \tag{9.7}$$

The radial function g can be expanded in multipoles as

$$g(|\mathbf{r} - \mathbf{r}_i|) = \sum_L \frac{4\pi}{2L+1} v_L(r, r_i) Y_L(\hat{\mathbf{r}}) \cdot Y_L(\hat{\mathbf{r}}_i), \tag{9.8}$$

where we use the dot product to denote

$$Y_L(\hat{\mathbf{r}}) \cdot Y_L(\hat{\mathbf{r}}_i) \equiv \sum_M Y_L^M(\hat{\mathbf{r}}) Y_L^{M*}(\hat{\mathbf{r}}_i). \tag{9.9}$$

The functions

$$v_L(r, r_i) = \tfrac{1}{2}(2L+1) \int g(|\mathbf{r} - \mathbf{r}_i|) P_L(\cos\omega)\, d\cos\omega \tag{9.10}$$

[where $\cos\omega = (\mathbf{r} \cdot \mathbf{r}_i)/(rr_i)$] can be obtained in closed form for some radial forms of the two-body potential, as is known from shell-model theory (see de-Shalit and Talmi, 1963).

To treat the spin-dependent and spin-independent terms in a unified way, introduce the spherical tensors (see Appendix)

$$T_{LSJ}^{\ M} = \begin{cases} Y_L^M(\hat{\mathbf{r}})\delta_{LJ}, & S = 0 \\ [Y_L(\hat{\mathbf{r}})\boldsymbol{\sigma}]_J^M, & S = 1, \end{cases} \tag{9.11}$$

where the square bracket denotes vector coupling. Now we can write

$$V = \sum_i V(\mathbf{r}, \mathbf{r}_i) = \sum_{LSJ} (-)^{L+S+J} V_S \mathscr{T}_{LSJ}(r, \mathbf{A}) \cdot T_{LSJ}(\hat{\mathbf{r}}, \boldsymbol{\sigma}), \tag{9.12}$$

where

$$\mathscr{T}_{LSJ}(r, \mathbf{A}) = \frac{4\pi}{2L+1} \sum_i v_L(r, r_i) T_{LSJ}(\hat{\mathbf{r}}_i, \boldsymbol{\sigma}_i). \tag{9.13}$$

This quantity carries the transformation properties of the nuclear part of the interaction, whereas T_{LSJ} does so for the free nucleon.

The matrix element of V can now be evaluated by the use of Racah algebra. The relevant results are contained in the Appendix. We find, by straightforward application of (A40)–(A46),

$$\begin{aligned} V^I_{\alpha'\alpha''}(r) &\equiv (\phi^M_{\alpha'\pi I}|V|\phi^M_{\alpha''\pi I}) \\ &\equiv ([\mathscr{Y}_{\ell' sj'}\Phi_{\alpha' J'}]^M_I|V|[\mathscr{Y}_{\ell'' sj''}\Phi_{\alpha'' J''}]^M_I) \\ &= \sum_{LSJ} V_S C^I_{LSJ}(J'J'')(J'\|\mathscr{T}_{LSJ}\|J''), \end{aligned} \tag{9.14}$$

where C is a purely geometrical factor

$$C^{I}_{LSJ} = (-)^{L+S+J+j''+J'+I} \begin{Bmatrix} I & J' & j' \\ J & j'' & J'' \end{Bmatrix} \langle j' \| T_{LSJ} \| j'' \rangle, \tag{9.15}$$

$$\begin{aligned} \langle j' \| T_{LSJ} \| j'' \rangle &\equiv \langle (\ell' \tfrac{1}{2}) j' \| T_{LSJ} \| (\ell'' \tfrac{1}{2}) j'' \rangle \\ &= (-)^{\ell'} \left[\frac{\hat{j}'\hat{j}''\hat{\ell}'\hat{\ell}''\hat{L}\hat{S}\hat{J}}{2\pi} \right]^{1/2} \begin{pmatrix} \ell' & L & \ell'' \\ 0 & 0 & 0 \end{pmatrix} \begin{Bmatrix} \ell' & \frac{1}{2} & j' \\ \ell'' & \frac{1}{2} & j'' \\ L & S & J \end{Bmatrix} \\ &= (-)^{j'-j''+L+S+J} \langle j'' \| T_{LSJ} \| j' \rangle. \end{aligned} \tag{9.16}$$

D. SELECTION RULES

The 3-j, 6-j, and 9-j symbols, as discussed in the Appendix and in the reference books on angular momentum, vanish unless their arguments obey certain relationships. They automatically enforce angular-momentum and parity conservation. These conservation laws, as has been emphasized in Chapter 2, are reflected in the angular distribution of reactions, and because direct reactions tend to be localized in the nuclear surface, the angular distribution is usually very characteristic of the angular momentum transfered in the reaction. At this stage, we can read from (9.15) and (9.16) the selection rules connecting the angular momentum. The 6-j symbol, among other relations, imposes the vector relation

$$\mathbf{J} = \mathbf{J}' + \mathbf{J}'', \tag{9.17a}$$

or, written more explicity,

$$|J' - J''| \le J \le J' + J''. \tag{9.17b}$$

The 3-j symbol implies a parity selection rule, because it vanishes unless

$$\ell' + \ell'' + L = \text{even} \tag{9.18a}$$

[see (A20)]. Because a transition of the scattered particle from the orbital angular momentum ℓ' to ℓ'' implies a parity change given by $(-)^{\ell'+\ell''}$, we see that L gives the change in parity

$$(-)^{\ell'+\ell''} = (-)^{L}. \tag{9.18b}$$

The 9-j symbol vanishes unless each row and each column satisfy a vector relationship. Most especially,

$$\ell' \ \ell'' = \mathbf{L}, \qquad \tfrac{1}{2} + \tfrac{1}{2} = \mathbf{S}, \qquad \mathbf{j}' + \mathbf{j}'' = \mathbf{J}, \qquad \mathbf{L} + \mathbf{S} = \mathbf{J} \tag{9.19}$$

which can be written more explicitly as in (9.17b).

Through the conservation of angular momentum and parity for the total system, the change in the nuclear state must compensate the change in the state of the scattered particle, or, more precisely, the state of relative motion. We have seen how the changes in angular momentum and parity of the scattered particle are connected to L, S, and J. Next we shall see how the change in the nucleonic motion in the nucleus is connected to these quantities.

E. NUCLEAR FORM FACTOR

The development of the theory has placed all information about the structure of the nucleus in the reduced matrix element of the operator of Eq. (9.13) that appears in Eq. (9.14). This is called the nuclear form factor:

$$\mathscr{F}_{LSJ}^{J'J''}(r) = (\Phi_{J'}(\mathrm{A})\|\mathscr{T}_{LSJ}\|\Phi_{J''}(\mathrm{A})). \qquad (9.20)$$

It involves an integration of all intrinsic nuclear coordinates and is a function of the separation between the scattered particle and nucleus.

If we neglect exchange effects arising from the Pauli principle, the structure of $\mathscr{F}$ is particularly simple. The neglect can be partially justified on the grounds that the exchange integrals may be smaller than the direct ones.‡ We make this simplifying assumption here. Relaxation of this approximation can be found in the literature (Petrovich *et al.*, 1977) and will be discussed again in Chapter 12.

With the neglect of exchange, the scattered particle behaves like a one-body operator acting on the nuclear coordinates. Thus, however complicated the nuclear states J' and J'' may be, the nuclear form factor can be expressed as combinations of single-particle form factors

$$F_{LSJ}^{ab}(r) = (\phi_{\mathrm{a}}(\mathbf{r}')\|\mathscr{T}_{LSJ}(r, r')\|\phi_{\mathrm{b}}(\mathbf{r}')), \qquad (9.21)$$

where ϕ_{a} and ϕ_{b} are shell-model wave functions. This is, of course, the form factor for the promotion of a single nucleon from b to a. We use the abbreviation $\mathrm{a} \equiv n_{\mathrm{a}}\ell_{\mathrm{a}}j_{\mathrm{a}}$ and write the linear combination in the form

$$\mathscr{F}_{LSJ}^{J'J''}(r) = \sum_{\mathrm{a,b}} S_J^{\mathrm{ab}}(J'J'')F_{LSJ}^{\mathrm{ab}}(r), \qquad (9.22)$$

‡ In the limit of an infinite-range force, they are in principle zero. In practice, because the bound and scattered waves are calculated in different potentials, they will not be precisely orthogonal but will at least have small overlap. For zero range, the direct and exchange integrals are equal. Thus both limits are simple. The actual situation lies between but more toward the short range. Therefore the principal effect of exchange is to amplify the cross section.

where the amplitudes depend on the particular structure of the nuclear states J' and J''. They are illustrated explicitly for a few simple situations later in this chapter.

F. SINGLE-PARTICLE FORM FACTORS

We have seen that the nuclear form factor is a linear combination of single-particle form factors that would be relevant to scattering from an odd nucleus. Let the shell-model wave function be written as

$$\phi_{\mathrm{a}}(\mathbf{r}) = u_{n_{\mathrm{a}}\ell_{\mathrm{a}}}(r)[Y_{\ell_{\mathrm{a}}}(\hat{\mathbf{r}})X_{1/2}(\sigma)]_{j_{\mathrm{a}}}, \tag{9.23}$$

where X is a spin function and u the radial wave function. From the definition of $\mathscr{T}$, the single-particle form factor can be written as

$$F_{LSJ}^{\mathrm{ab}}(r) = (\phi_{\mathrm{a}}(\mathbf{r}')\|\mathscr{T}_{LSJ}\|\phi_{\mathrm{b}}(\mathbf{r}')) = 4\pi\, R_{\mathrm{ab}}^{L}(r)\langle j_{\mathrm{a}}\|T_{LSJ}\|j_{\mathrm{b}}\rangle. \tag{9.24}$$

The radial factor appearing in (9.24) denotes

$$R_{\mathrm{ab}}^{L}(r) = \frac{1}{2L+1}\int u_a(r')v_L(r, r')u_{\mathrm{b}}(r')r'^2\, dr'. \tag{9.25}$$

This can be evaluated in closed form as a polynomial if the potential $V(r, r')$ is Gaussian and the radial functions of the shell model are taken to be harmonic oscillator functions (Glendenning and Veneroni, 1966; Glendenning, 1959, 1968a).

The short range of the force suggests that the form of R is strongly governed by the product of the wave functions. For example, if the force range is zero, then

$$\begin{aligned} v_L(r, r') &= \frac{2L+1}{4\pi}\,\delta(r - r')\,\frac{1}{rr'}, \\ R_L^{\mathrm{ab}}(r) &= \frac{1}{4\pi}\, u_{\mathrm{a}}(r)u_{\mathrm{b}}(r). \end{aligned} \tag{9.26}$$

The radial form factor R_{ab} is illustrated for several transitions and force ranges in Fig. 9.1.

Because of the appearance of the reduced matrix element of T_{LSJ} in the single-particle form factor (9.24) the same discussion given earlier in connection with (9.16) applies to the connection of the excited nuclear particle in its initial $(l_{\mathrm{a}}\frac{1}{2}j_{\mathrm{a}})$ and final $(l_{\mathrm{b}}\frac{1}{2}j_{\mathrm{b}})$ states, with the transfer angular momenta LSJ. In particular,

$$\begin{gathered} (-)^{\ell_{\mathrm{a}}+\ell_{\mathrm{b}}} = (-)^{L} \\ \boldsymbol{\ell}_{\mathrm{a}} + \boldsymbol{\ell}_{\mathrm{b}} = \mathbf{L}, \qquad \tfrac{1}{2} + \tfrac{1}{2} = \mathbf{S}, \qquad \mathbf{j}_{\mathrm{a}} + \mathbf{j}_{\mathrm{b}} = \mathbf{J}. \end{gathered} \tag{9.27}$$

This makes clear why we refer to LSJ as the transfer angular momenta.

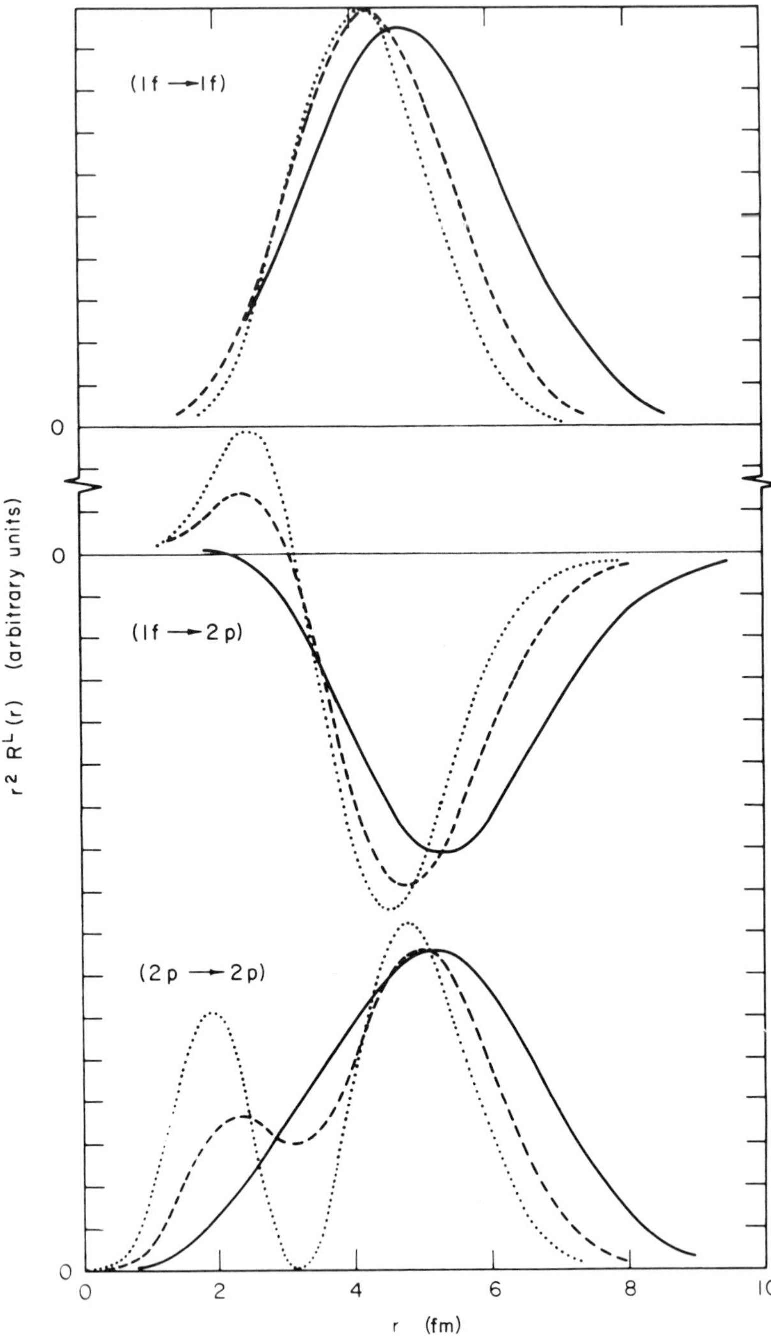

Fig. 9.1. Shapes of several single-particle form factors that contribute to excitation of 2^+ states in nickel isotopes are shown for several force ranges of the direct interaction. The oscillator parameter $v = m\omega/\hbar = 0.25\ \text{fm}^{-2}$. Force range: (· · ·) 0 fm; (– – – –) 1 fm; (———) 1.85 fm. (From Glendenning and Veneroni, 1966.)

Moreover, because the change in parity $(-)^{\ell_a+\ell_b}$ of the nucleon motion must indicate the change in parity between the initial and final nuclear states, we have

$$\pi'\pi'' = (-)^L, \tag{9.28}$$

where π' denotes the parity of the initial nuclear state.

We give this very detailed discussion of the selection rules imposed by angular momentum and parity conservation because these rules are central to the application of direct reactions to nuclear spectroscopy. This is, of course, only one aspect of the role that is played by direct reactions in the study of nuclear properties. We proceed now to others.

G. NUCLEAR STRUCTURE AMPLITUDES

It has been shown that the nuclear form factor can be written as a linear combination of single-particle form factors. The particular linear combination, expressed by the amplitudes S_J^{ab} defined implicitly by (9.22), depends on the structure of the nuclear states. They are the distillation of the nuclear structure information as it is needed to describe inelastic scattering and electromagnetic transitions. (The difference between these processes enters in the definition of the single-particle form factors.) We illustrate with a few examples how they can be calculated if the structure of the nuclear wave functions is assumed to be known (Glendenning, 1968a, 1969a).

H. SHELL-MODEL CONFIGURATIONS

For an even–even nucleus with two nucleons beyond closed shells, the simplest assumptions that can be made for the wave functions are

$$\Phi_J = |(j^2)J\rangle \tag{9.29}$$

or

$$\begin{aligned}\Phi_J &= \mathscr{A}|(jj')J\rangle \\ &= 1/\sqrt{2}[|(j(1)j'(2))J\rangle - |(j(2)j'(1))J\rangle],\end{aligned} \tag{9.30}$$

where $\mathscr{A}$ is an antisymmetrization operator. Note that the first wave function is automatically antisymmetric if J is even, as can be proved by directly writing out the Clebsch–Gordan sum and examining its properties. According to the notation introduced earlier, the nuclear form factor corresponding to a transition between these two states is

$$\begin{aligned}
\mathscr{F}_{LSJ}(r) &= (\Phi_{(j^2)J'}\|\mathscr{T}_{LSJ}\|\Phi_{(jj')J''}) \\
&= 1/\sqrt{2}\left\{\left((j^2)J'\left\|\sum_{i=1}^{2}\mathscr{T}_{LSJ}(r,\mathbf{r}_i)\right\|(j(1)j'(2))J''\right)\right. \\
&\quad \left.-\left((j^2)J'\left\|\sum_{i=1}^{2}\mathscr{T}_{LSJ}(r,\mathbf{r}_i)\right\|(j(2)j'(1))J''\right)\right\} \\
&= 1/\sqrt{2}\{((j^2)J'\|\mathscr{T}_{LSJ}(r,\mathbf{r}_2)\|(j(1)j'(2))J'') \\
&\quad -((j^2)J'\|\mathscr{T}_{LSJ}(r,\mathbf{r}_1)\|(j(2)j'(1))J'')\} \\
&= \sqrt{2}((j^2)J'\|\mathscr{T}_{LSJ}(r,\mathbf{r}_2)\|(j(1)j'(2))J'') \\
&= (-)^{j+j'+J+J'}[2\hat{J}'\hat{J}'']^{1/2}\begin{Bmatrix} j & j' & J \\ J'' & J' & j \end{Bmatrix} F_{LSJ}^{jj'}(r).
\end{aligned} \tag{9.31}$$

The last line is obtained by use of Racah algebra and the definition of the single-particle form factor (9.21.)

I. PARTICLE CREATION AND DESTRUCTION OPERATORS

It is useful to introduce at this point the notation of particle creation and destruction operators. Let $\beta_{jm}^\dagger$ create a particle in the shell-model state jm. The adjoint is a destruction operator. For fermions, they obey the anti-commutation relations

$$\begin{aligned}
\beta_k\beta_\ell^\dagger + \beta_\ell^\dagger\beta_k \equiv \{\beta_k, \beta_\ell^\dagger\} &= \delta_{k\ell}, \\
\{\beta_k^\dagger, \beta_\ell^\dagger\} = \{\beta_k, \beta_\ell\} &= 0.
\end{aligned} \tag{9.32}$$

It can be demonstrated that

$$\beta_{jm}^\dagger \quad \text{and} \quad (-)^{j-m}\beta_{j-m} \tag{9.33}$$

behave like irreducible tensors T_j^m. Recall that such a tensor is defined as one that satisfies the commutation relations (Appendix)

$$\begin{aligned}
[J_z, T_j^m] &= mT_j^m, \\
[J_\pm, T_j^m] &= \sqrt{(j\mp m)(j\pm m+1)}\,T_j^{m\pm 1},
\end{aligned} \tag{9.34}$$

where $\mathbf{J}$ is the angular momentum operator and $J_\pm = J_x \pm iJ_y$.

The import of the preceding statements is that

$$\tilde{\beta}_{jm} \equiv (-)^{j-m}\beta_{j-m} \tag{9.35}$$

carries angular momentum j and projection m (not $-m$).

In terms of the particle creation and destruction operators, one can write a general one-body operator as

$$\begin{aligned} T_J^M &= \sum_{jmj'm'} \langle jm|T_J^M|j'm'\rangle \beta_{jm}^\dagger \beta_{j'm'} \\ &= \sum \frac{(-)^{j'-m'}}{\sqrt{2J+1}} C_{m\,-m'\,M}^{j\;\,j'\;\,J} \langle j\|T_J\|j'\rangle \beta_{jm}^\dagger \beta_{j'm'} \\ &= -\frac{1}{\sqrt{2J+1}} \sum_{jj'} \langle j\|T_J\|j'\rangle [\beta_j^\dagger \tilde{\beta}_{j'}]_j^M, \end{aligned} \tag{9.36}$$

where we have used Edmonds's definition of the reduced matrix element $\langle j\|T_J\|j'\rangle$. The square brackets denote, as always, vector coupling

$$[\beta_j^\dagger \tilde{\beta}_{j'}]_J^M = \sum_{mm'} C_{mm'M}^{j\,j'\,J} \beta_{jm}^\dagger \tilde{\beta}_{j'm'}. \tag{9.37}$$

From the defining equation (9.22) for the structure amplitudes it follows that

$$S_J^{\mathrm{ab}}(J'J'') = -\frac{1}{\sqrt{2J+1}} \langle J'\|[\beta_{\mathrm{a}}^\dagger \tilde{\beta}_{\mathrm{b}}]_J\|J''\rangle \tag{9.38}$$

is the way that S can be defined in terms of creation operators.

J. PARTICLE-HOLE CONFIGURATIONS

In lowest approximation, the vibrational states in closed-shell nuclei may be described as a superposition of particle-hole excitations. The closed-shell ground state, in lowest approximation (i.e., no ground-state correlations), can be written as the so-called vacuum state of the A particle system:

$$|\omega\rangle = \beta_1^\dagger \cdots \beta_{\mathrm{A}}^\dagger |0\rangle, \tag{9.39}$$

where $|0\rangle$ is the particle vacuum. The state $|\omega\rangle$ is intended to be one in which all shell-model states are filled in order of their energy and in which all shells are closed. Excited states consist of one-particle, one-hole states and then at higher excitation $2p$, $2h$ states, etc. The former can be written as

$$\Phi_J^M = |JM\rangle = \sum_{p,h} C_{ph} [\beta_p^\dagger \tilde{\beta}_h]_J^M |\omega\rangle, \tag{9.40}$$

where h is summed over levels j, m below, and p over levels above those that are occupied by the ground state $|\omega\rangle$. The amplitudes C_{ph} are obtained as a solution to the nuclear problem.

The structure amplitude for transitions between the ground state (spin 0) and one of these multipole vibrational states is to be obtained from our general expression (9.38) as

$$S_J^{ab}(0, J) = -\frac{1}{\sqrt{2J+1}} \sum_{p,h} C_{ph} \langle \omega \| [\beta_a^\dagger \tilde{\beta}_b]_J \| [\beta_p^\dagger \tilde{\beta}_h]_J | \omega \rangle. \qquad (9.41)$$

The reduced matrix element of the operators must be evaluated:

$$\begin{aligned} \langle \omega | [\beta_a^\dagger \tilde{\beta}_b]_J^M | [\beta^\dagger \tilde{\beta}_h]_J^M | \omega \rangle &= \sum_{m's} C^{a\ b\ J}_{m_a m_b M} C^{p\ h\ J}_{m_p m_h M} \langle \omega | \beta^\dagger_{am_a} \tilde{\beta}_{bm_b} \beta^\dagger_{pm_p} \tilde{\beta}_{hm_h} | \omega \rangle \\ &= \sum C^{a\ b\ J}_{m_a m_b M} C^{p\ h\ J}_{m_p m_h M} \{ \delta_{pb} \delta_{ah} - \langle \omega | \beta_a^\dagger \beta_p^\dagger \tilde{\beta}_b \tilde{\beta}_h | \omega \rangle \} \\ &= \delta_{pd} \delta_{ah} (-)^{ja + jb - J}. \end{aligned} \qquad (9.42)$$

The last term obviously gives zero. Also from the definition of reduced matrix element,

$$\delta_{pb} \delta_{ah} (-)^{j_a + j_b - J} = \frac{1}{\sqrt{2J+1}} C^{O\ J\ J}_{OMM} \langle \omega \| \cdots \| \omega \rangle = \frac{1}{\sqrt{2J+1}} \langle \omega \| \cdots \| \omega \rangle, \qquad (9.43)$$

from which

$$S_J^{hp}(0, J) = (-)^{j_h + j_p - J + 1} C_{ph}. \qquad (9.44)$$

Similarly one finds

$$S_J^{ph}(0, J) = 0. \qquad (9.45)$$

The amplitude for transitions between one p–h state and another are much more involved to derive but can be found in Glendenning (1968a).

K. QUASI-PARTICLE CONFIGURATIONS

One method of approximately solving the nuclear many-body problem consists of isolating the pairing effects of the residual interaction by the BCS method. This corresponds to transforming from the real nucleons of the system to quasi-particles which, as far as the pairing effect of the interaction is concerned, are approximately diagonalized in the quasi-particle representation. This means that the quasi-particles behave almost like noninteracting particles. The transformation between particles $\beta^\dagger$ and quasi-particles $\alpha^\dagger$ is

$$\beta_{jm}^\dagger = U_j \alpha_{jm}^\dagger + V_j \tilde{\alpha}_{jm}, \qquad (9.46)$$

where U and V are coefficients of the Bogolyubov–Valatin transformation. With the assumption of phases for spherical harmonics that are known as the Condon–Shortley phases (used also by Edmonds) such that

$$Y_{\ell -m} = (-)^m Y^*_{\ell m}, \qquad (9.47)$$

then, for the state of minimum energy,

$$UV(-)^{\ell} \geq 0, \tag{9.48}$$

and the normalization of the BCS wave function implies also that

$$U^2 + V^2 = 1. \tag{9.49}$$

The remaining part of the interaction is treated approximately by a diagonalization within the subspace of two-quasi-particle configurations. In this approximation, excited states are of the form

$$|JM\rangle = \tfrac{1}{2} \sum_{\mathrm{ab}} \eta^J_{\mathrm{ab}} \mathrm{A}^\dagger_{JM}(\mathrm{ab})|\omega\rangle, \tag{9.50}$$

where $|\omega\rangle$ denotes the BCS vacuum or ground state and

$$\mathrm{A}^\dagger_{JM}(\mathrm{a, b}) \equiv -[\alpha^\dagger_{\mathrm{a}}\alpha^\dagger_{\mathrm{b}}]^M_J = (-)^{j_{\mathrm{a}} - j_{\mathrm{b}} + J}\mathrm{A}^\dagger_{JM}(\mathrm{b, a}) \tag{9.51}$$

is a pair-creation operator. In the preceding expression for the wave function, the sum on a, b is unrestricted as to order (i.e., terms ab and ba occur).

The expression of a one-body operator S in quasi-particle terms is obtained from (9.38) by use of the transformation between particles and quasi-particles and is found, after manipulation, to be Glendenning and Veneroni, 1966)

$$\begin{aligned} S^{\mathrm{ab}}_J = (1/\sqrt{2J+1})\{ & V_{\mathrm{a}}\sqrt{2j_{\mathrm{a}}+1}\,\delta_{\mathrm{ab}}\delta_{JO} \\ & +\tfrac{1}{2}[U_{\mathrm{a}}U_{\mathrm{b}} - (-)^{J+\sigma}V_{\mathrm{a}}V_{\mathrm{b}}][N^\dagger_{JM}(\mathrm{a, b}) + (-)^{M+\sigma}N_{J-M}(\mathrm{a, b})] \\ & -\tfrac{1}{2}[U_{\mathrm{a}}V_{\mathrm{b}} + (-)^{J+\sigma}V_{\mathrm{a}}U_{\mathrm{b}}][\mathrm{A}^\dagger_{JM}(\mathrm{a, b}) + (-)^{M+\sigma}\mathrm{A}_{J-M}(\mathrm{a, b})]\}, \end{aligned} \tag{9.52}$$

where the phase σ depends on the property of the particular operator T^M_J and is defined by

$$\langle \mathrm{a}\|T_J\|\mathrm{b}\rangle = (-)^{j_{\mathrm{a}} - j_{\mathrm{b}} + \sigma}\langle \mathrm{b}\|T_J\|\mathrm{a}\rangle. \tag{9.53}$$

In the present application to the amplitudes S of (9.38), we find from (9.16) that

$$\sigma = L + S + J. \tag{9.54}$$

The operator N is

$$N^\dagger_{JM}(\mathrm{a, b}) \equiv -[\alpha^\dagger_{\mathrm{a}}\tilde{\alpha}_{\mathrm{b}}]^M_J = (-)^{j_{\mathrm{a}} - j_{\mathrm{b}} + M}N_{J-M}(\mathrm{b, a}). \tag{9.55}$$

The nuclear structure amplitudes appearing in expression (9.22) are the reduced matrix elements of the S^{ab}_J operator. In the case of a transition from the BCS ground-state vacuum to a two-quasi-particle state,

$$S^{\mathrm{ab}}_J(J, 0) = -\tfrac{1}{4}[U_{\mathrm{a}}V_{\mathrm{b}} + (-)^{L+S}V_{\mathrm{a}}U_{\mathrm{b}}] \sum_{\mathrm{cd}} \eta^J_{\mathrm{cd}}\langle\omega|\mathrm{A}_{JM}(\mathrm{c, d})\mathrm{A}^\dagger_{JM}(\mathrm{a, b})|\omega\rangle.$$

It can be readily proven that

$$\langle \omega | \mathrm{A}_{\lambda\mu}(\mathrm{ab}) \mathrm{A}^{\dagger}_{\lambda'\mu'}(\mathrm{a'b'}) | \omega \rangle = \delta_{\lambda\lambda'}\delta_{\mu\mu'}(\delta_{\mathrm{aa'}}\delta_{\mathrm{bb'}} + (-)^{j_{\mathrm{a}}-j_{\mathrm{b}}+\lambda}\delta_{\mathrm{ab'}}\delta_{\mathrm{ba'}}),$$

from which

$$S^{\mathrm{ab}}_{J}(J, 0) = -\tfrac{1}{2}[U_{\mathrm{a}}V_{\mathrm{b}} + (-)^{L+S}V_{\mathrm{a}}U_{\mathrm{b}}]\eta^{J}_{\mathrm{ab}}. \tag{9.56}$$

For transition between a pair of two-quasi-particle states, by reference to (9.52), one sees that it is the $UU - VV$ term that contributes. The algebra is involved; the result is in Glendenning, 1968.

We have seen how the nuclear structure information enters the coupled equations for scattering through the reduced matrix elements of an operator S^{ab}_{LSJ}, which is implicitly defined by (9.22) or explicitly by (9.38) in terms of particle operators, or (9.52) in terms of BCS quasi-particles. Its calculation is several specific examples was illustrated, and other cases can be found in the literature. No exhaustive list can be made because we cannot anticipate the novel forms that future theories of nuclear structure may take. In any case, the corresponding amplitudes can be calculated from the general definition of S. The examples were chosen to illustrate a variety of cases.

L. RECAPITULATION

The coupled equations representing the elastic and inelastic scattering of two nuclei a and A are given by (9.4). Once the physical ingredients, the optical potential U, and the off-diagonal matrix elements of the interaction causing transitions $(\phi_{\alpha}|V|\phi_{\alpha'})$ are defined, the problem is then merely a computational one.

The parameters of the optical potential are frequently closely related to those that describe the elastic cross section alone in the absence of coupling to excited states (Chapter 4). However, if the coupling to some of the excited states is particularly strong, as for collective rotational levels especially, and to a lesser degree, vibrational levels, then the optical potential needed in the coupled problem will have different parameters. This can be understood in terms of the structure of the effective interaction of (8.18) and will be discussed in greater detail in Chapter 13.

The other, and most interesting, physical ingredients are the matrix elements of the direct interaction between the *nuclear* wave functions (9.14). Any scalar two-body operator or sum of two-body operators acting on different parts of a physical system (e.g., the projectile and target) can be written as a sum of scalar products of tensor operators whose transformation properties refer to the two parts of the system,

$$V(1, 2) = \sum_{J} \mathscr{T}_{J}(1) \cdot T_{J}(2). \tag{9.57}$$

This simply means that the powerful angular-momentum algebra developed by Racah can be used in the evaluation of the matrix element. For our special case, the preceding representation of the interaction was explicitly obtained and given in detail in (9.5)–(9.13). The matrix elements (9.14), with respect to product states of the two parts of the system, can be written immediately, by application of Racah algebra (see Appendix), in terms of purely geometrical coefficients ($3-j$, $6-j$ symbols, etc.) and the reduced matrix elements of $\mathcal{T}_J(1)$ and $T_J(2)$. In the example treated (nucleon–nucleus scattering), the reduced matrix element of the nucleon part is itself purely geometrical (9.20) and therefore of no particular interest. The interesting physics is in the *nuclear* reduced matrix element referred to as the form factor (9.17). Still seeking the very heart of the matter, this was expressed in terms of a *superposition* of shell-model single-particle reduced matrix elements, or single-particle form factors (9.21). Shell-model states are presumed to be known. They are the eigenstates of some appropriate shell-model Hamiltonian. The actual detailed structure of the *nuclear* states enters the scattering theory through the *structure amplitudes* $S_J^{\mathrm{ab}}(J'J'')$ of (9.22), which specifies what superposition of shell-model form factors is needed. This was written in terms of creation and destruction operators in (9.38), namely,

$$S_J^{\mathrm{ab}} = [\beta_{\mathrm{a}}^{\dagger}\tilde{\beta}_{\mathrm{b}}]_J, \qquad \mathrm{a} = n_{\mathrm{a}}\ell_{\mathrm{a}}j_{\mathrm{a}}. \tag{9.58}$$

There are a number of approximate solutions to the nuclear structure problem beginning with the shell model with a residual interaction and more general extended shell-model theories, such as those that employ the Bogolyubov–Valatin transformation to isolate the pairing effect of the nucleon–nucleon interaction followed by a diagonalization of the residual quasi-particle interaction in the so-called Tamm–Dancoff or random-phase approximation. The point is that *whatever* theory is proposed to describe a particular set of nuclear states—any one of the preceding or some as yet undiscovered theory—all that is needed to test such a theory concerning scattering data, is to provide the reduced matrix elements of S_J^{ab}. All the rest is geometry and the numerical solution of coupled differential equations. The scattering theory here presented (Glendenning, 1968a, 1969a) has isolated the content of any nuclear structure theory in the specification of the preceding amplitudes!

Strictly speaking, this statement is true only if exchange arising from the Pauli principle and the nature of the interaction is neglected. Even so, approximate ways of accounting for the exchange terms preserve the preceding structure (Petrovich *et al.*, 1977), as will be shown in detail in Chapter 12.

M. THE DIRECT INTERACTION AND ITS SPIN DEPENDENCE

The central part of the interaction between the scattered particle and the nucleons in the nucleus can be written as in Eqs. (9.6) and (9.7):

$$V(r_1, r_2) = V_{00} + V_{01}(\boldsymbol{\tau}_1 \cdot \boldsymbol{\tau}_2) + [V_{10} + V_{11}(\boldsymbol{\tau}_1 \cdot \boldsymbol{\tau}_2)](\boldsymbol{\sigma}_1 \cdot \boldsymbol{\sigma}_2). \quad (9.59)$$

There are other ways of writing this same interaction, which for some purposes are more convenient. Because

$$\boldsymbol{\sigma}_1 \cdot \boldsymbol{\sigma}_2 = 4\mathbf{s}_1 \cdot \mathbf{s}_2 = 2(S^2 - s_1^2 - s_2^2) = 2S^2 - 3 = \begin{cases} -3, & S = 0, \\ +1, & S = 1, \end{cases}$$

we can construct projection operators of spin-singlet and -triplet states as

$$P_0^\sigma = \tfrac{1}{4}(1 - \boldsymbol{\sigma}_1 \cdot \boldsymbol{\sigma}_2), \qquad P_1^\sigma = \tfrac{1}{4}(3 + \boldsymbol{\sigma}_1 \cdot \boldsymbol{\sigma}_2), \quad (9.60)$$

and similarly for isospin-singlet and -triplet. Because the two-nucleon wave function involving spin, isospin, and space parts must be antisymmetric, we can construct out of products of the two spin and two isospin singlet and triplet projection operators, projection operators for states that we can refer to as singlet or triplet in spin and even or odd under interchange of the space coordinates $\mathbf{r}_1 \leftrightarrow \mathbf{r}_2$. Referring to these operators as P^{SE}, P^{SO}, P^{TE}, P^{TO} for singlet-even, singlet-odd, etc., we can write the interaction as

$$V(r_1, r_2) = V_{\mathrm{SE}}P^{\mathrm{SE}} + V_{\mathrm{SO}}P^{\mathrm{TO}} + V_{\mathrm{TE}}P^{\mathrm{TE}} + V_{\mathrm{TO}}P^{\mathrm{TO}}, \quad (9.61)$$

and a little algebra yields the connection among the four potential strengths in each representation. The latter form is useful in the analysis of two-nucleon scattering data.

Depending on the circumstances in a nucleon–nucleus scattering problem, various pieces of the interaction will be effective in one case and others in another. Consider, for example, the scattering of protons from a nucleus like nickel. The valence nucleons are neutrons, and the low-lying states will be composed mainly of neutron excitations, the proton magic shell of 28 being closed. The isospin operator $\boldsymbol{\tau}_1 \cdot \boldsymbol{\tau}_2$ yields the value $+1$ when acting on two like nucleons and -1 when acting on a neutron–proton pair. Recall that we can write

$$\boldsymbol{\tau}_1 \cdot \boldsymbol{\tau}_2 = \tfrac{1}{2}(\tau_1^+ \tau_2^- + \tau_1^- \tau_2^+) + \tau_1^Z \tau_2^Z, \quad (9.62)$$

where

$$\tau^{\pm} = \tau^x \pm i\tau^y, \quad (9.63)$$

and it is the Z component that has diagonal expectation values. Consequently, in terms of the first representation of the interaction (9.59), the proton–neutron interaction for inelastic scattering takes the form

$$V_{\mathrm{p-n}} = (V_{00} - V_{01}) + (V_{10} - V_{11})\boldsymbol{\sigma}_1 \cdot \boldsymbol{\sigma}_2 \tag{9.64}$$

On the other hand, if neutrons were scattered from the same nucleus, we would need the neutron–neutron interaction, which is

$$V_{\mathrm{n-n}} = (V_{00} + V_{01}) + (V_{10} + V_{11})\boldsymbol{\sigma}_1 \cdot \boldsymbol{\sigma}_2. \tag{9.65}$$

Thus, according to the strengths of these combinations, protons and neutrons will interact differently with respect both to the (spin) scalar and vector parts of the interaction.

Carrying out the algebra connecting the two representations of the interaction, we find

$$\left.\begin{aligned} V_0 &= V_{00} - V_{01} = \tfrac{1}{8}(3\mathrm{TE} + 3\mathrm{TO} + \mathrm{SE} + \mathrm{SO}) \\ V_1 &= V_{10} - V_{11} = \tfrac{1}{8}(\mathrm{TE} + \mathrm{TO} - \mathrm{SE} - \mathrm{SO}) \end{aligned}\right\} (\mathrm{p-n}), \tag{9.66}$$

$$\left.\begin{aligned} V_0 &= V_{00} + V_{01} = \tfrac{1}{4}(3\mathrm{TO} + \mathrm{SE}) \\ V_1 &= V_{10} + V_{11} = \tfrac{1}{4}(\mathrm{TO} - \mathrm{SE}) \end{aligned}\right\} \mathrm{n-n \ or \ p-p}. \tag{9.67}$$

At the time that the microscopic theory of the inelastic scattering was developed, the knowledge and treatment of the effective interaction was rudimentary. Consistent with low-energy scattering data and shell-model calculations for certain nuclei, the interaction has the approximate character (as shown in Table I)

$$\begin{aligned} \mathrm{SE} &\simeq \tfrac{2}{3}\mathrm{TE} = \text{attractive}, \\ \mathrm{TO} &\simeq \mathrm{SO} \simeq 0, \end{aligned} \tag{9.68}$$

TABLE I

Comparison of Strengths of Various Real 1F Range Yukawa "Equivalent" Interactions[a]

Force	V_{00}	V_{10}	V_{01}	V_{11}	V_{pp}^0	V_{pp}^1	V_{pn}^0	V_{pn}^1	Ref.
A	−36.2	6.30	17.8	12.1	−18.4	18.4	−54	−5.75	Kalio and Koltveit, (1964)
B	−43.8	4.30	18.3	13.6	−25.5	18.0	−59.1	−9.20	Slanina and McManus, 1968
C	|30–40|	|11–27|	|21|	|17|	—	—	—	—	Ball and Cerny, 1969
D	−40.5	6.80	20.2	13.5	−20.3	20.3	−60.7	−6.70	Glendenning and Veneroni, 1966
E	−41.1	7.40	20.0	13.7	−21.1	21.1	−61.1	−6.30	True, 1963

[a] All values are in mega electron volts. A is the K–K force, B the impulse at $E_{\mathrm{LAB}} = 60$ MeV, C the interaction determined by Ball and Cerny, D the interaction of Glendenning and Veneroni, and E the interaction of True. (From Petrovich, 1970.)

which yields, in units of the strengths of the triplet-even attraction,

$$V_0 \simeq \tfrac{11}{24}, \qquad V_1 \simeq \tfrac{1}{24}, \qquad \text{(p–n)}, \tag{9.69}$$

$$V_0 \simeq \tfrac{1}{6}, \qquad V_1 \simeq -\tfrac{1}{6}, \qquad \text{(n–n or p–p)}. \tag{9.70}$$

Based on these results one can conclude that protons and neutrons can have different cross sections for exciting the same nuclear state and that by comparison, the role of the spin-independent and spin-dependent parts of the interaction can be studied (Glendenning and Veneroni, 1966). Note that a different nuclear form factor $S = 0$ or 1 in (9.20) is involved in the two cases.

In recent years, the role of correlations in nuclear states induced by various pieces of the interaction and their excitation by various probes has become a very fruitful area of study (Goodman *et al.*, 1980; Petrovich and Love, 1981). This was made possible by important advances in understanding and handling the interaction. Spin–orbit and tensor components can now be handled, and contact has been established with the free nucleon–nucleon scattering data (Bertsch *et al.*, 1977; Love and Franey, 1980). We shall look at those important topics in Chapters 11 and 12.

N. SELECTION RULES AND THE DIRECT INTERACTION

The general expression for the nuclear form factors was discussed earlier and is defined by

$$\mathscr{F}^{J'J''}_{LSJ}(r) = (\Phi_{J'}(\mathbf{A})\|\mathscr{T}_{LSJ}\|\Phi_{J''}(\mathbf{A})) \tag{9.71}$$

Clearly, for transitions between any pair of nuclear states J' and J'', several such form factors are involved, corresponding to possible combinations of L, S, and J. Their values are constrained by angular momentum and parity conservation:

$$\begin{gathered} |J' - J''| \le J \le J' + J'', \\ \mathbf{J} = \mathbf{L} + \mathbf{S}, \\ (-)^L = \pi'\pi'', \\ S = 0 \quad \text{or} \quad 1. \end{gathered} \tag{9.72}$$

As an example, consider explicitly transitions from the ground to a state of spin J and parity $(-)^J$:

$$0^+ \to J(-)^J. \tag{9.73}$$

(Such states $J(-)^J$ are referred to as natural parity states, $0^+, 1^-, 2^+, 3^-, \ldots.$) Then one readily confirms that two combinations of LSJ are possible for natural parity states, namely,

$$LSJ = \begin{cases} J & 0 & J \\ J & 1 & J \end{cases} \tag{9.74}$$

It is convenient to refer to the form factors as being scalar if $S = 0$ and vector if $S = 1$. The latter is responsible for spin–flip transitions, according to Eq. (9.6).

On the other hand, another set of selection rules apply to LSJ for transitions to unnatural parity states in an even–even nucleus

$$0^+ \to J(-)^{J+1}. \tag{9.75}$$

Then, according to (9.72), the allowed combinations of LSJ are

$$LSJ = \begin{cases} J-1 & 1 & J \\ J+1 & 1 & J \end{cases} \tag{9.76}$$

The excitation of an unnatural parity state requires that $S = 1$. Thus, according to (9.6) and (9.11), the spin-dependent part of the interaction alone is responsible for exciting such states. Thus, we begin to see how various pieces of the interaction are responsible for exciting specific types of nuclear states. It is worthwhile emphasizing this more strongly by quoting results obtained in the distorted-curve Born approximation for several probes (Glendenning and Veneroni, 1966). In particular, for spinless particles (alpha), nucleons, and spin 1 (deuteron), the cross sections for exciting natural parity states have the form

$$\frac{d\sigma}{d\Omega}[0^+ \to J(-)^J] = \begin{cases} V_0^2\sigma_{J0J}, & \text{alpha} \\ V_0^2\sigma_{J0J} + V_1^2\sigma_{J1J}, & \text{nucleon} \\ V_0^2\sigma_{J0J} + \frac{2}{3}V_1^2\sigma_{J1J}, & \text{deuteron} \end{cases} \tag{9.77}$$

In the case of the nucleon, V_0 and V_1 are related through (9.64) and (9.65) to the various spin and isospin pieces of the interaction. In the case of the composite particles, V_0 and V_1 can be related to the interaction (9.50) after its diagonal matrix element with respect to the wave function of the composite particle is performed (Glendenning and Veneroni, 1966). As one might expect, the ranges of the resulting potentials are somewhat larger than the original nucleon–nucleon potential because of the finite size of the composite. Details can be found in the Appendix to Glendenning and Veneroni, 1966.

For the excitation of unnatural parity states, the corresponding results are

$$\frac{d\sigma}{d\Omega}[0^+ \to J(-)^{J+1}] = \begin{cases} 0, & \text{alpha} \\ V_1^2(\sigma_{J-1,1,J} + \sigma_{J+1,1,J}), & \text{nucleon} \\ \frac{2}{3}V_1^2(\sigma_{J-1,1,J} + \sigma_{J+1,1,J}), & \text{deuteron.} \end{cases} \tag{9.78}$$

These results are written here in DWBA to display the parts of the interaction and selection rules that come into play in the direct (lowest order) transition. They were obtained in Chapter 7 for nucleons. The results for alphas and deuterons can be inferred from the development there. In any case, the definition of σ_{LSJ} in (9.77) and (9.78) is

$$\sigma_{LSJ} = \frac{k'}{k}\left(\frac{m}{2\pi\hbar^2}\right)^2 \sum_M |\mathrm{B}_{LSJ}^M|^2, \tag{9.79}$$

where m is the reduced mass of the projectile, k and k' are its initial and final wave numbers, and B, according to (7.103), is

$$\mathrm{B}_{LSJ}^M = i^{-L}(2L+1)^{-1/2} \int \chi^{(-)*}(\mathbf{k}', \mathbf{r}) \mathscr{F}_{LSJ}^{J'J}(r) Y_L^{M*}(\hat{\mathbf{r}}) \chi^{(+)}(\mathbf{k}, \mathbf{r})\, d\mathbf{r}, \tag{9.80}$$

where the form factor $\mathscr{F}$ is given by (9.20) in general and, for particular examples of nuclear wave functions, was worked out in subsequent sections.

O. APPLICATION OF THE THEORY

The theory has been applied to various nickel and tin isotopes (Glendenning and Veneroni, 1966; Glendenning, 1969a. Faessler *et al.*, 1967). These are vibrational-like nuclei having a closed shell of protons and an open shell of neutrons. The structure of these nuclei has been the subject of several theoretical studies. In one of them, the Bogolyubov–Valatin transformation was calculated to extract the pairing effects of a phenomenological residual interaction in the open shells of neutrons (Arvieu and Veneroni, 1963; Arvieu *et al.*, 1963, 1964). In the second step the remaining interaction was diagonalized on a basis of two-quasi-particle states. The structure amplitudes for just such a theory were among those illustrated earlier.

The scalar form factor for the transition from the ground to the first 2^+ state is shown in Fig. 9.2 based on such wave functions. The single-particle transitions, together with their phases and amplitudes, are also shown. For this lowest 2^+ state each contributes constructively to the nuclear form factor yielding the large (compared with single particle) peaked function shown. This has become a well-known property of *collective* transitions.

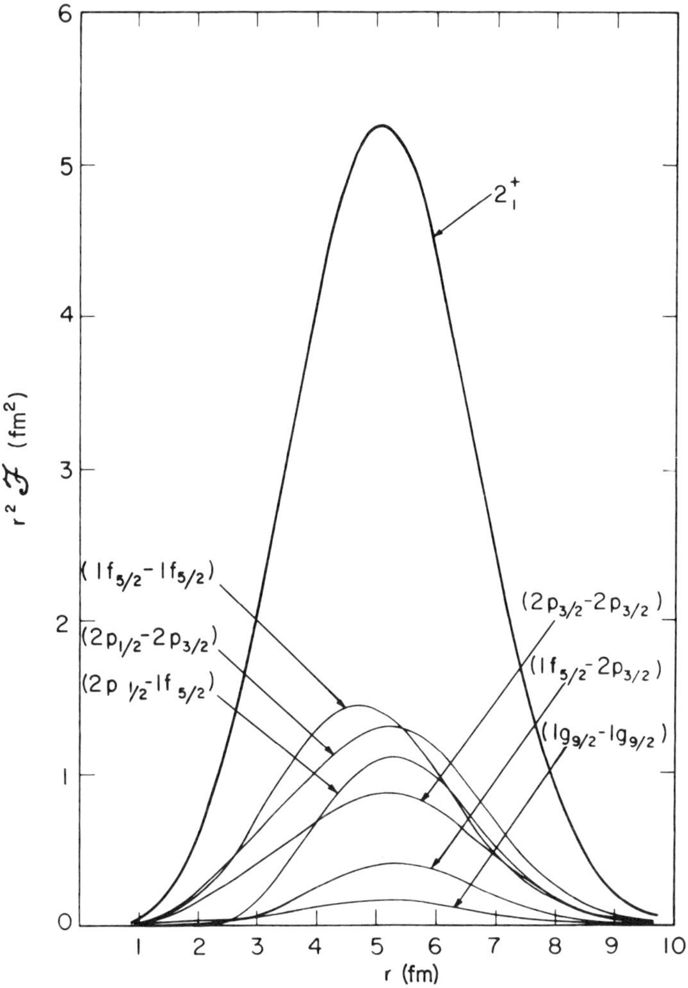

Fig. 9.2. Scalar form factor for the lowest 2^+ state in ^{60}Ni is shown together with the single-particle form factors that contribute to it. They all have such phases that they add constructively for the lowest 2^+ but must therefore be destructive for all other 2^+ states. (From Glendenning and Veneroni, 1966.)

The scalar form factor for the transition to the second 0^+ state is shown in Fig. 9.3 and is an example of a *noncollective* transition. The interference between the various single-particle transitions is destructive, leading to a small-amplitude oscillatory form factor. Clearly the incoherent transitions are much more sensitive to the details of the nuclear wave functions than the coherent or collective transitions.

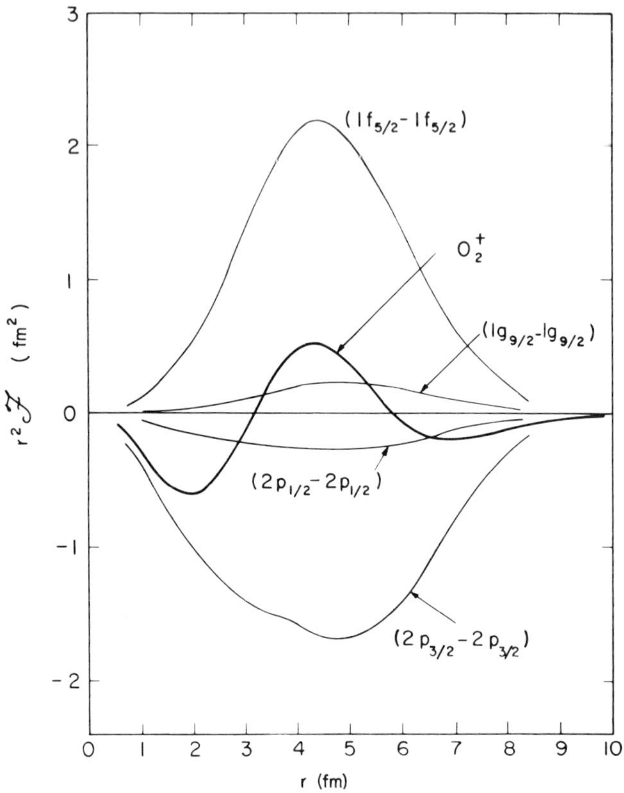

Fig 9.3. The monopole single-particle form factors that contribute to the 0^+ states of ^{60}Ni are shown with magnitudes and signs appropriate to the first excited 0^+ stated labeled 0_2. Here they interfere destructively to give the small form factor for the nuclear state shown by the heavy line labeled 0_2^+ (Glendenning and Veneroni, 1966).

In Fig. 9.4 the scalar and vector form factors for transitions to three 2^+ states in ^{60}Ni are shown together with cross sections calculated in DWBA at two energies. One sees again the larger form factor of the lowest 2^+ state in comparison with the incoherent transitions that lead to inhibited form factors. The cross sections reflect the difference in the form factor in magnitude as they must, and for the higher energy, they also more sharply reflect differences in shape. This is so because the higher the projectile energy, the shorter its wavelength and therefore the greater its spatial resolution.

It is worth commenting on the relative magnitudes of the scalar and vector form factors. For the lowest 2^+ state the scalar form factor is about five times larger than the vector. This reflects the dominant role of the singlet-even part of the interaction in inducing the coherent correlations in this state. Recall

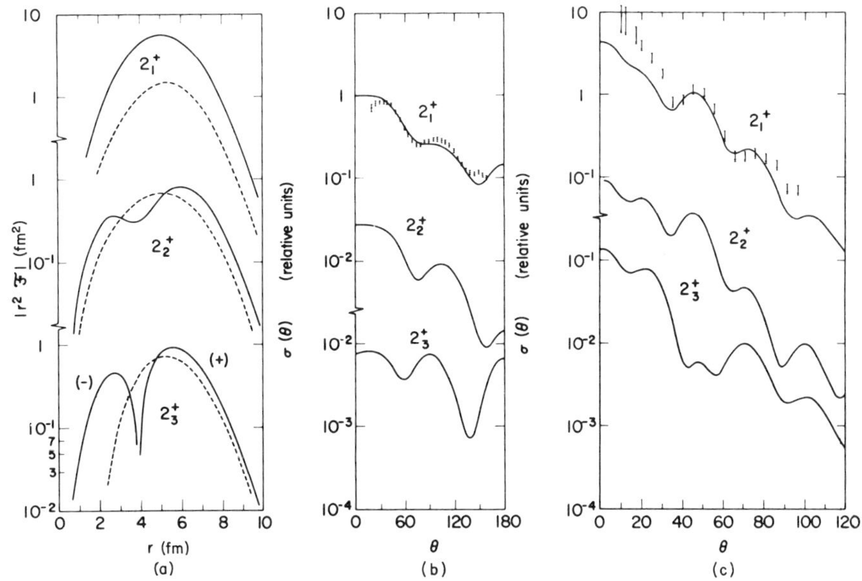

Fig. 9.4. Scalar (———) and vector (---) form factors for 2^+ states in ^{60}Ni are shown in (a). The absolute values are plotted, and oscillations are indicated by (+) and (−). Corresponding cross sections at 11 MeV in (b) and at 40 MeV in (c) are shown for proton scattering calculated in the distorted-wave approximation ((a) from Glendenning and Veneroni (1966), (b) from Dickens, *et al.*, 1963, and (c) from Stovall and Hintz, 1964).

that, according to (9.68), the interaction is attractive in the even states and weak or repulsive in the odd states and, for like nucleons, which comprise the outer shell nucleons in Ni, the available states are the antisymmetric space–spin combinations SE and TO. In terms of the alternative representation of the interaction (9.70), V_0 is attractive and V_1 repulsive, so that in a low-lying state the correlations will be such as to exploit V_0 and to minimize the effect of $V_1 \boldsymbol{\sigma}_1 \cdot \boldsymbol{\sigma}_2$.

On the other hand, one can anticipate that other parts of the interaction will tend to induce correlations in levels lying elsewhere in the spectrum. After all, the wave functions must be orthogonal. Indeed, these qualitative predictions have turned out to be valid (Glendenning and Veneroni, 1966).

The distorted-wave Born approximation may be fairly reliable for moderately collective states of a nucleus, for which the direct transition from the ground state dominates, but is not too strong. However, for weakly excited states, for which the direct transition is hindered, indirect transitions through intermediate states may be of equal or more importance. In such cases the interference between several transition amplitudes may lead to major corrections to DWBA. This is illustrated in Fig. 9.5, where DWBA and coupled channel calculations are compared (Glendenning, 1969a).

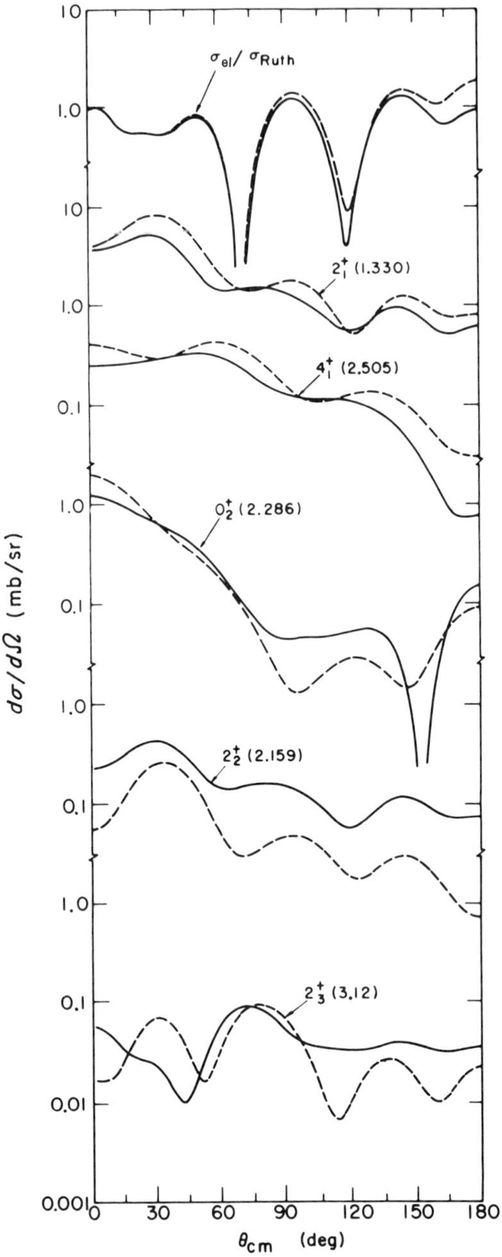

Fig. 9.5. Coupled-channel (———) is compared with the optical-model calculation of elastic and distorted-wave calculation of inelastic proton scattering (– – –) for ^{60}Ni. Note that the elastic cross section is barely affected by the additional coupling. Although several cross sections are only slightly changed others are significantly modified (From Glendenning, 1969).

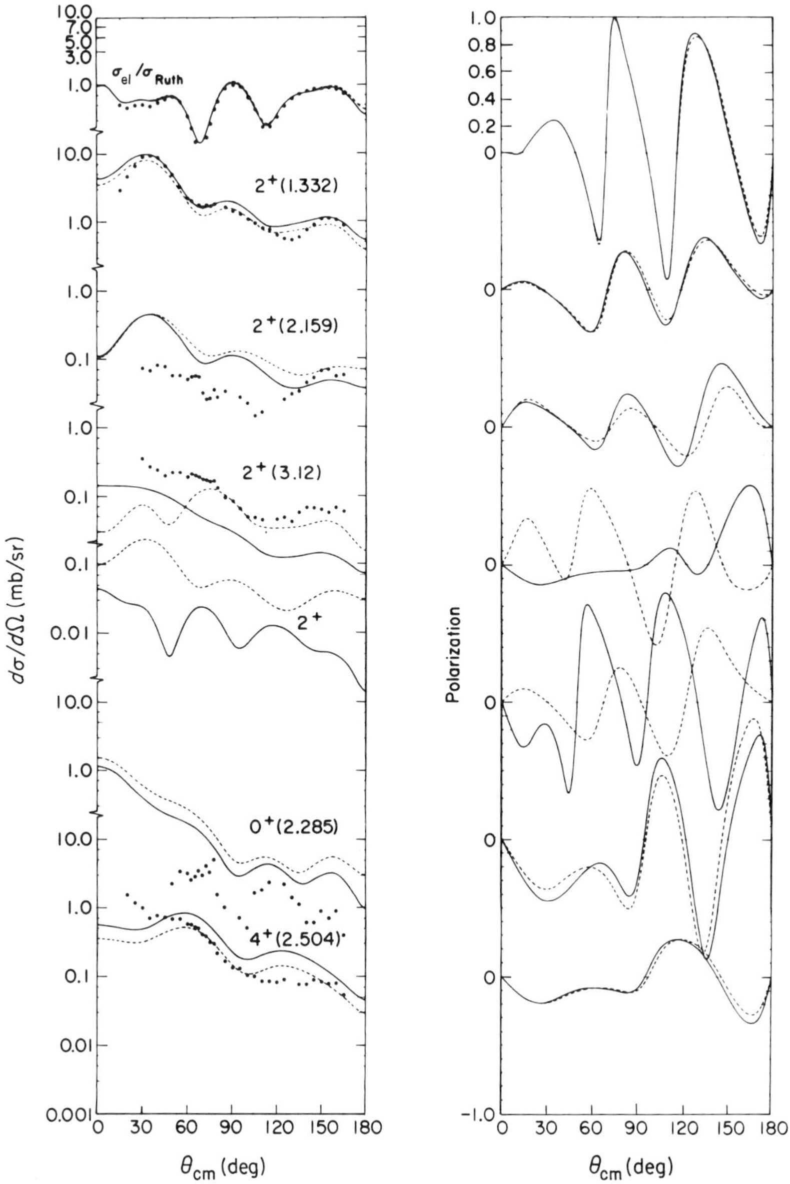

Fig. 9.6. Cross sections and polarizations for 17.8 MeV protons from ^{60}Ni. Curves are coupled-equation calculations using microscopic description employing a surface delta interaction (———) and a volume finite-range interaction (- - -). Direct interaction parameters are $V_\sigma = -55$ MeV, $V_1 = 0$, and $\sigma = 1.85$ fm. (Data from Jarvis, *et al.*, 1967; calculations by Faessler *et al.*, 1967.)

Because of the greater sensitivity of noncollective states to the detailed accuracy of the nuclear wave functions, they are not as well accounted for by the theory as the collective states, as can be seen in Fig. 9.6, which compares a coupled-channel calculation with experiment (Faessler *et al.*, 1967). The calculated polarizations are also shown, though none were measured at

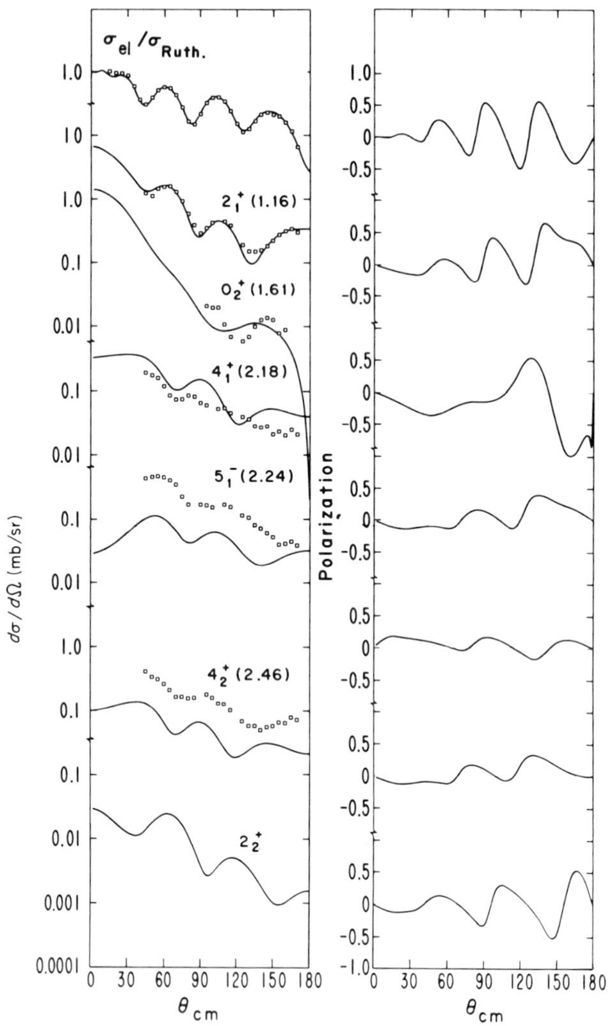

Fig. 9.7. Cross sections and polarizations for 17.8 MeV protons from ^{120}Sn, measured at Berkeley. Solid lines represent coupled-channel calculation based on microscopic theory with $V_0 = -45$ MeV, $V_1 = 0$, and $\sigma = 1.85$ fm (Data from Jarvis *et al.*, 1967; calculations by Glendenning (1969)).

the time that the calculations were performed. The same situation holds in the tin region as Fig. 9.7 suggests.

A common problem in all cases concerned the strength of the transitions. Although relative magnitudes and shapes, especially for the collective transitions, were reasonably well reproduced, the absolute magnitudes were underestimated by a factor of ~ 2 by a direct interaction having a strength that seemed reasonable in terms of a phenomenological representation of the vacuum interaction. Recalling the discussion at the beginning of this chapter of the bare and renormalized G matrix, we are reminded of the possibility that the nuclear structure calculations were performed on a basis that was too severely truncated. In the next chapter we shall discuss how corrections referred to as core polarization can be made for effects of the truncation.

Notes to Chapter 9

General references to the microscopic theory of nuclear structure can be found in various texts and review articles such as those by Bohr and Mottelson (1969, 1975), Preston and Bhaduri (1975), Brown (1971), de Shalit and Feshbach (1974), Jean (1969), Baranger (1963) and Lane (1964).

Chapter 10

Core Polarization

In the introduction to Chapter 9, it was mentioned that a discussion of nuclear reactions may, in general, require corrections arising both from a truncation of the channel space in which the scattering problem is solved as well as from the truncation of the configuration space in which the nuclear problem is solved. The general formulation of effective operators within a truncated shell-model space would parallel the development in Chapter 8. However, instead we now discuss a simple model, introduced by Love and Satchler (1967), that is a particular case of the general theory.

It is well known that spherical nuclei possess certain collective states for which electromagnetic and inelastic transitions are enhanced. Such states have often been regarded as consisting of surface vibrations of the nucleus. This is the Bohr–Mottelson collective model (Bohr and Mottelson, 1969 and 1975). The shape of the nucleus is defined to be

$$R(\theta, \phi) = R_0\left[1 + \sum_{LM} \alpha_{LM} Y^*_{LM}(\theta, \phi)\right], \tag{10.1}$$

where the α_{LM} are dynamical variables describing the oscillations of the surface. For small-amplitude oscillations, they obey the quantum-mechanical oscillator equation. The spectrum of such a system consists of equally spaced energy levels characterized as having 0, 1, 2, ... phonons of excitation $\hbar\omega_L$.

Consider now an external particle with coordinate **r** in interaction with such a nucleus. We have already postulated that the average effect of the nucleus on the scattering of such a particle can be represented by an average potential, which in the case of bound nucleons is the shell-model potential, whereas for free particles it is the optical potential. Furthermore, it is reasonable to assume, as follows from the short range of the nucleon–nucleon interaction, that the shape of the potential follows closely the nuclear density shape. If $V(r)$ denotes the equipotentials in the case of a spherical nucleus, then the equipotentials for a nucleus having the shape as in Eq. (10.1) are

$$V(\alpha, \mathbf{r}) = V\left(\frac{r}{1 + \sum_{LM} \alpha_{LM} Y^*_{LM}(\hat{\mathbf{r}})}\right). \tag{10.2}$$

Supposing that the vibrations always have small amplitude, this can be expanded in a Taylor series:

$$\begin{aligned} V(\alpha, \mathbf{r}) &= V(0, r) + \sum_{LM} (\partial V/\partial \alpha_{LM})_{\alpha=0}\alpha_{LM} + \cdots \\ &= V(r) - r(\partial V/\partial r) \sum_{LM} \alpha_{LM} Y^*_{LM}(\hat{\mathbf{r}}) + \cdots. \end{aligned} \tag{10.3}$$

It is a property of the oscillator that the matrix elements of α_{LM} between states of the oscillator Hamiltonian exists only if the states differ by a single phonon. Thus the interaction of a particle with the oscillator can, through the second term, change the state of the oscillator by one phonon.

Now consider a nucleus with, say, two particles beyond a spherical closed shell of nucleons, referred to as the core. Suppose further that the nucleus corresponding to the core has a vibrational-like spectrum. Then, in the absence of any interaction between the particles and the core, the spectrum of states of the nucleus would have the form

$$|j^2, N_L\rangle, \qquad \text{and} \qquad |jj', N_L\rangle,$$

where N_L denotes the number of phonons of multipolarity L. In this case the spectrum would look like that shown in Fig. 10.1. We do not mention the angular momentum to which the particles are coupled, nor the vibrational angular momentum, for simplicity of notation. Otherwise we would write, for example,

$$|(j^2)J, N_L R; I\rangle,$$

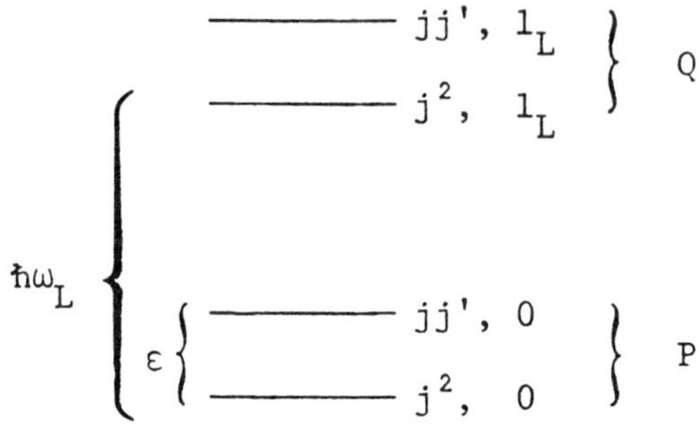

Fig. 10.1. A shell-model truncation to the states P that omits the collectivity carried by states in the omitted space of states Q.

where

$$\mathbf{I} = \mathbf{J} + \mathbf{R}, \qquad \mathbf{J} = \mathbf{j} + \mathbf{j}$$

(and $R = 0$ if $N_L = 0$, $R = L$ if $N_L = 1$, and in general R is the angular momentum carried by the phonons).

Suppose now that we wish to describe the scattering of a nucleon from such a nucleus as that described; one with two valence particles outside a vibrational core. Suppose further that we consider *explicitly* only nuclear states having zero phonons present, such as j^2, jj', jj'', etc., all with $N_L \equiv 0$. We recognize that this is a drastic truncation because the interaction term in Eq. (10.3) acting between the *valence* nucleons and the core will admix one-phonon components into the zero-phonon states and vice versa. Yet for ease of handling the scattering problem we want to insist on dealing with only the simple states. As in Chapter 8, the effect of the truncation can be embedded in the effective interaction, acting between the scattered nucleon and the valence nucleons.

To do this we first note, in accord with Eq. (10.3), that the interaction between the scattered particle and the core V_{pc} and between the valence nucleon and the core V_{vc} can be written as

$$V_{\text{pc}}(r) = -k_{\text{p}}(r) \sum_{LM} \alpha_{LM} Y^*_{LM}(\hat{\mathbf{r}}),$$
$$V_{\text{vc}} = -\sum_{i=1}^{2} k_{\text{v}}(r_i) \sum_{LM} \alpha_{LM} Y^*_{LM}(\hat{\mathbf{r}}_i), \tag{10.4}$$

where

$$k_{\text{p}}(r) = r\,\partial U/\partial r, \qquad k_{\text{v}}(r) = r\,\partial V/\partial r. \tag{10.5}$$

Here U is the optical potential for nucleon–nucleus scattering and V the shell-model central potential.

In first-order perturbation theory it is simple to write the corrections to the wave functions in the subspace P. To this order the corrected wave functions are, respectively (see Fig. 10.1),

$$\begin{aligned}
|\alpha\rangle &= |j^2, 0\rangle + \sum_L |j^2, L\rangle \frac{\langle j^2, L|V_{vc}|j^2, 0\rangle}{(-\hbar\omega_L)} \\
&\quad + \sum_L |jj', L\rangle \frac{\langle jj', L|V_{vc}|j^2, 0\rangle}{(-\hbar\omega_L - \varepsilon)}, \\
|\alpha'\rangle &= |jj', 0\rangle + \sum_L |jj', L\rangle \frac{\langle jj', L|V_{vc}|jj', 0\rangle}{(-\hbar\omega_L)} \\
&\quad + \sum_L |j^2, L\rangle \frac{\langle j^2, L|V_{vc}|jj', 0\rangle}{(\varepsilon - \hbar\omega_L)}.
\end{aligned} \tag{10.6}$$

Notice that we are using a very abbreviated notation in which $|j^2, 0\rangle$ means that no phonon is present and $|j^2, L\rangle$ means that one phonon of multipolarity L is present. In this example no higher phonon states are considered. Notice also that, for brevity, we do not explicitly mention the angular momentum to which the particles are coupled. Clearly all three components of $|\alpha\rangle$ have the same total angular momentum. All of these details can be simply filled in, and must be at the end to actually carry out the numerical work. Finally, to simplify the notation and make clearest the physics arising from the admixture of core collective motion into the valence nuclear wave functions, we ignore the splitting of j^2 and jj' in their different angular momentum states 0, 2, . . . , etc.

The interaction between the scattered particle and the total nucleus is

$$V_{pA} = V_{pv} + V_{pc}. \tag{10.7}$$

Corresponding to a transition between the two correlated states, we want the matrix element (to lowest order in the admixed configurations),

$$\begin{aligned}
&\langle\alpha'|V_{pv} + V_{pc}|\alpha\rangle \\
&\quad = \langle jj', 0|V_{pv}|j^2, 0\rangle \\
&\qquad - \sum_L \langle jj', 0|V_{pv} + V_{pc}|j^2, L\rangle\langle j^2, L|V_{vc}|j^2, 0\rangle/(\hbar\omega_L) \\
&\qquad - \sum_L \langle jj', 0|V_{pv} + V_{pc}|jj', L\rangle\langle jj', L|V_{vc}|j^2, 0\rangle/(\hbar\omega_L + \varepsilon) \\
&\qquad - \sum_L \langle jj', L|V_{pv} + V_{pc}|j^2, 0\rangle\langle jj', L|V_{vc}|jj', 0\rangle^\dagger/(\hbar\omega_L) \\
&\qquad - \sum_L \langle j^2, L|V_{pv} + V_{pc}|j^2, 0\rangle\langle j^2, L|V_{vc}|jj', 0\rangle^\dagger/(\hbar\omega_L - \varepsilon).
\end{aligned} \tag{10.8}$$

In the last four terms, the V_{pv} term vanishes identically because the core configuration is different on opposite sides of the matrix elements and V_{pv} does not act on the core. Similarly the V_{pc} term vanishes in the second and fourth terms because the valence configurations are changed, but V_{pc} does not act on valence nucleons. Therefore

$$\begin{aligned}\langle\alpha'|V_{\text{pv}} + V_{\text{pc}}|\alpha\rangle = \langle jj'|V_{\text{pv}}|j^2\rangle &- \sum_L \langle jj', 0|V_{\text{pc}}|jj', L\rangle\langle jj', L|V_{\text{vc}}|j^2, 0\rangle/(\hbar\omega_L + \varepsilon) \\ &- \sum_L \langle jj', 0|V_{\text{vc}}^\dagger|j^2, L\rangle\langle j^2, L|V_{\text{pc}}|j^2, 0\rangle/(\hbar\omega_L - \varepsilon). \quad (10.9)\end{aligned}$$

For the four matrix elements we have

$$\begin{aligned}\langle jj', 0|V_{\text{pc}}|jj', L\rangle &= -k_{\text{p}}(r)\langle 0|\alpha_{LM}|LM\rangle Y_{LM}^*(\hat{\mathbf{r}}), \\ \langle jj', L|V_{\text{vc}}|j^2, 0\rangle &= -\sum_i \langle jj'|k_{\text{v}}(r_i)Y_{LM}(\hat{\mathbf{r}}_i)|j^2\rangle\langle LM|\alpha_{LM}^\dagger|0\rangle, \\ \langle jj', 0|V_{\text{vc}}^\dagger|j^2, L\rangle &= -\sum_i \langle jj'|k_{\text{v}}(r_i)Y_{LM}^*(\hat{\mathbf{r}}_i)|j^2\rangle \sum_{LM} \langle 0|\alpha_{LM}|LM\rangle, \\ \langle j^2L|V_{\text{pc}}|j^2 0\rangle &= -k_{\text{p}}(r) \sum_{LM} \langle LM|\alpha_{LM}^\dagger|0\rangle\, Y_{LM}(\hat{\mathbf{r}}).\end{aligned} \quad (10.10)$$

(We have used the fact that $\sum_M \alpha_{LM} Y_{LM}^*$ is real in several of these equations.)

From the vibrational properties of the core (Bohr and Mottelson, 1969 and 1975),

$$\langle 0|\alpha_{LM}|LM\rangle = \langle LM|\alpha_{LM}^\dagger|0\rangle = \sqrt{(\hbar\omega_L/2C_L)}, \quad (10.11)$$

where C_L is a constant and $\hbar\omega_L$ is the phonon energy. Putting this together we find

$$\begin{aligned}\langle\alpha'|V_{\text{pv}} + V_{\text{pc}}|\alpha\rangle = \langle jj'|V_{\text{pv}}|j^2\rangle &- \sum_L \frac{1}{C_L}\frac{(\hbar\omega_L)^2}{(\hbar\omega_L)^2 - \varepsilon^2} k_{\text{p}}(r) \sum_M Y_L^{M*}(\hat{\mathbf{r}}) \\ &\times \langle jj'|\sum_i k_{\text{v}}(r_i) Y_L^M(\hat{\mathbf{r}}_i)|j^2\rangle. \quad (10.12)\end{aligned}$$

This is the desired result. It is a matrix element between the two simple states j^2 and jj' of an effective operator differing from the original interaction V_{pv} by the second term. Each term V_{p1} and V_{p2} of V_{pv} is replaced by

$$V_{\text{eff}}(\mathbf{r} - \mathbf{r}_i) = V(\mathbf{r} - \mathbf{r}_i) - \sum_L \frac{1}{C_L}\frac{(\hbar\omega_L)^2}{(\hbar\omega_L)^2 - \varepsilon^2} k_{\text{p}}(r)k_{\text{v}}(r_i)Y_L^{M*}(\hat{\mathbf{r}})Y_L^M(\hat{\mathbf{r}}_i). \quad (10.13)$$

Moreover, because V is negative for an attractive interaction and the second term is also, the effect of the polarization of the core is to *increase* the strength

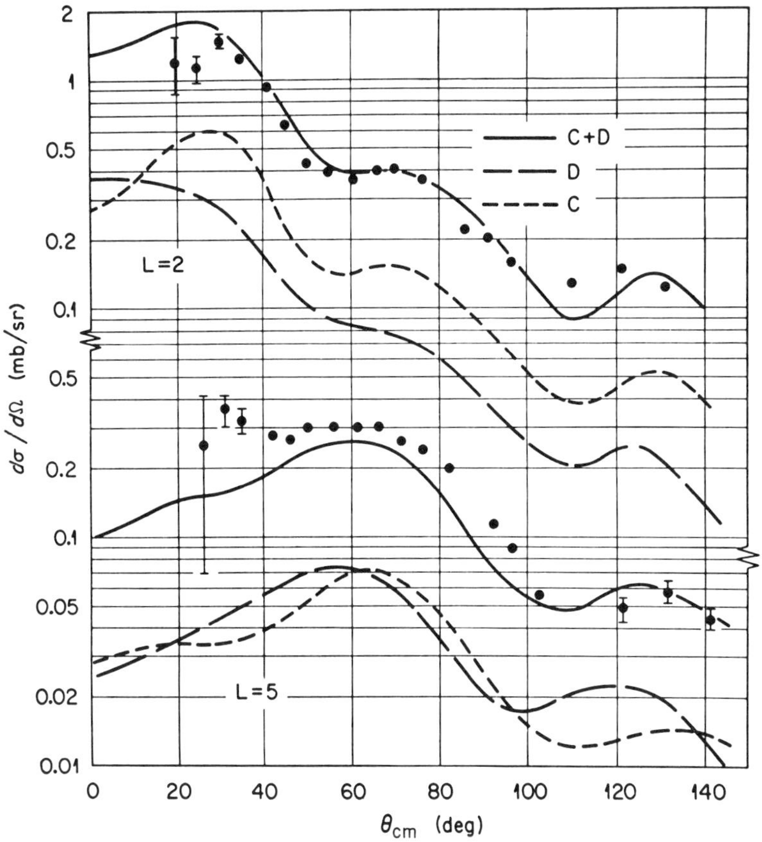

Fig. 10.2. Cross sections for 18.8-MeV protons inelastically scattered to produce the 2^+ and 5^- levels in ^{90}Zr. The effect of core polarization (C) on the direct (D) cross section computed in DWBA is shown. [From Love and Satchler (1967).]

of the direct interaction. This is, of course, entirely analogous to the treatment of effective charges for electromagnetic transitions.

An example is shown in Fig. 10.2 of the large enhancement of the cross section due to the polarization of the core. The effect is so large because of the severe truncation of the shell-model space, which fails to describe the collectivity of the state. In Chapter 9 we learned that collective enhancement can be very large because it corresponds to the constructive superposition of many single-particle transitions. Of course the alternative to calculating core polarization effects is to enlarge the shell model basis, and this model calculation can be interpreted as providing a motivation for doing so. This is not always practicable, however. The fact that neutrons carry an effective charge that is appreciable indicates that even in single closed-shell nuclei, it is necessary to incorporate the polarization of both neutrons and protons to properly account for such effects.

The core polarization, which was calculated here in the framework of the collective model as a simple illustration, can also be carried out in a microscopic scheme, and much progress has been made in recent years on this problem (Petrovich *et al.*, 1977; Brown and Madsen, 1975).

Chapter 11

Effective Interactions and the Free Nucleon–Nucleon Force

A. INTRODUCTION

The two-nucleon interaction is known to have a strong repulsion at short range. Consequently, a perturbative approach to scattering theory or the bound-state problem cannot be based directly on the free interaction V. Early applications of the theory of direct reactions therefore substituted a nonsingular phenomenological interaction. One justification for this approach can be understood as follows: Recall from Chapter 6 that Watson (1953) and Frances and Watson (1953) achieved a reordering of the Born series for the T operator so that each nucleon–nucleon interaction is represented by the τ matrix rather than V. The τ matrix is just the two-body t-matrix defined for scattering in the nuclear medium. It describes the complete scattering sequence, iterating the effect of V to all orders. As such it is well behaved even though V may be singular. In Chapter 8 we carried out a further development in which the space of target nuclear states was divided into two classes, denoted by P and Q. The P space includes the ground and other low-lying levels of interest, but especially collective states. The Q space refers to all the other states. We saw that a formally exact solution can be obtained in the truncated space P, provided that the interaction V is replaced by a new interaction operator $\mathscr{V}$. This new interaction is a many-body operator and does not have the form of a sum of two-body operators, such as the formulation of the theory of rearrangement collisions assumes. However, it can be developed in a Watson-type series in which the τ operator satisfies the integral equation,

$$\tau_{ij} = V_{ij} + V_{ij}\frac{Q}{E - H_N - T_i + i\varepsilon}\tau_{ij}, \tag{11.1}$$

where H_N is the nuclear Hamiltonian and T_i the relative kinetic energy of the projectile. This has a possible practical advantage over Watson's original formulation, in that the presence of the projection operator Q may allow approximations to the spectrum of intermediate states that enter the second term of (11.1) that would not be tenable if the full spectrum, including low-lying and collective states, were needed. The multiple scattering series for $\mathscr{V}$ is

$$\mathscr{V} = \sum_j \tau_{ij} + \sum_{j,\, k \neq j} \tau_{ij} g^{(+)} \tau_{ik} + \cdots, \tag{11.2}$$

where we denote the Green's functions in (11.1) by $g^{(+)}$. The series is arranged in a hierarchy of ever more complex interactions. Alternatively, a series based on a generalization of the Brueckner G matrix can be developed. By the very nature of direct reactions, an early truncation of these series is plausible. If the series is truncated at the first term, we obtain a theory based on a sum of nonsingular two-body interactions, just as has been assumed throughout the development of the theory of direct reactions.

Therefore, the phenomenological interactions that have been employed so frequently in applications of the theory can be regarded as a substitute for τ, which in a more basic approach would have to form part of the calculation. Although, from its defining equation, τ is complex, in practice a real phenomenological interaction is usually substituted.

In early applications, the parametrization of the interaction was chosen to reproduce the effective range and scattering length of the low-energy nucleon data and to account for some of the energy levels in selected nuclei near closed shells (Glendenning and Veneroni, 1965, 1966). In some other work of the same era, the parameters of the direct interaction were simply adjusted to individual transitions, with the hope of establishing some systematics (Satchler, 1967).

The choice of a simple phenomenological interaction was adequate for confirming the broad outline of the theory of inelastic scattering and for illuminating certain aspects of nuclear structure, such as how the coherent contribution of single-particle excitations could build up collective form factors for transitions to vibrational states (Glendenning and Veneroni,

1966). It was also perceived in the early work that the nucleon–nucleon interaction, being different in the various states of relative motion, could give rise to many types of collective states in nuclei and that comparisons between neutron and proton scattering from nuclei ought to contribute to the identification and elucidation of their character.

However, to address these and other interesting questions quantitatively and to place the theory on a secure foundation, a quantitative link between the direct interaction and the free nucleon–nucleon interaction is essential. In this respect, much more progress has been made on the bound-state problem than on the scattering problem. The theory of Brueckner *et al.* (1954, 1955), Bethe (1956), and Goldstone (1957) for nuclear matter and its applications to finite nuclei has established a quantitative link between nuclear properties and the two-nucleon interaction. Therefore, we now briefly discuss the free interaction, the Brueckner method for deriving an effective interaction for nuclei, and the corresponding problem for nuclear scattering. At sufficiently high energy, the effective interaction for scattering should approach the free t matrix, so that two energy regimes can be identified, one in which the medium effects are predominant and one in which they are negligible. Finally, we shall touch on recent progress in relating the optical potential to the free interaction. In the following Chapter we shall then take up the technical matters involved in incorporating the realistic interactions derived from the free interaction into the theory of inelastic scattering.

B. THE FREE-NUCLEON–NUCLEON INTERACTION

It has been understood since the work of Yukawa that nuclear forces arise through the exchange of quanta. However, Lagrangian field theories, which attribute the force to the exchange of mesons with various quantum numbers, have never yielded unambiguous results for the nonrelativistic potential. Currently, the basic exchange quanta are believed to be the quarks and gluons. Because of the unsolved problem of confinement, the derivation of a nucleon–nucleon potential from this theory is even more remote. Nevertheless, over distances that are not too small, the exchange of the basic quanta can be simulated by the exchange of the lighter mesons, especially the pion; and the theory can then be used to suggest the form of the potential, if not all its details. The one-pion exchange potential has long been valuable as a constraint on the long-range part, whereas the data are used to determine the phase shifts or a phenomenological potential corresponding to the shorter range part. The work of the Yale (Lassela *et al.*, 1962) and Livermore (MacGregor *et al.*, 1969) groups is important in connection with phase shift analysis of data. Such work is devoted to characterizing the scattering in terms of a set of phase shifts which, of course, are functions of energy. The work of Hamada and Johnston (1962) and of Ried (1968) are early examples of potential models that are designed to yield the energy dependence of the phase shifts as best as is possible within the limitations of the potential model.

Recent research of the Paris group is based on a more fundamental approach to the potential problem (Vinh Mau, p. 151, 1979). It avoids the perturbative calculation of the potential from a specific Lagrangian model and instead relies on unitarity, dispersion relations, and crossing symmetries to calculate the contributions of one-pion, two-pion, and omega exchange, while treating the short-range part phenomenologically. It has been quite a fruitful approach, which reproduces the empirical phase shifts very well.

Invariance principles can be used to derive the most general form that the free two-nucleon scattering amplitude can take (Wolfenstein, 1956). A potential, linear in velocity, can have only central, spin–orbit, and tensor components, which can act differently in the four states of relative motion. The fact that the S wave phase shift in nucleon–nucleon scattering changes sign around 250 MeV, signaling a change from attractive to repulsive interaction, is interpreted as indicating that the two-nucleon potential is repulsive at short distance. The phenomenological analysis of the scattering phase shifts indicates that it has to be strong. In some work, such as that of Hamada and Johnston (1962), it is represented by an infinitely strong repulsion, referred to as a hard core. Ried (1968), on the other hand, employs a soft core, meaning that the repulsive potential has a finite magnitude.

C. THE BRUECKNER G MATRIX

The means of solving the nuclear many-body problem was found by Brueckner and elaborated and clarified by Brueckner *et al.*, (1954, 1955), Bethe (1956), and Goldstone (1957). The theory is based on an analogy with the multiple scattering theory of Watson (1953) and Frances and Watson (1953). Watson achieved a reordering of the Born series so that each nucleon–nucleon collision is represented by the finite τ matrix rather than the singular free interaction. The τ matrix contains the nuclear Hamiltonian. Because the scattering theory of the earlier chapters was developed in terms of model nuclear wave functions for the purpose of exposing them to the test of consistency with scattering and reaction data, the appearance of H_N in τ is appropriate. On the other hand, the fundamental problem of nuclear structure is to calculate the ground-state energy from the free interaction. From this perspective, the nuclear wave function is unknown. Therefore it is convenient to introduce, as in the shell model, a one-body potential that defines unperturbed single-particle states in the nucleus. The nuclear Hamiltonian

$$H_N = \sum_{i=1}^{A} T_i + \sum_{i<j} V_{ij} \tag{11.3}$$

is rewritten as

$$H_N = H_0 + v,$$
$$H_0 = \sum_{i=1}^{A} (T_i + U_i), \qquad v = \sum_{i<j} v_{ij} - \sum_i U_i. \tag{11.4}$$

The single-particle Hamiltonians $T_i + U_i$ define the standard independent shell-model wave functions, a typical one of which we can denote by $\phi_p(r_1)$, where p denotes the single-particle quantum numbers. The unperturbed nuclear ground-state wave function is

$$\Phi = \mathscr{A}[\phi_p(\mathbf{r}_,)\phi_q(\mathbf{r}_2)\cdots\phi_s(\mathbf{r}_A)], \tag{11.5}$$

where the levels $p, q, \ldots$ are filled in the order of their energies. Let us denote the sum of single-particle eigenvalues by E_0, so that

$$(E_0 - H_0)\Phi = 0. \tag{11.6}$$

A solution to the Schrödinger equation

$$(E - H)\Psi = 0 \tag{11.7a}$$

can be written in terms of Φ by rewriting (11.7a) as

$$(E_0 - H_0)\Psi = (v - E + E_0)\Psi, \tag{11.7b}$$

from which we infer that

$$\Psi = \Phi + \frac{Q_0}{E_0 - H_0}(v - E + E_0)\Psi. \tag{11.8a}$$

Here the operator

$$Q_0 = 1 - |\Phi\rangle\langle\Phi| \tag{11.8b}$$

projects out the ground state. The solution (11.8a) can be verified by acting on it with $E_0 - H_0$. After rearranging terms and using (11.7a) we find that

$$\langle\Phi|v - E + E_0|\Phi\rangle = 0.$$

However, from (11.8) we also find

$$\langle\Phi|\Psi\rangle = 1, \tag{11.9}$$

so that

$$E = E_0 + \langle\Phi|v|\Psi\rangle. \tag{11.8c}$$

Iterating (11.8a), we obtain the Rayleigh–Schrödinger series. For brevity, call

$$g = Q_0/(E_0 - H_0). \tag{11.10}$$

Then

$$\begin{aligned}\Phi &= \sum_{n=0}^{\infty} [g(v - E + E_0)]^n \Phi \\ E - E_0 &= \langle \Phi | v | \Phi \rangle + \langle \Phi | v g (v - E + E_0) | \Phi \rangle \\ &\quad + \langle \Phi | v g (v - E + E_0) g (v - E + E_0) | \Phi \rangle + \cdots \\ &= \langle \Phi | v | \Phi \rangle + \langle \Phi | v g v | \Phi \rangle \\ &\quad + \langle \Phi | v g v g v | \Phi \rangle - (E - E_0) \langle \Phi | v g^2 v | \Phi \rangle + \cdots . \end{aligned} \tag{11.11}$$

In the last term on the right, substitute the series for $E - E_0$ to find

$$\begin{aligned} E - E_0 &= \langle \Phi | v | \Phi \rangle + \langle \Phi | v g v | \Phi \rangle \\ &\quad + \langle \Phi | v g v g v | \Phi \rangle - \langle \Phi | v | \Phi \rangle \langle \Phi | v g^2 v | \Phi \rangle + \cdots . \end{aligned} \tag{11.12}$$

The wave function Φ is the A body unperturbed ground-state wave functions, whose occupied particle states constitute the Fermi sea. The potential v is the sum (11.4) over pairs. The expansion (11.12) is not useful as it stands because of the singular character of V. A reordering of the summation is therefore desirable, so that each two-body scattering is represented by a nonsingular operator that sums the effect of V to all orders. In Watson's theory, the free particle was treated as distinguishable from the bound ones. Here we explicitly deal with antisymmetrized states (11.5). The reduction is most easily accomplished by introducing a graphical method. The graphs are a convenient way of cataloging the two-body matrix elements of V_{ij}, and the one-body matrix elements of U_i, that actually occur in the matrix elements between the unperturbed many-body ground-state wave functions in (11.12). A very clear exposition on the use of diagrams is given by Day (1967), and we will not go into it here.

However, several essential elements can be mentioned. Notice that there are two types of third-order terms in (11.12). The second of these can be shown by graphical methods to cancel certain types of graphs in the other, which are unlinked. One can appreciate from the derivation that it would become increasingly difficult to recognize this cancellation in higher orders. Goldstone has proven a theorem that shows that the cancellation does indeed occur at each order. Another important element is the choice of one-body potential. The choice is arbitrary. However, if U is chosen to be a Hartree–Fock potential, defined in terms of the G matrix discussed later, then certain higher-order graphs in G are cancelled by graphs containing a U interaction.

By the graphical method one can see how, in the expansion (11.12) for E, whole classes of graphs in V can be summed to yield an expression for the energy that is given in terms of a quantity called G instead of V. The G matrix is defined by the Bethe–Goldstone equation

$$G_{ij}(E) = V_{ij} + V_{ij} \frac{Q_0}{E - H_0} G_{ij}(E). \tag{11.13}$$

This has a formal similarity to the equation defining the τ matrix but differs in several important respects. First, for the bound state problem, G is real. Second, the Hamiltonian in (11.13) is the unperturbed Hamiltonian (11.4) for the pair (i, j). Third, the expression for the ground-state energy in terms of G is not the same as Watson's series for the T operator.

To get some insight into G_{ij}, consider the problem defined by two particles embedded in a nuclear medium in which they each experience a potential U and interact through V_{ij} (Preston and Bhaduri, p. 253, 1975). Let the wave function defining this pair be ψ_{pg}, where p and q denote the states of H_0 to which ψ_{pg} is asymptotic. The wave function for the A particle system so defined is

$$\Psi_{pq} = \mathscr{A}[\psi_{pq}(\mathbf{r}_1, \mathbf{r}_2)\phi_r(\mathbf{r}_3) \cdots \phi_s(\mathbf{r}_A)]. \tag{11.14}$$

Of course $\psi_{pq}(1, 2)$ is orthogonal to all of the other occupied states, and it satisfies

$$(\varepsilon_{pq} - T_1 - U_1 - T_2 - U_2 - V_{12})\psi_{pq} = 0 \tag{11.15}$$

We can write its solution

$$\psi_{pq} = \phi_p\phi_q + \frac{Q_0}{\varepsilon_{pq} - H_0(1) - H_0(2)} V_{12}\psi_{pq}, \tag{11.16}$$

where Q_0 rejects any states $r, s, \ldots$ that are occupied in (11.12) except p and q. Now define G_{12} by

$$G_{12}\phi_p\phi_q = V_{12}\psi_{pq}. \tag{11.17}$$

Multiply (11.16) by V_{12} and use (11.17) to find

$$G_{12}\phi_p\phi_q = \left(V_{12} + V_{12}\frac{Q_0}{\varepsilon_{pq} - H_0} G_{12}\right)\phi_p\phi_q,$$

where $H_0 = H_0(1) + H_0(2)$. The preceding equation shows that the operator G that accomplishes (11.17) is the same as (11.13). Now (11.17) is a familiar type of equation. The T matrix in scattering problems satisfies such a relation. The wave function ψ_{pq} is the solution to (11.15). Suppose that V_{12} has a strong short-range repulsion. Then ψ_{pq} will be small in the region of the repulsion. The product $V_{12}\psi_{pq}$ is well behaved even though V_{12} may be singular. The product wave function $\phi_p\phi_q$, of course, has no such compensating property. Therefore G_{12} does not possess the singularity of V_{12}. Hence a series expansion for the energy, which diverges term by term, can be converted into one, all of whose terms are finite. Consequently, such a series constitutes a viable means of calculating an approximate energy and systematically improving it.

We can exhibit the interactions that G sums by iterating the Bethe–Goldstone equation (11.13):

$$G(E) = V + V\frac{Q_0}{E - H_0}V + V\frac{Q_0}{E - H_0}V\frac{Q_0}{E - H_0}V + \cdots. \tag{11.18}$$

If we take two-particle matrix elements of G and insert a complete set between operators, then

$$\langle k_1k_2|G|k_3k_4\rangle = \langle k_1k_2|V|k_3k_4\rangle + \sum_{i>k_F}\frac{\langle k_1k_2|V|i_1i_2\rangle\langle i_1i_2|V|k_3k_4\rangle}{E - \varepsilon_{i_1} - \varepsilon_{i_2}} + \cdots. \tag{11.19}$$

For states k above the Fermi sea, this could be drawn as:

$\langle k_1k_2|G|k_3k_4\rangle =$ 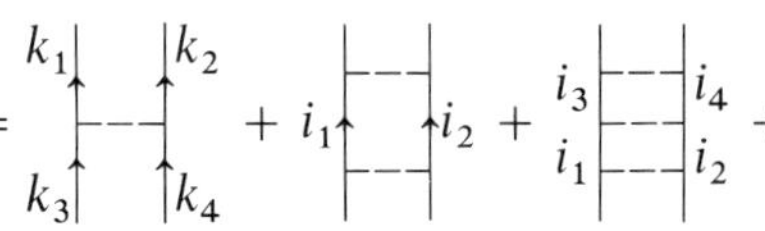 (11.20)

Thus G is often said to be the sum of ladder diagrams, where each dashed line represents V. Each term diverges by itself for a singular interaction. However, the sum is finite as already remarked in connection with (11.17).

The left side of equation (11.20) is often represented diagrammatically as

$\langle k_1k_2|G|k_3k_4\rangle =$ 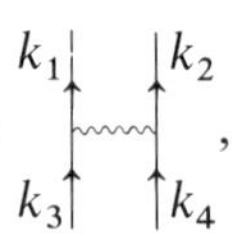 , (11.21)

with the G interaction represented by the wavy line. The contribution to the energy of the system, represented by diagonal elements of G summed over occupied states, is

$W = \sum_{k<k_F}\langle k_1k_2|G|k_3k_4\rangle \equiv \sum$

$= \sum\{$ 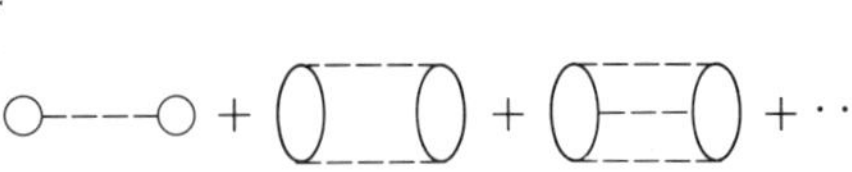 $+ \cdots\}$. (11.22)

(In this case the ends of the diagram in (11.22) are closed because it is a diagonal matrix element with respect to the unperturbed ground state.) Of course, W is only the lowest-order estimate of the binding energy in terms of G. However, the essential point is that the series is in terms of the nonsingular G, so that it can be evaluated term by term. In principle the energy of the many-body system can thus be calculated to any desired accuracy.

However, the actual calculation of G is very difficult. For example, there is a double self-consistency condition imposed by the choice of the one-body potential U as a Hartree–Fock potential based on G. The action of the Pauli operator also is difficult to implement in finite-nucleus calculations. For this reason, G is sometimes calculated for infinite nuclear matter so that the intermediate states are plane waves. A very interesting and important development called the separation method was published by Moszkowski and Scott (1961, 1960). It provides a simple estimate of G and a prescription for systematic improvements. They divide the free two-nucleon potential into short and long-range parts,

$$V = V_s + V_l \tag{11.23}$$

so as to achieve a particular end. The short-range part contains the repulsive core and some of the attraction. The separation distance d at which V is divided is chosen so that the wave function ψ of a pair interacting through V_s will join smoothly at the distance d to the free wave function ϕ. At distances greater than d, it is as if V_s were absent. Of course, the separation distance must depend on the relative momentum of the pair. Even so, it is a fortunate feature of the N–N interaction that the short-range part involves strong attractive and repulsive pieces such that, in the region where ϕ and ψ differ, the latter has a high curvature. Its Fourier components lie automatically, to good approximation, in the space Q, so that the Pauli operator is essentially unity with respect to V_s. At the same time, V_l is smooth and nonsingular and consequently its contribution can be calculated perturbatively. To see the form that G takes under this separation, return to the series (11.18), writing for brevity,

$$g = Q_0/(E - H_0). \tag{11.24}$$

Rearrange the terms of

$$\begin{aligned} G = {} & (V_s + V_l) + (V_s + V_l)g(V_s + V_l) \\ & + (V_s + V_l)g(V_s + V_l)g(V_s + V_l) + \cdots \end{aligned} \tag{11.25}$$

as

$$\begin{aligned} \mathrm{G} &= V_l + V_l g V_l + V_l g V_l g V_l + \cdots + V_s + V_s g V_s + V_s g V_s g V_s + \cdots \\ &\quad + V_s g V_l + V_s g V_s g V_l + \cdots + V_l g V_s + V_l g V_s g V_s + \cdots \\ &= G_l + G_s + G_s g V_l + V_l g G_s + \cdots. \end{aligned} \tag{11.26}$$

We have selected terms from the series in an obvious order. The G matrices with the subscripts l represent the series

$$\begin{aligned} G_l &= V_l + V_l g V_l + V_l g V_l g V_l + \cdots \\ &= V_l + V_l \frac{Q_0}{E - H_0} G_l. \end{aligned} \tag{11.27}$$

Because V_l is a smooth function dominated by low Fourier components, the matrix elements connecting low-lying states to those of Q will be small. Consequently, the series will converge rapidly, and to lowest approximation,

$$G_l \simeq V_l. \tag{11.28}$$

A similar argument suggests that the cross terms linking l and s in (11.26) are small.

The series definition of G_s is analogous to that of G:

$$\begin{aligned} G_s &= V_s + V_s g V_s + V_s g V_s g V_s + \cdots \\ &= V_s + V_s \frac{Q_0}{E - H_0} G_s. \end{aligned} \tag{11.29}$$

Now employ the preceding claim that V_s generates Fourier components that lie dominantly in the subspace Q, so that, to good approximation, the Pauli operator can be replaced by unity. In such a case,

$$G_s \simeq G_s^0 = V_s + V_s \frac{1}{E - H_0} G_s^0. \tag{11.30}$$

With this approximate result, it can be shown that, with the choice of separation distance previously defined (Moszkowski and Scott, 1961, 1960),

$$G_s^0 = 0. \tag{11.31}$$

In that case we have, for a rough approximation to the complete G matrix,

$$G \simeq V_l \tag{11.32a}$$

or, in next approximation,‡

$$G \simeq V_l + V_l g V_l. \tag{11.32b}$$

This is a remarkable result, which shows that to first approximation the G matrix, which is a complicated operator defined by the Bethe–Goldstone equation (11.13), can be represented simply by the long-range part of the potential. Many authors have calculated G to better approximation. However, the Moszkowski–Scott separation method shows how the approximation can be systematically improved.

For a more complete description of the theory and its applications, the interested reader is referred to some of the excellent reviews and references therein (Baranger, 1969, p. 511; Brown, 1971; and MacFarlane, 1969).

‡ The tensor force makes a large contribution in the second-order term of (11.32b) for triplet states (see Brown, 1971).

D. BARE G MATRIX AND CORE POLARIZATION

The calculation of the G matrix is the first step in relating the free interaction to the properties of nuclei. It provides a well-behaved operator in place of the singular free interaction.

Needless to say, all calculations of G are approximate. The many-body problem cannot be solved exactly. For nuclear matter the sum over intermediate states in the Bethe–Goldstone equation is over plane-wave states with momenta above the Fermi momentum. The density of nuclear matter is, of course, characterized by the Fermi momentum k_F because the density is

$$\rho = 4 \int_0^{k_F} \frac{d\mathbf{k}}{(2\pi)^3} = \frac{2}{3\pi^2} k_F^3. \tag{11.33}$$

Here 4 is the spin–isospin degeneracy of each momentum state. In finite, nuclei, the problem is much more difficult. Frequently the calculation of G is done as though each region of the nucleus was locally like nuclear matter. Because the density varies with distance from the center, such a calculation would, if this variation were taken into account, provide a density-dependent G matrix. Sometimes the separated long-range part of the potential is used as the G matrix (11.32). In any case, the G matrix is defined with respect to a free propagator. This makes sense from a calculational point of view. Otherwise one would need to know the nuclear spectrum to calculate G. However, as a consequence the long-range correlations in the nuclear wave function that are induced by the force must enter in higher order in G. The corresponding large-basis shell-model problem with G as interaction could be truncated as in Chapter 8 for the scattering problem. In the truncated space, a new effective interaction or renormalized G matrix would have to be used. This alteration of G is sometimes referred to as core polarization (Kuo and Brown, 1966), and the original G can be referred to as the bare G matrix.

E. EFFECTIVE INTERACTION IN SCATTERING

In lowest order, the effective interaction that is appropriate for a scattering problem might be τ from (11.1) or the generalization of the G matrix to the case where one particle is free. In principle this means that the outgoing-wave boundary condition must be inserted into (11.13) in the form of $+i\varepsilon$. The two approaches can be related through the identity that relates two operators obeying integral equations like (11.1), namely,

$$\begin{aligned} T_a &= T_b + T_b(G_a - G_b)T_a \\ &= T_b + T_b(G_a - G_b)T_b + \cdots . \end{aligned} \tag{11.34}$$

The first alternative, in the form presented in Chapter 6 has been the basis for applications to high-energy scattering (Kerman *et al.*, 1956; Haybron and McManus, 1964; Kawai *et al.*, 1964). So far this alternative in the form (11.2) has not been pursued. At low energies only the second alternative has been considered in any detail. The first important steps in this direction were taken by Slanina and McManus (1968), Petrovich *et al.* (1969), and Love and Satchler (1970). The separation prescription (11.32) was used in these works. Of course, the renormalization of G due to long-range correlations must also be considered for the scattering problem (Petrovich *et al.*, 1977). The proper treatment of the many-body aspects of this approach has been carried out by Mahaux (1979) and Rook, Brieva, and von Geramb (1979).

F. LOW-ENERGY DOMAIN

At low bombarding energy, the scattering of the incident nucleon by a bound one inside the nuclear environment will be strongly effected by the presence of other nucleons. The de Broglie wavelength is not small compared with the average nucleon spacing. Moreover, the possible final scattering states of the interacting pair are severely restricted by the Pauli principle. In the low-energy domain, these difficulties cannot be evaded. The appropriate interaction appears to be the G matrix, which incorporates the Pauli principle correctly. It should be calculated with one particle in the continuum (i.e., with the outgoing wave condition inserted in the denominator). Frequently, however, a simpler interaction has been employed, such as the bare bound-state G matrix, calculated in one way or another (Slanina and McManus, 1968; Petrovich *et al.*, 1969; Love and Satchler, 1970). One approximation is provided by (11.32), according to which the long-range part of, say, the Hamada–Johnston potential is used to approximate G. This is not quite as straightforward as it may seem because, in the odd states of relative motion, the bare interaction is only weakly attractive or even repulsive, in which case the separation method is not applicable. One alternative is to derive the matrix elements in various states of relative motion directly from the N–N phase shifts, which is claimed to be similar to the separation method (Elliott *et al.*, 1968). Alternatively, the bare G matrix can be calculated for a less singular potential, like the Reid soft-core potential. Bertsch *et al.* have fitted Yukawa forms to bare G matrix elements calculated by all three methods (Bertsch *et al.*, 1977). These forms can be easily used in a computer program for calculating the transition amplitude $\mathscr{T}$. Such programs are generally constructed to accept the effective interaction as a superposition of local Yukawa potentials in configuration space.

G. HIGH-ENERGY DOMAIN ($E > 100$ MeV)

In the high-energy domain, the bare G matrix is frequently approximated as the free two-body scattering amplitude. This is known as the impulse approximation (Kerman *et al.* 1956) and when the transition amplitude $\mathcal{T}$ for nucleon–nucleus scattering is calculated using this approximation for the effective interaction and distorted waves are used for the scattering states of the reaction, the combined approximations go under the name of distorted-wave impulse approximation (DWIA) (Haybron and McManus, 1964; Kawai *et al.* 1964), which was discussed in Chapter 7.

H. MICROSCOPIC CALCULATION OF THE OPTICAL POTENTIAL

In the high-energy domain, the optical potential might be approximated by "folding" the nuclear density distribution with the interaction

$$U(r) = \int \bar{t}(\mathbf{r} - \mathbf{r}')\rho(r')\, d\mathbf{r}', \tag{11.35}$$

where $\rho(r')$ is the ground-state density distribution of the target nucleus and $\bar{t}$ is $t(r)(1 - P_{12})$ averaged over the spin and isospins of the target nucleons (Kerman *et al.* 1956; Slanina and McManus, 1968). This approximation attributes the complexity of the optical potential solely to the complexity of the free-scattering matrix, whereas the rationale for the imaginary part of the optical potential (Chapter 4) attributed it to the absorption out of the elastic channel into other reaction channels. The derivation in Chapter 8 of the effective interaction appropriate for a truncated space was quite explicit in this respect. That discussion was given in the context of a nonsingular two-body interaction, which we now understand to be related to the bare τ or G matrix. So far, a unified calculation of the optical potential has not been performed, although this very difficult problem has been attacked in two extremes. The work of Vinh Mau and collaborators and Madsen *et al.* emphasizes the structure of the low-lying nuclear states that belong to the excluded space of the optical potential (i.e., all excited states) but employs a purely phenomenological interaction (Vinh Mau, 1970; Vinh Mau and Bouyssey, 1974, 1976; Bernard and van Giai, 1979; Osterfeld *et al.*, 1981). Two other groups have concentrated on the calculation of the bare G matrix in the case in which one particle is in the continuum at positive energy and the others lie below the Fermi sea (Mahaux, 1979; von Geramb, 1979). This is implemented in finite nuclei through the local density approximation and accounts for the decay of the original state (particle in continuum + nuclear matter) into a multitude of states of the same energy but with the incident particle scattered out of its initial state. Use of the local density approximation for nuclear matter neglects the particular structure of individual nuclear states, the collective ones of which can give rise to appreciable contributions to the optical potential (Glendenning, 1969a; Glendenning *et al.*, 1968). According to our earlier argument, it is an approximation that is expected to improve at higher energy. Otherwise, it could be regarded as the smooth average optical potential on which corrections arising from the particular structure of individual nuclei should be superimposed (Glendenning, 1969a; Glendenning *et al.*, 1968). A calculation, in the local density approximation, of the real and imaginary parts of the central part of the average optical potential is shown in Figures 11.1 and 11.2, respectively, for ^{58}Ni (von Geramb, 1979). At low energy the real part has a profile similar to the nuclear density, whereas the imaginary part has a tendency to peak on the surface, corresponding to a volume + derivative term in the phenomenological optical potential. At high energy the imaginary absorption goes over to the shape of the nuclear density, but the real part takes on a rather unusual form, becoming very weak at about half the nuclear radius. All this behavior is attributable to the density dependence of the nuclear matter G matrix and not to the struc-

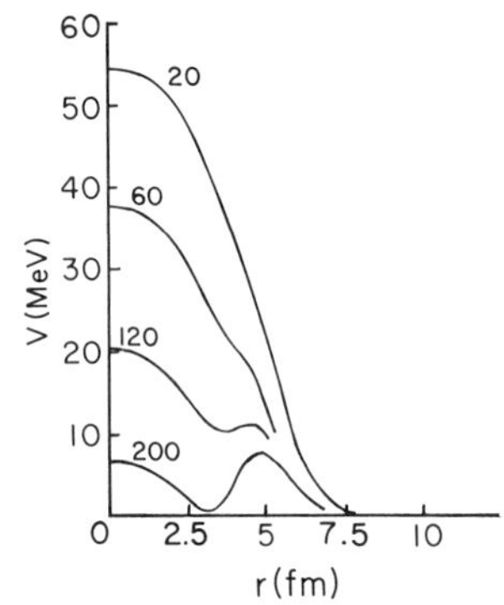

Fig. 11.1. Real part of the optical potential of ^{58}Ni calculated from a local density approximation to the G matrix and based on the Hamada—Johnston potential. (From von Geramb, *et al.*, 1979).

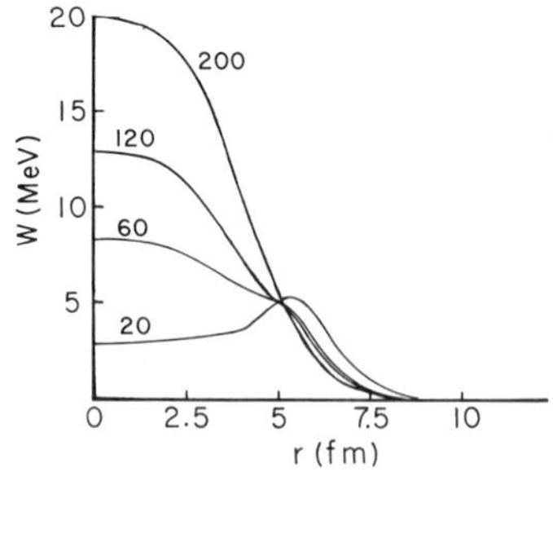

Fig. 11.2. Imaginary part of the optical potential of ^{58}Ni calculated from the local density approximation to the G matrix (From von Geramb *et al.*, 1979).

ture of individual nuclear states, which is neglected in this calculation. Clearly, a consistent treatment of the high momentum states, which are an essential ingredient of the Bethe–Goldstone equation as concerns the treatment of the singular part of the interaction and the peculiar aspects of the structure of the nucleus under consideration, is a very difficult problem. A consistent treatment must, of course, address the nonorthogonality of the two sets of intermediate states that are presently emphasized in the two extreme approaches previously mentioned (Osterfeld *et al.* 1981).

Chapter 12

Further Developments in the Theory of Inelastic Scattering

A. INTRODUCTION

In the preceding chapter the relationship between the effective and the free nucleon–nucleon interaction was discussed. The relationship is complicated by the strong repulsion of the free interaction at short distance and by the severe truncation of configuration space in the shell model. Breuckner's method of handling a singular interaction in the many-body problem is to reorder the perturbation series to replace the free interaction by the non-singular G matrix. At sufficiently high energy, a calculation of the G matrix can be avoided because it should approach the free nucleon–nucleon t matrix, about which there is considerable empirical information. At lower energy, the G matrix must be used if contact with the free interaction is to be made. In practice, the G matrix is often approximated by the bare G matrix, corresponding to, say, a nuclear-matter local-density approximation in the intermediate states. Sometimes it is simply approximated by the long-range part of the separated potential. In principle, the bare G matrix should be renormalized if the configuration space in which the nuclear wave functions are calculated is a severely truncated one. A model for such a renormalization was discussed in Chapter 10 and references to more recent approaches were cited.

The free t matrix $t(\mathbf{k}, \mathbf{k}')$ is derived from the data as a momentum-space operator, whose from follows. It is nonlocal, as can be seen from

$$\begin{aligned} t(\mathbf{r}, \mathbf{r}') &\equiv \langle \mathbf{r}|t|\mathbf{r}'\rangle = \sum_{kk'} \langle \mathbf{r} \mid \mathbf{k}\rangle \langle \mathbf{k}|t|\mathbf{k}'\rangle \langle \mathbf{k}' \mid \mathbf{r}'\rangle \\ &= (2\pi)^{-6} \int d\mathbf{k}\, d\mathbf{k}'\, e^{i\mathbf{k}\cdot\mathbf{r}} t(\mathbf{k}, \mathbf{k}') e^{-i\mathbf{k}'\cdot\mathbf{r}'}. \end{aligned} \tag{12.1}$$

The G matrix, in principle, is also nonlocal unless it is approximated by the separated long-range part of a potential, such as the Ried or Hamada–Johnston potentials. The nonlocal potentials can be employed in a distorted-wave calculation of the amplitude for nucleon–nucleus scattering. However, it is more convenient from several points of view to have a local approximation to the interaction. First, it is much easier to gain insights from a local interaction. Second, a local representation can be employed more easily in computer programs that calculate the amplitude for nucleon–nucleus scattering in the distorted-wave approximation. These are both strong motivations for finding an equivalent local potential.

Bertsch *et al.* (1977) and Love and Franey (1981) have implemented a method for matching the antisymmetrized Fourier transform of a local potential to the momentum-space representation of the free t matrix or G matrix, as the case may be. The direct interaction, when so related to the data (t matrix) or a calculation of the G matrix, is often referred to as a realistic interaction or as an effective interaction. We shall show in detail how the preceding relationship can be made in the case of the t matrix.

Such a local realistic interaction can be easily used in calculations of the distorted-wave amplitude for nucleon–nucleus scattering. Early work using phenomenological interactions frequently neglected the exchange contribution arising through the exchange nature of the force and the antisymmetrization of the wave functions. In general, such contributions are not negligible‡ and should be accounted for in a quantitative theory, which is what we here pursue. In addition to the central components, the realistic interaction has tensor and spin–orbit pieces. These also need to be incorporated into the calculation (Raynal, 1967; Love and Parish, 1970; Love, 1971).

There are computer programs available that can calculate the distorted-wave amplitude using the full complexity of the realistic interaction (For

‡ See Amos *et al.*, 1967; Agassi and Schaeffer, 1968; Atkinson and Madsen, 1968, 1970; Schaeffer, 1969; Satchler and Love, 1970.

example, R. Schaeffer and J. Raynal, computer code DWBA 70, unpublished). Not surprisingly, because of the exchange integrals, the general formalism is not suited to gaining insights into the selectivity for particular nuclear excitations possessed by various pieces of the interaction and the evolution of this selectivity with energy or momentum transfer. It is, of course, extremely valuable to develop such insights both for the interpretation of data and the design of new experiments. We therefore display a formalism developed by Petrovich and collaborators (Petrovich, pers. comm.; Petrovich *et al.*, 1983) that, through the use of momentum-space techniques (Petrovich, 1975) and a factorization approximation for exchange, achieves the desired transparency of the theory. Following these developments, a few applications that illustrate the power available in the theory will be discussed.

B. GENERAL FORM OF THE TWO-BODY AMPLITUDE

The two–nucleon scattering matrix (related by a factor to the amplitude) has a most general form allowed by invariance principles (Wolfenstein, 1956). There are several possible representations of the general form. One of them is

$$t_{12}(\mathbf{k}_f, \mathbf{k}_i) = a + b\boldsymbol{\sigma}_1 \cdot \hat{\mathbf{n}}\boldsymbol{\sigma}_2 \cdot \hat{\mathbf{n}} + c(\boldsymbol{\sigma}_1 + \boldsymbol{\sigma}_2) \cdot \hat{\mathbf{n}} + e\boldsymbol{\sigma}_1 \cdot \hat{\mathbf{q}}\boldsymbol{\sigma}_2 \cdot \hat{\mathbf{q}} + f\boldsymbol{\sigma}_1 \cdot \hat{\mathbf{Q}}\boldsymbol{\sigma}_2 \cdot \hat{\mathbf{Q}}, \quad (12.2)$$

where the coefficients *a*, *b*, *c*, *e*, *f* are themselves functions of the c.m. energy, scattering angle, and isospin,

$$a = a_0 + a_1\boldsymbol{\tau}_1 \cdot \boldsymbol{\tau}_2, \qquad \text{etc.} \quad (12.3)$$

The three *unit* vectors, $\hat{\mathbf{Q}}$, $\hat{\mathbf{n}}$ and $\hat{\mathbf{q}}$ form a right-handed Cartesian coordinate system, oriented with respect to the initial and final momenta of one of the nucleons, $\mathbf{k}_i$ and $\mathbf{k}_f$ in the c.m., so that

$$\mathbf{Q} = \mathbf{k}_i + \mathbf{k}_f, \qquad \mathbf{q} = \mathbf{k}_i - \mathbf{k}_f, \qquad \mathbf{n} = \mathbf{q} \times \mathbf{Q}, \qquad \text{where } |\mathbf{k}_i| = |\mathbf{k}_f|. \quad (12.4)$$

The preceding form of *t* can also be written in terms of the projection operators P_0^σ and P_1^σ for singlet and triplet spin states and the tensor operators

$$P_0^\sigma = \tfrac{1}{4}(1 - \boldsymbol{\sigma}_1 \cdot \boldsymbol{\sigma}_2), \qquad P_1^\sigma = \tfrac{1}{4}(3 + \boldsymbol{\sigma}_1 \cdot \boldsymbol{\sigma}_2) \quad (12.5)$$

$$S_{12}(\hat{\mathbf{u}}) = 3\boldsymbol{\sigma}_1 \cdot \hat{\mathbf{u}}\boldsymbol{\sigma}_2 \cdot \hat{\mathbf{u}} - \boldsymbol{\sigma}_1 \cdot \boldsymbol{\sigma}_2, \quad (12.6)$$

namely,

$$t_{12}(\mathbf{k}_f, \mathbf{k}_i) = a'P_0^\sigma + b'P_1^\sigma + c(\boldsymbol{\sigma}_1 + \boldsymbol{\sigma}_2) \cdot \mathbf{n} + e'S_{12}(\hat{\mathbf{q}}) + f'S_{12}(\hat{\mathbf{Q}}). \quad (12.7)$$

The relationship between the coefficients is

$$a' = a - b - e - f, \qquad b' = a + \tfrac{1}{3}(b + e + f),$$
$$c' = c, \qquad e' = \tfrac{1}{3}(e - b), \qquad f' = \tfrac{1}{3}(f - b). \tag{12.8}$$

The coefficients a' and b' are connected with the central part of the interaction, as in earlier chapters, the c is associated with the spin–orbit, and the e' and f' coefficients are connected with the tensor parts of the interaction.

C. LOCAL COORDINATE-SPACE REPRESENTATION OF G OR t

The G matrix is, in general, nonlocal in coordinate space, as is the coordinate-space representation of $t(\mathbf{k}_f, \mathbf{k}_i)$, which we wish to use in the high-energy domain (according to the discussion of the previous chapter).

The nucleon–nucleus scattering problem has generally been treated by seeking a local coordinate-space representation for the interaction, which can then be used in standard computer programs (R. Schaeffer and J. Raynal, computer code DWBA, unpublished) for calculating the transition amplitude in the distorted-wave approximation. Bertsch *et al.* (1977) and Love and Franey (1981) have carried out a program of representing the nonlocal t or G matrix by a local operator in each N–N channel (singlet-even, singlet-odd, . . .), having the form

$$t(\mathbf{r}, \mathbf{p}) = t_C(r) + t_{LS}(r)\mathbf{L} \cdot \mathbf{S} + t_T(r)S_{12}(\hat{\mathbf{r}})$$
$$\mathbf{L} = \mathbf{r} \times \mathbf{p}, \qquad \mathbf{r} = \mathbf{r}_1 - \mathbf{r}_2, \qquad \mathbf{p} = \tfrac{1}{2}(\mathbf{p}_1 - \mathbf{p}_2), \tag{12.9a}$$
$$\mathbf{S} = \tfrac{1}{2}\boldsymbol{\sigma}, \qquad \boldsymbol{\sigma} = \boldsymbol{\sigma}_1 + \boldsymbol{\sigma}_2,$$

and referring to central, spin–orbit, and tensor components. The central part itself has four components due to the spin–isospin dependence, and t_{LS} and t_T are both isospin dependent,

$$t_C(r) = t_0(r) + t_\sigma \boldsymbol{\sigma}_1 \cdot \boldsymbol{\sigma}_2 + t_\tau(r)\boldsymbol{\tau}_1 \cdot \boldsymbol{\tau}_2 + t_{\sigma\tau}(r)\boldsymbol{\sigma}_1 \cdot \boldsymbol{\sigma}_2 \boldsymbol{\tau}_1 \cdot \boldsymbol{\tau}_2,$$
$$t_x(r) = t_x^0(r) + t_x^1(r)\boldsymbol{\tau}_1 \cdot \boldsymbol{\tau}_2, \qquad x = \mathrm{LS} \quad \text{or} \quad \mathrm{T}. \tag{12.9b}$$

The longest range part of the potential is constrained (see Bertsch *et al.*, 1977; Love and Franey, 1981) to have the behavior of the one-pion exchange potential. Otherwise, the potential functions $t(r)$ are represented as the sum of Yukawa potentials of various ranges for the central and spin–orbit components, for example,

$$t_0(r) = \sum_n V_n^0 e^{-(r/r_n)}/(r/r_n), \tag{12.9c}$$

or sums of $r^2 \times$ Yukawa functions for the tensor components. The strengths and ranges for each component can then be adjusted so that the momentum-space representation of the local interaction is equal to t or G within the accuracy allowed by the local representation and its parametrization.

We now exhibit in more detail the transformation of the interaction (12.9) into momentum space and the identification of $t(\mathbf{k}_f, \mathbf{k}_i)$ with the form (12.7). This will be very useful for appreciating the connections among the interaction, its dependence on energy and on momentum transfer, and the type of nuclear states that can be excited by various pieces of the interaction.

An interaction of the form (12.9) is intended for use as the direct interaction in a distorted-wave transition amplitude with properly antisymmetrized wave functions between projectile and bound nucleons. It is easy to see that the use of antisymmetrized wave functions and the interaction $t(\mathbf{r}, \mathbf{p})$ is equivalent to using an unsymmetrized wave function with the interaction $t(1 - P_{12})$, where P_{12} is the operator that interchanges all coordinates of particles 1 and 2, so denoted

$$\hat{t}(\mathbf{r}, \mathbf{p}) = t(\mathbf{r}, \mathbf{p})(1 - P_{12}). \tag{12.10a}$$

The antisymmetrization of the spatial coordinates achieved by the space exchange part P^x of P_{12},

$$P_{12} = P^x P^\sigma P^\tau, \tag{12.11}$$

causes the most difficulty in calculations. The other two operators are the spin and isospin exchange operators,

$$P^\sigma = (1 + \boldsymbol{\sigma}_1 \cdot \boldsymbol{\sigma}_2)/2, \qquad P^\tau = (1 + \boldsymbol{\tau}_1 \cdot \boldsymbol{\tau}_2)/2. \tag{12.12}$$

It is easy to verify that (12.11) can be rewritten as

$$P_{12} = P^x \sum_{ST} (-)^{S+T} P_S^\sigma P_T^\tau, \tag{12.13}$$

where P_S^σ is the spin-singlet ($S = 0$) or-triplet ($S = 1$) projection operator (12.5), with an analogous definition of P_T^τ.

Therefore, if we separate t into pieces that act on singlet or triplet states in spin and isospin

$$t(\mathbf{r}, \mathbf{p}) \equiv \sum_{ST} t^{ST}(\mathbf{r}, \mathbf{p}) P_S^\sigma P_T^\tau, \tag{12.14}$$

then

$$\hat{t}(\mathbf{r}, \mathbf{p}) = \sum_{ST} t^{ST}(\mathbf{r}, \mathbf{p})[1 + (-)^{S+T+1} P^x] P_S^\sigma P_T^\tau. \tag{12.10b}$$

Define t^D and t^E by

$$t^D(\mathbf{r}, \mathbf{p}) = t(\mathbf{r}, \mathbf{p}) = \sum_{S,\,T} t^{ST}(\mathbf{r}, \mathbf{p}) P_S^\sigma P_T^\sigma, \tag{12.15a}$$

$$t^E(\mathbf{r}, \mathbf{p}) = -\sum_{S,\,T} (-)^{S+T} t^{ST}(\mathbf{r}, \mathbf{p}) P_S^\sigma P_T^\tau. \tag{12.15b}$$

Then

$$\hat{t}(\mathbf{r}, \mathbf{p}) = t^D(\mathbf{r}, \mathbf{p}) + t^E(\mathbf{r}, \mathbf{p})P^x. \tag{12.15c}$$

So far this is nothing but a trivial definition of t^{TS} and a rewriting of $\hat{t}$. For a two-particle spin–isospin state $|S, T\rangle$, the projection operators simply select out the relevant part of $\hat{\mathbf{t}}$,

$$\hat{t}\,|S, T\rangle = [t^{\mathrm{ST}}(\mathbf{r}, \mathbf{p}) + (-)^{S+T+1}t^{\mathrm{ST}}(\mathbf{r}, \mathbf{p})P^x]\,|S, T\rangle. \tag{12.16}$$

We can now concentrate on P^x, which introduces the greater difficulty.

Our problem at this stage is to identify the components of $\hat{\mathbf{t}}(\mathbf{r}, \mathbf{p})$ with the free-scattering matrix $t(\mathbf{k}_{\mathrm{f}}, \mathbf{k}_{\mathrm{i}})$ given by (12.7) through the correspondence

$$\begin{aligned} t(\mathbf{k}_{\mathrm{f}}, \mathbf{k}_{\mathrm{i}}) &\doteq \int \exp(-i\mathbf{k}_{\mathrm{f}} \cdot \mathbf{r})\hat{\mathbf{t}}(\mathbf{r}, \mathbf{p}) \exp(i\mathbf{k}_{\mathrm{i}} \cdot \mathbf{r})\, d\mathbf{r} \\ &= \int \exp(-i\mathbf{k}_{\mathrm{i}} \cdot \mathbf{r})t(\mathbf{r}, \mathbf{p})(1 - P_{12}) \exp(i\mathbf{k}_{\mathrm{i}} \cdot \mathbf{r})\, d\mathbf{r} \\ &= \int \exp(-i\mathbf{k}_{\mathrm{f}} \cdot \mathbf{r})[t^D(\mathbf{r}, \mathbf{p}) + t^E(\mathbf{r}, \mathbf{p})P^x] \exp(i\mathbf{k}_{\mathrm{i}} \cdot \mathbf{r})\, d\mathbf{r}. \end{aligned} \tag{12.17}$$

The symbol $\doteq$ denotes that $t(\mathbf{r}, \mathbf{p})$ is to be chosen to best reproduce the experimentally determined left side. As remarked in connection with (12.1), there is no local operator that satisfies a strict equality. Through such an identification, the local operator $t(\mathbf{r}, \mathbf{p})$ can be constructed so that its antisymmetrized Fourier transform approximates the free t matrix.

From the last line of (12.17) we define the Fourier components of direct and exchange parts of $\hat{\mathbf{t}}(\mathbf{r}, \mathbf{p})$,

$$\begin{aligned} t^D(\mathbf{q}, \mathbf{k}_i) &\equiv \int \exp(-i\mathbf{k}_{\mathrm{f}} \cdot \mathbf{r})t^D(\mathbf{r}, \mathbf{p}) \exp(i\mathbf{k}_{\mathrm{i}} \cdot \mathbf{r})\, d\mathbf{r} \\ &= \int \exp(i\mathbf{q} \cdot \mathbf{r})t^D(\mathbf{r}, \mathbf{k}_{\mathrm{i}})\, d\mathbf{r}, \end{aligned} \tag{12.18a}$$

$$\begin{aligned} t^E(\mathbf{Q}, -\mathbf{k}_{\mathrm{i}}) &\equiv \int \exp(-i\mathbf{k}_{\mathrm{f}} \cdot \mathbf{r})t^E(\mathbf{r}, \mathbf{p})P^x \exp(i\mathbf{k}_{\mathrm{i}} \cdot \mathbf{r})\, d\mathbf{r} \\ &= \int \exp(-i\mathbf{Q} \cdot \mathbf{r})t^E(\mathbf{r}, -\mathbf{k}_{\mathrm{i}})\, d\mathbf{r}, \end{aligned} \tag{12.18b}$$

so that (12.17) becomes

$$t(\mathbf{k}_{\mathrm{f}}, \mathbf{k}_{\mathrm{i}}) \doteq t^D(\mathbf{q}, \mathbf{k}_{\mathrm{i}}) + t^E(\mathbf{Q}, -\mathbf{k}_{\mathrm{i}}). \tag{12.19}$$

We now evaluate the Fourier transforms of each of the three components of $\hat{t}(\mathbf{r}, \mathbf{p})$ separately.

Central

Because the central interaction has no operators that act in the coordinate space and depend on $r = |\mathbf{r}|$ only, it is very easy to evaluate the preceding

Fourier transforms. Use the plane-wave expansion (2.8) for the exponential factor in

$$\int \exp(i\mathbf{q}\cdot\mathbf{r})t_C^D(r)\,dr = 4\pi \sum_L i^L \int j_L(qr)Y_L(\hat{\mathbf{r}})\cdot Y_L(\hat{\mathbf{q}})t_C^D(r)\,d\mathbf{r}$$

$$= 4\pi\sqrt{4\pi}\,Y_0(\hat{\mathbf{q}}) \int j_0(qr)t_C^D(r)r^2\,dr, \tag{12.20}$$

where we have introducted unity into the r integral in the form of $\sqrt{4\pi}\,Y_0(\hat{\mathbf{r}}) \equiv 1$, in order to do the $\hat{\mathbf{r}}$ integral. In an entirely analogous fashion, the exchange transform can be evaluated. The results are

$$t_C^D(\mathbf{q}, \mathbf{k}_i) \equiv t_C^D(q) = 4\pi \int j_0(qr)t_C^D(r)r^2\,dr, \tag{12.21a}$$

$$t_C^E(\mathbf{Q}, -\mathbf{k}_i) \equiv t_C^E(Q) = 4\pi \int j_0(Qr)t_C^E(r)r^2\,dr. \tag{12.21b}$$

For a representation of $t_C(r)$ as a sum of Yukawa potentials (12.9c), the momentum functions in the preceding equations are

$$t_C^D(q) = 4\pi \sum_n V_n r_n^3/[1 + (qr_n)^2], \tag{12.22}$$

with a similar expression for $t^E(Q)$. It is through correspondence (12.17) that the parameters V_n and r_n for the various components of the force are determined, for the central force the correspondence is

$$t_C(\mathbf{k}_f, \mathbf{k}_i) \doteq t_C^D(q) + t_C^E(Q)$$
$$= \sum_{ST} [t_C^{ST}(q) + (-)^{S+T+1} t_C^{ST}(Q)] P_S^\sigma P_T^\tau, \tag{12.23}$$

where we used (12.15) in the last step. Referring to (12.7), we have the correspondence

$$\sum_T [t_C^{0T}(q) + (-)^{T+1} t_C^{0T}(Q)] P_T^\tau \doteq a' = a_0' + a_1'(\boldsymbol{\tau}_1\cdot\boldsymbol{\tau}_2),$$
$$\sum_T [t_C^{1T}(q) + (-)^{T} t_C^{1T}(Q)] P_T^\tau \doteq b' = b_0' + b_1'(\boldsymbol{\tau}_1\cdot\boldsymbol{\tau}_2). \tag{12.24}$$

These are the equations through which the central part of the local $t(\mathbf{r})$ can be related to the scattering amplitude.

Tensor

For the tensor components of (12.17) it is convenient to notice that the tensor operator can be written as the scalar product of two spherical tensors,

$$S_{12}(\hat{\mathbf{r}}) = 3\,\boldsymbol{\sigma}_1\cdot\hat{\mathbf{r}}\,\boldsymbol{\sigma}_2\cdot\hat{\mathbf{r}} - \boldsymbol{\sigma}_1\cdot\boldsymbol{\sigma}_2 = \sqrt{24\pi/5}\,Y_2(\hat{\mathbf{r}})\cdot T_2(\boldsymbol{\sigma}_1, \boldsymbol{\sigma}_2), \tag{12.25}$$

where

$$T_2^m(\boldsymbol{\sigma}_1, \boldsymbol{\sigma}_2) = [\boldsymbol{\sigma}_1, \boldsymbol{\sigma}_2]_2^m \equiv \sum_\mu \sigma_1^\mu \sigma_2^{m-\mu} C^{1\ \ 1\ \ \ 2}_{\mu\ m-\mu\ m} \tag{12.26}$$

is the vector-coupled product of the two vectors $\boldsymbol{\sigma}_1$ and $\boldsymbol{\sigma}_2$ to form a rank-2 tensor (see Appendix for the spherical components of a vector). Again making the plane-wave expansion, the angle integration can be performed. For example,

$$\begin{aligned}(24\pi/5)^{1/2} &\int e^{i\mathbf{q}\cdot\mathbf{r}} t_{\mathrm{T}}^D(\hat{\mathbf{r}}) Y_2(\hat{\mathbf{r}}) \cdot T_2(\boldsymbol{\sigma}_1, \boldsymbol{\sigma}_2)\, d\mathbf{r} \\ &= (24\pi/5)^{1/2}\, 4\pi \sum i^L \int j_{\mathrm{L}}(qr) t^D(r) Y_{\mathrm{L}}(\hat{\mathbf{r}}) \cdot Y_{\mathrm{L}}(\hat{\mathbf{q}}) Y_2(\hat{\mathbf{r}}) \cdot T_2(\boldsymbol{\sigma}_1, \boldsymbol{\sigma}_2)\, d\mathbf{r} \\ &= -t_{\mathrm{T}}^D(q) S_{12}(\hat{\mathbf{q}}),\end{aligned} \tag{12.27}$$

where

$$t_{\mathrm{T}}^D(q) = 4\pi \int j_2(qr) t_{\mathrm{T}}^D(r) r^2\, dr, \tag{12.28a}$$

$$t_{\mathrm{T}}^E(Q) = 4\pi \int j_2(Qr) t_{\mathrm{T}}^E(r) r^2\, dr. \tag{12.28b}$$

Thus,

$$t_{\mathrm{T}}^D(\mathbf{q}, \mathbf{k}_{\mathrm{i}}) = -t_{\mathrm{T}}^D(q) S_{12}(\hat{\mathbf{q}}), \tag{12.29a}$$

$$t_{\mathrm{T}}^E(\mathbf{Q}, -\mathbf{k}_{\mathrm{i}}) = -t_{\mathrm{T}}^E(Q) S_{12}(\hat{\mathbf{Q}}). \tag{12.29b}$$

The tensor force acts only in spin triplet states. Referring to (12.15), (12.29), and (12.7), the correspondence with t is

$$\begin{aligned} t_{\mathrm{T}}^D(q) &= \sum_T t_{\mathrm{T}}^{1T}(q) P_{\mathrm{T}}^\tau \doteq -e' = -(e_0' + e_1'\, \boldsymbol{\tau}_1 \cdot \boldsymbol{\tau}_2) \\ t_{\mathrm{T}}^E(Q) &\equiv \sum_T (-)^T t_{\mathrm{T}}^{1T}(Q) P_T^\tau \doteq -f' = -(f_0' + f_1'\, \boldsymbol{\tau}_1 \cdot \boldsymbol{\tau}_2)\end{aligned} \tag{12.30}$$

Spin–Orbit

We come now to the calculation of the spin–orbit part of (12.17). Here we have

$$t_{\mathrm{LS}}^D(\mathbf{q}, \mathbf{k}_{\mathrm{i}}) = \int \exp(i\mathbf{q}\cdot\mathbf{r})\, t_{\mathrm{LS}}^D(r)\, \mathbf{r} \times \mathbf{k}_i \cdot \mathbf{S}\, d\mathbf{r}, \tag{12.31a}$$

$$t_{\mathrm{LS}}^E(\mathbf{Q}, -\mathbf{k}_{\mathrm{i}}) = -\int \exp(-i\mathbf{Q}\cdot\mathbf{r})\, t_{\mathrm{LS}}^E(r)\, \mathbf{r} \times \mathbf{k}_i \cdot \mathbf{S}\, d\mathbf{r}. \tag{12.31b}$$

The integrals contain the vector $\mathbf{r}$. Using the spherical representation of $\mathbf{r}$, whose mth component (see Appendix) is

$$r_m = (4\pi/3)^{1/2} r Y_1^m(\hat{\mathbf{r}}). \tag{12.32}$$

We note

$$\int \exp(i\mathbf{q}\cdot\mathbf{r})\, t_{\mathrm{LS}}^{D}(r) r_m \, d\mathbf{r} = (4\pi/3)^{1/2} \int \exp(i\mathbf{q}\cdot\mathbf{r})\, t_{\mathrm{LS}}^{D}(r) r\, Y_1^m(\hat{\mathbf{r}})\, d\mathbf{r}$$

$$= (4\pi/3)^{1/2}\, 4\pi \sum_L i^L \int j_L(qr) Y_L(\hat{\mathbf{r}}) \cdot Y_L(\hat{\mathbf{q}}) t_{\mathrm{LS}}^{D}(r) r\, Y_1^m(\hat{\mathbf{r}})\, d\mathbf{r}$$

$$= (4\pi/3)^{1/2} 4\pi i \int j_1(qr) t_{\mathrm{LS}}^{D}(r) r^3\, dr\, Y_1^m(\hat{\mathbf{q}})$$

$$= 4\pi i (q_m/q) \int j_1(qr) t_{\mathrm{LS}}^{D}(r) r^3\, dr. \qquad (12.33)$$

Similarly, the exchange integral can be calculated. Let us denote the transforms of $t_{\mathrm{LS}}(r)$ by

$$t_{\mathrm{LS}}^{D}(q) = -4\pi q \int j_1(qr) t_{\mathrm{LS}}^{D}(r) r^3\, dr, \qquad (12.34a)$$

$$t_{\mathrm{LS}}^{E}(Q) = -4\pi Q \int j_1(Qr) t_{\mathrm{LS}}^{E}(r) r^3\, dr. \qquad (12.34b)$$

These are defined to have the dimension MeV-fm^3 when $t(r)$ is in MeV. Then

$$t_{\mathrm{LS}}^{D}(\mathbf{q}, \mathbf{k}_i) = -(i/q^2) t_{\mathrm{LS}}^{D}(q) \mathbf{q} \times \mathbf{k}_i \cdot \mathbf{S}, \qquad (12.35a)$$

$$t_{\mathrm{LS}}^{E}(\mathbf{Q}, -k_i) = -(i/Q^2) t_{\mathrm{LS}}^{E}(Q) \mathbf{Q} \times \mathbf{k}_i \cdot \mathbf{S}$$
$$= (i/Q^2) t_{\mathrm{LS}}^{E}(Q) \mathbf{q} \times \mathbf{k}_i \cdot \mathbf{S}. \qquad (12.35b)$$

For the spin–orbit interaction, correspondence (12.17) therefore reads

$$t_{\mathrm{LS}}(k_f, k_i) \doteq -i \left\{ \frac{1}{q^2} t_{\mathrm{LS}}^{D}(q) - \frac{1}{Q^2} t_{\mathrm{LS}}^{E}(Q) \right\} \mathbf{q} \times \mathbf{k}_i \cdot \mathbf{S}$$

$$= -\frac{iQ}{q^4} \left\{ t_{\mathrm{LS}}^{D}(q) - \left(\frac{q}{Q}\right)^2 t_{\mathrm{LS}}^{E}(Q) \right\} \hat{\mathbf{n}} \cdot (\boldsymbol{\sigma}_1 + \boldsymbol{\sigma}_2), \qquad (12.36)$$

or in terms of (12.7),

$$-\frac{iQ}{q^4} \left\{ t_{\mathrm{LS}}^{D}(q) - \left(\frac{q}{Q}\right)^2 t_{\mathrm{LS}}^{E}(Q) \right\} \doteq c'. \qquad (12.37)$$

Complete Amplitude

Gathering together the preceding results, we can write, for the Fourier transform of a local coordinate-space representation of the interaction,

$$t^D(\mathbf{q}, \mathbf{k}_i) \equiv \int \exp(i\mathbf{q}\cdot\mathbf{r}) t^D(\mathbf{r}, \mathbf{k}_i)\, d\mathbf{r}$$

$$= t_{\mathrm{C}}^{D}(q) - (i/q^2) t_{\mathrm{LS}}^{D}(q) \mathbf{q} \times \mathbf{k}_i \cdot \mathbf{S} - t_{\mathrm{T}}^{D}(q) S_{12}(\hat{\mathbf{q}}), \qquad (12.38a)$$

$$t^E(\mathbf{Q}, -\mathbf{k}_i) \equiv \int \exp(-i\mathbf{Q}\cdot\mathbf{r})t^E(\mathbf{r}, \mathbf{k}_i)\,d\mathbf{r}$$
$$= t_C^E(Q) + (i/Q^2)t_{LS}^E(Q)\mathbf{q}\times\mathbf{k}_i\cdot\mathbf{S} - t_T^E(Q)S_{12}(\hat{\mathbf{Q}}). \quad (12.38b)$$

The momentum functions on the right side can be expressed in terms of the local potential (12.9) through such expressions as (12.22). The potential parameters V_n and r_n are then varied to achieve a best fit to the free t matrix (or a calculated G matrix, depending on the energy domain) through the correspondence (12.17).

D. ENERGY AND MOMENTUM DEPENDENCE OF THE EFFECTIVE INTERACTION IN THE HIGH-ENERGY DOMAIN

As discussed in Chapter 11, an approximation to the momentum-space representation of the bare G matrix in the high-energy domain is the free nucleon–nucleon scattering matrix t. For the convenience of inserting this free-scattering data into the nucleon–nucleus scattering problem, a local coordinate-space representation was sought in the previous section.

Love and Franey (1981) have carried out such a program of representing the currently available free data by constraining the long-range part of $t(\mathbf{r})$ by the one-pion exchange potential and searching on a Yukawa parametrization of the inner region to satisfy correspondence (12.17). They have tabulated their results for the real and imaginary parts of $t(\mathbf{r})$ at selected energies from 100 to 800 MeV. This tabulation of the parameters representing $t(r)$ would be used in the calculation of nucleon–nucleus scattering in this energy range in the approximation $G \simeq t$.

To provide a qualitative impression of the energy and the momentum-transfer dependence of these results, Petrovich and Love (1981) and Love *et al.* (1983) have plotted the absolute values of the various components of $t(q)$. In Fig. 12.1, the volume integrals $[t(q=0)]$ of t_C, as given in (12.9b), are plotted as a function of energy. Figure 12.2 shows the spin–orbit and tensor forces corresponding to the decomposition into isospin-independent and isospin-dependent components (12.9b) at a typical value of momentum transfer $q = 1.5\ \text{fm}^{-1}$. Finally, at one particular center-of-mass energy $E =$ 140 MeV, the momentum dependence of all components of the force are shown in Fig. 12.3. Figure 12.3a corresponds to the parts that can give rise to spin transfer (i.e., the central components with the operator $\boldsymbol{\sigma}_1 \cdot \boldsymbol{\sigma}_2$ and the tensor and spin–orbit forces). In Fig. 12.3b are shown the spin-independent components of the central force. The spin–orbit interaction $\mathbf{L}\cdot\mathbf{S} = \mathbf{L}\cdot\mathbf{S}_1 + \mathbf{L}\cdot\mathbf{S}_2$ has one piece that can transfer spin to the target nucleus and one that cannot. For this reason, the coefficient of $\mathbf{L}\cdot\mathbf{S}$ is shown on both sides of the

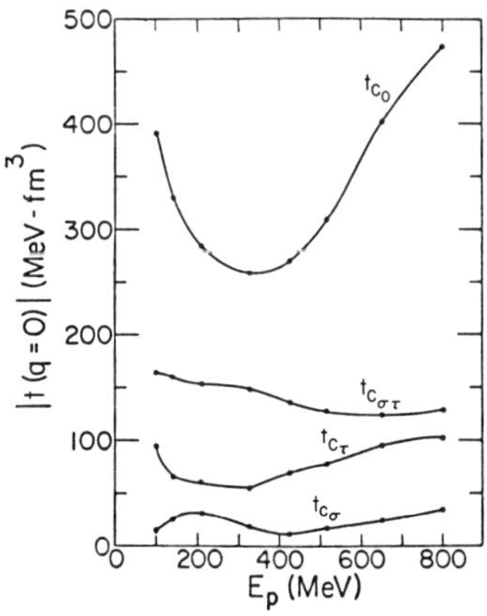

Fig. 12.1. Energy dependence of the magnitude of the of the volume integral of the central components of the N–N t matrix (Love *et al.*, 1983.)

figure. The selection rules for natural and unnatural parity transitions have been discussed in Chapter 9. Recall that for normal parity, the $S = 0$ and $S = 1$ components can contribute, although the former dominate. Only the $S = 1$ components are responsible for the excitation of unnatural parity states. The two halves of the figure therefore correspond to unnatural and natural parity excitations.

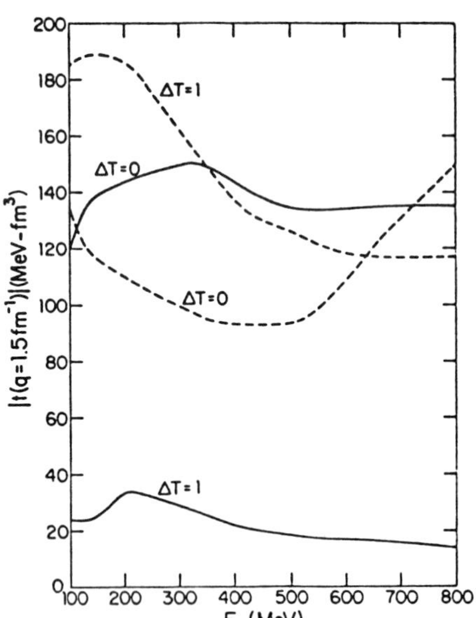

Fig. 12.2. Energy dependence of the magnitude of the spin–orbit (——) and tensor (– – –) components of the N–N t matrix at $q = 1.5\ \text{fm}^{-1}$ (Love *et al.*, 1983.)

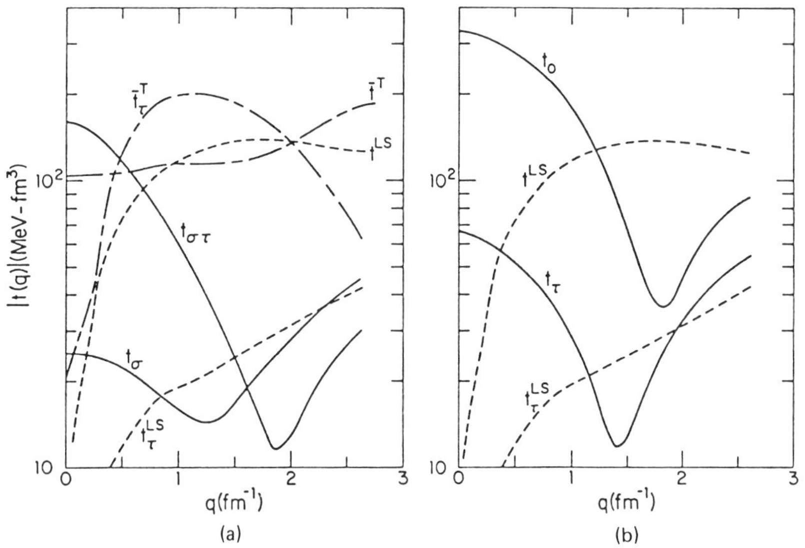

Fig. 12.3. At a fixed energy of 140 MeV all components of the N–N t matrix are shown as a function of momentum transfer $q = 2k_i \sin \theta/2$, where C(———), LS(- - -), and T(——-). (From Love *et al.*, 1983.)

For the central and spin–orbit terms in Figs. 12.1–12.3, the direct and exchange contributions as prescribed by (12.38) are shown together with the exchange contributions evaluated at the fixed value of Q corresponding to forward knockout ($\mathbf{Q} = \mathbf{k}_i + \mathbf{k}_f$). Because the direct and exchange parts of the tensor force are associated with different operators, only the direct tensor part is shown.

A discussion of the way in which these various components and their energy and momentum transfer dependence manifest themselves in specific nuclear excitations will be deferred to the end of this chapter. For the present we merely comment on the strong dependence on E and q exhibited by some of the components.

E. EFFECTIVE OPERATOR FOR THE EXCHANGE CONTRIBUTION

In the formalism for inelastic scattering that was developed in Chapter 9, the exchange integrals arising through the exchange nature of the force and the Pauli principle with regard to projectile and target nucleons were neglected. In this section, following Petrovich *et al.* (1983), we seek an operator whose *direct* matrix element will approximate these exchange contributions.

Once this has been achieved, the interaction between projectile and target will once again have the form of a sum of one-body operators in the target space. As a consequence, a general nuclear transition can be written as a superposition of single-particle transitions, and all the methods discussed in Chapter 9 for calculating the nuclear form factors will apply.

In addition to incorporating the exchange contribution approximately, the full complexity of the force, including the central, spin–orbit, and tensor components, will be retained. In the preceding sections we have discussed how a local coordinate-space representation of the force can be matched to the free t matrix or the G matrix, depending on the energy domain in question. The momentum dependence of the exchange operator will turn out to be given by $t^E(Q, -k_i)$, derived earlier. We may therefore regard it as a known operator.

According to Chapter 5, the distorted-wave approximation to the transition amplitude is a matrix element of the interaction between initial and final states in which the relative motion between the scattering partners is described by distorted waves that are solutions of a one-body Schrödinger equation containing an optical potential. We do not refer to the approximation as a Born approximation because the interaction is now taken to be the complex G matrix or the t matrix, which iterates the two-body interaction at each scattering site to all orders [see (11.18)].

The following notation can be used for the initial-state wave functions, with the appropriate analogs for the final state:

$$\begin{aligned} |\mathrm{i}\rangle &= \chi_{\mathrm{i}}^{m_\mathrm{i}\mu_\mathrm{i}(+)}(\mathbf{k}_\mathrm{i}, \mathbf{r}_1)\phi_{\mathrm{i}}^{M_\mathrm{i}\mathcal{M}_\mathrm{i}}(\mathbf{r}_2), \\ \chi_{\mathrm{i}}^{m_\mathrm{i}\mu_\mathrm{i}(+)} &= \chi_{\mathrm{i}}^{(+)}(\mathbf{k}_\mathrm{i}, \mathbf{r}_1)X_{1/2}^{m_\mathrm{i}}(\boldsymbol{\sigma}_1)X_{1/2}^{\mu_\mathrm{i}}(\boldsymbol{\tau}_1), \\ \phi_{\mathrm{i}}^{M_\mathrm{i}\mathcal{M}_\mathrm{i}} &= \phi_{n_\mathrm{i}\ell_\mathrm{i}j_\mathrm{i}}^{M_\mathrm{i}}(\mathbf{r}_2, \boldsymbol{\sigma}_2)X_{1/2}^{\mathcal{M}_\mathrm{i}}(\boldsymbol{\tau}_2). \end{aligned} \tag{12.39}$$

The $\chi^{(+)}$ is the distorted wave of Chapters 4 and 5, and $\phi_{n\ell j}$ a single-particle spin–orbit state such as that encountered in Chapter 9 [e.g., (9.23)]. The $X_{1/2}$ are spin and isospin functions. We can simplify the problem to one that corresponds to an excitation of a single target nucleon $j_\mathrm{i} \rightarrow j_\mathrm{f}$ because, as previously discussed, once we have found the operator whose *direct* integral approximates the exchange integrals, then a general nuclear excitation is just the superposition of single-particle excitations.

The antisymmetrization of the amplitude is achieved by using the reduced form of (12.10) contained in (12.15) for the effective interaction in the amplitude, that is,

$$\mathscr{T}_{\mathrm{fi}} = -\int \chi_{\mathrm{f}}^{(-)*}(\mathbf{r}_1)\phi_{\mathrm{f}}^{*}(\mathbf{r}_2)\{t^D(\mathbf{r}, \mathbf{p}) + t^E(\mathbf{r}, \mathbf{p})P^x\}\phi_\mathrm{i}(\mathbf{r}_2)\chi_\mathrm{i}^{(+)}(\mathbf{r}_1)\, d\mathbf{r}_1\, d\mathbf{r}_2, \tag{12.40}$$

where we have additionally streamlined the notation by suppressing reference to spin and isospin coordinates in χ_f^*, ϕ_f^*, ϕ_i, and χ_i. To see the approximation of Petrovich and co-workers (1983) for the exchange amplitude in the most transparent way, it is best to transform the amplitude to momentum space.

Recall the conventions of Chapter 6 for Fourier transforms,

$$f(\mathbf{r}) = (2\pi)^{-3} \int \exp(i\mathbf{k} \cdot \mathbf{r}) f(\mathbf{k})\, d\mathbf{k},$$
$$f(\mathbf{k}) = \int \exp(-i\mathbf{k} \cdot \mathbf{r}) f(\mathbf{r})\, d\mathbf{r}, \tag{12.41}$$

and the representation of the delta function,

$$\delta(\mathbf{r} - \mathbf{r}') = (2\pi)^{-3} \int \exp[\pm i\mathbf{k} \cdot (\mathbf{r} - \mathbf{r}')]\, d\mathbf{k}. \tag{12.42}$$

Introducing these into the integral in (12.40), we obtain

$$\mathcal{T}_{\rm fi} = (2\pi)^{-12} \int \chi_{\rm f}^*(\mathbf{k}_1)\phi_{\rm f}^*(\mathbf{k}_2)\phi_{\rm i}(\mathbf{k}_3)\chi_{\rm i}(\mathbf{k}_4)$$
$$\times \left\{ \int \exp[-i(\mathbf{k}_1 \cdot \mathbf{r}_1 + \mathbf{k}_2 \cdot \mathbf{r}_2)][t^D(\mathbf{r}, \mathbf{p}) + t^E(\mathbf{r}, \mathbf{p})P^x] \right.$$
$$\left. \times \exp[i(\mathbf{k}_3 \cdot \mathbf{r}_2 + \mathbf{k}_4 \cdot \mathbf{r}_1)]\, d\mathbf{r}_1\, d\mathbf{r}_2 \right\} d\mathbf{k}_1\, d\mathbf{k}_2\, d\mathbf{k}_3\, d\mathbf{k}_4. \tag{12.43}$$

To do the coordinate-space integral, transform to the coordinates

$$\mathbf{r} = \mathbf{r}_1 - \mathbf{r}_2 \qquad \text{and} \qquad \mathbf{R} = 1/2(\mathbf{r}_1 + \mathbf{r}_2). \tag{12.44}$$

Then,

$$\mathcal{T}_{\rm fi} = (2\pi)^{-12} \int \chi_{\rm f}^*(\mathbf{k}_1)\phi_{\rm f}^*(\mathbf{k}_2)\phi_{\rm i}(\mathbf{k}_3)\chi_{\rm i}(\mathbf{k}_4)$$
$$\times \left\{ \int \exp[-i(\mathbf{k}_1 - \mathbf{k}_2) \cdot \mathbf{r}/2][t^D(\mathbf{r}, \mathbf{p}) + t^E(\mathbf{r}, \mathbf{p})P^x] \right.$$
$$\left. \times \exp[i(\mathbf{k}_4 - \mathbf{k}_3) \cdot \mathbf{r}/2]\, d\mathbf{r} \right\}$$
$$\times \left\{ \int \exp[-i(\mathbf{k}_3 + \mathbf{k}_4 - \mathbf{k}_1 - \mathbf{k}_2) \cdot \mathbf{R}\, d\mathbf{R} \right\} d\mathbf{k}_1\, d\mathbf{k}_2\, d\mathbf{k}_3\, d\mathbf{k}_4, \tag{12.45}$$

which reduces to

$$\mathcal{T}_{\rm fi} = (2\pi)^{-9} \int \chi_{\rm f}^*(\mathbf{k}_3 + \mathbf{k}_4 - \mathbf{k}_2)\phi_{\rm f}^*(\mathbf{k}_2)\phi_{\rm i}(\mathbf{k}_3)\chi_{\rm i}(\mathbf{k}_4)$$
$$\times \left\{ \int \exp(-i\mathbf{k}_{\rm f} \cdot \mathbf{r})[t^0(\mathbf{r}, \mathbf{p}) + t^E(\mathbf{r}, \mathbf{p})P^x] \right.$$
$$\left. \times \exp(i\mathbf{k}_{\rm i} \cdot \mathbf{r})\, d\mathbf{r} \right\} d\mathbf{k}_2\, d\mathbf{k}_3\, d\mathbf{k}_4, \tag{12.46}$$

with

$$\mathbf{k}_f = \tfrac{1}{2}(\mathbf{k}_1 - \mathbf{k}_2), \qquad \mathbf{k}_i = \tfrac{1}{2}(\mathbf{k}_4 - \mathbf{k}_3),$$
$$\mathbf{q} = \mathbf{k}_i - \mathbf{k}_f, \qquad \mathbf{Q} = \mathbf{k}_i + \mathbf{k}_f. \tag{12.47}$$

The integral over r for the direct and exchange pieces can therefore be written as

$$t^D(\mathbf{q}, \mathbf{k}_i) = \int \exp(-i\mathbf{k}_f \cdot \mathbf{r})\, t^D(\mathbf{r}, \mathbf{p}) \exp(i\mathbf{k}_i \cdot \mathbf{r})\, d\mathbf{r}$$
$$= \int \exp(i\mathbf{q} \cdot \mathbf{r})\, t^D(\mathbf{r}, \mathbf{k}_i)\, d\mathbf{r}, \tag{12.48}$$

where the momentum operator conjugate to r, namely, $\mathbf{p} = -i\boldsymbol{\nabla}$, acting on the plane wave, replaces $\mathbf{p}$ by $\mathbf{k}_i$. Similarly

$$t^E(\mathbf{Q}, -\mathbf{k}_i) = \int \exp(-i\mathbf{k}_f \cdot \mathbf{r})\, t^E(\mathbf{r}, \mathbf{p}) \exp(-i\mathbf{k}_i \cdot \mathbf{r}) d\mathbf{r}$$
$$= \int \exp(-i\mathbf{Q} \cdot \mathbf{r})\, t^E(\mathbf{r}, -\mathbf{k}_i)\, d\mathbf{r}. \tag{12.49}$$

In this case, the space-exchange operator P^x changes the sign of the exponential on the right in the first line. Recall that these Fourier transforms were worked out in detail in an earlier section of this chapter, with the object of choosing the local coordinate-space operator so that its Fourier transform agrees with the G or t matrix according to the energy domain of interest.

At this point we have

$$\mathcal{T}_{fi} = (2\pi)^{-9} \int \chi_f^*(\mathbf{k}_3 + \mathbf{k}_4 - \mathbf{k}_2)\phi_f^*(\mathbf{k}_2)\phi_i(\mathbf{k}_3)\chi_i(\mathbf{k}_4)$$
$$\times \left[t^D(\mathbf{q}, \mathbf{k}_i) + t^E(\mathbf{Q}, -\mathbf{k}_i)\right] d\mathbf{k}_2\, d\mathbf{k}_3\, d\mathbf{k}_4, \tag{12.50}$$

where of course $\mathbf{q}$, $\mathbf{Q}$, and $\mathbf{k}_i$ are given in terms of $\mathbf{k}_2$, $\mathbf{k}_3$, and $\mathbf{k}_4$, as described previously. In particular, $\mathbf{q} = \mathbf{k}_2 - \mathbf{k}_3$. We can change the variables of integration from $(\mathbf{k}_2, \mathbf{k}_3)$ to $(\mathbf{q}, \mathbf{k}_3)$. Then

$$\mathcal{T}_{fi} = (2\pi)^{-9} \int \chi_f^*(\mathbf{k}_4 - \mathbf{q})\phi_f^*(\mathbf{k}_3 + \mathbf{q})\phi_i(\mathbf{k}_3)\chi_i(\mathbf{k}_4)$$
$$\times \left[t^D(\mathbf{q}, \mathbf{k}_i) + t^E(\mathbf{Q}, -\mathbf{k}_i)\right] d\mathbf{q}\, d\mathbf{k}_3\, d\mathbf{k}_4, \tag{12.51}$$

where now $\mathbf{Q} = \mathbf{k}_4 - \mathbf{q} - \mathbf{k}_3$.

Now we make an approximation that $t^E(\mathbf{Q}, -\mathbf{p})$ varies slowly with $\mathbf{Q}$. We could think of making a Taylor expansion, for example, around a chosen constant value. At high energies, $\mathbf{k}_4$ is approximately the incident momentum, which is much greater than $\mathbf{q}$ and $\mathbf{k}_3$ for angles at which cross sections are typically measured. We look only at the constant term with $Q \simeq k_4$. With this approximation, $\mathcal{T}_{fi}$ will look like a direct matrix element in coordinate space.

To transform the amplitude back to coordinate space, for the central and tensor parts of t, which do not have $\mathbf{k}_i$ in them, we can use

$$\int \phi_f^*(\mathbf{k}_3 + \mathbf{q})\phi_i(\mathbf{k}_3)\, d\mathbf{k}_3 = (2\pi)^3 \int \phi_f^*(\mathbf{r}_2)\phi_i(\mathbf{r}_2) \exp(i\mathbf{q}\cdot\mathbf{r}_2)\, d\mathbf{r}_2, \tag{12.52a}$$

$$\int \chi_f^*(\mathbf{k}_4 - \mathbf{q})\chi_i(\mathbf{k}_4)\, d\mathbf{k}_4 = (2\pi)^3 \int \chi_f^*(\mathbf{r}_1)\chi_i(\mathbf{r}_1) \exp(-i\mathbf{q}\cdot\mathbf{r}_1)\, d\mathbf{r}_1. \tag{12.52b}$$

For the spin–orbit part of t, which has the vector $\mathbf{k}_i$ and depends on $\mathbf{k}_3$ and $\mathbf{k}_4$, we need the integrals

$$\begin{aligned}
&\int \phi_f^*(\mathbf{k}_3 + \mathbf{q})\phi_i(\mathbf{k}_3)\mathbf{k}_3\, d\mathbf{k}_3 \\
&\quad = \int \phi_f^*(\mathbf{r}_2)\phi_i(\mathbf{r}_3) \exp[i(\mathbf{k}_3 + \mathbf{q})\cdot\mathbf{r}_2] \exp(i\mathbf{k}_3\cdot\mathbf{r}_3)\mathbf{k}_3\, d\mathbf{k}_3\, d\mathbf{r}_2\, d\mathbf{r}_3 \\
&\quad = \int \phi_f^*(\mathbf{r}_2)\phi_i(\mathbf{r}_3) \exp[i(\mathbf{k}_3 + \mathbf{q})\cdot\mathbf{r}_2] i\nabla_3 \exp(-i\mathbf{k}_3\cdot\mathbf{r}_3)\, d\mathbf{k}_3\, d\mathbf{r}_2\, d\mathbf{r}_3 \\
&\quad = -\int \phi_f^*(r_2)[i\nabla_3\, \phi_i(\mathbf{r}_3)] \exp[i\mathbf{k}_3\cdot(\mathbf{r}_2 - \mathbf{r}_3)] \exp(i\mathbf{q}\cdot\mathbf{r}_2)\, d\mathbf{k}_3\, d\mathbf{r}_2\, d\mathbf{r}_3 \\
&\quad = (2\pi)^3 \int \phi_f^*(\mathbf{r}_2)[\mathbf{p}_2\phi_i(\mathbf{r}_2)] \exp(i\mathbf{q}\cdot\mathbf{r}_2)\, d\mathbf{r}_2.
\end{aligned} \tag{12.53}$$

An integration by parts was done in going from line 3 to line 4 in the right-hand side. Likewise, we need

$$\int \chi_f^*(\mathbf{k}_4 - \mathbf{q})\chi_i(\mathbf{k}_4)\mathbf{k}_4\, d\mathbf{k}_4 = (2\pi)^3 \int \chi_f^*(\mathbf{r}_1)[\mathbf{p}_1\chi_i(\mathbf{r}_1)] \exp(-i\mathbf{q}\cdot\mathbf{r}_1)\, d\mathbf{r}_1. \tag{12.54}$$

Using these results in $\mathscr{T}_{fi}$ we now have

$$\mathscr{T}_{fi} = \int \chi_f^*(\mathbf{r}_1)\phi_f^*(\mathbf{r}_2)\hat{\mathbf{t}}(\mathbf{r}, \mathbf{p})\phi_i(\mathbf{r}_2)\chi_i(\mathbf{r}_1)\, d\mathbf{r}_1\, d\mathbf{r}_2, \tag{12.55a}$$

$$\hat{t}(\mathbf{r}, \mathbf{p}) = (2\pi)^{-3} \int d\mathbf{q} \exp[-i\mathbf{q}\cdot(\mathbf{r}_1 - \mathbf{r}_2)][t^D(\mathbf{q}, \mathbf{p}) + t^E(\mathbf{Q}, -\mathbf{p})], \tag{12.55b}$$

where the momentum operator acts on $\phi_i\chi_i$ because $\mathbf{p} = \frac{1}{2}(\mathbf{p}_1 - \mathbf{p}_2)$.

By executing the integral in (12.55b) we obtain a coordinate-space representation of the interaction that includes exchange in an approximate manner. Recalling (12.38b) for the central and tensor parts of the exchange contribution to $\hat{t}$ we obtain an r dependence $\delta(\mathbf{r})$, whereas for the spin–orbit part, because of the $\mathbf{q} \times \mathbf{k}_i \cdot \mathbf{S}$ dependence on $\mathbf{q}$, we obtain for the r dependence

$$(2\pi)^{-3} \int d\mathbf{q} \exp(i\mathbf{q}\cdot\mathbf{r})\mathbf{q} = i\nabla\,\boldsymbol{\delta}(\mathbf{r}). \tag{12.56}$$

Therefore the q integral in $\hat{t}$ yields for the exchange contribution, under the constant-Q approximation, an effective operator,

$$\bar{t}^E(\mathbf{r}, \mathbf{p}) = [t_C^E(Q) - t_T^E(Q)S_{12}(\hat{\mathbf{Q}})]\,\delta(\mathbf{r}) - (1/Q^2)t_{LS}^E(Q)\,\nabla\,\delta(\mathbf{r}) \times \mathbf{p}\cdot\mathbf{S}, \tag{12.57a}$$

whose *direct* matrix element approximates the exchange contribution to $\mathscr{T}_{fi}$.

The direct terms under the q integration, of course, yield the original $t(\mathbf{r}, \mathbf{p})$ of Eqs. (12.9a) and (12.9b). Therefore the complete effective operator $\bar{t}$, which includes all terms in the interaction and exchange in an approximate way, is

$$\bar{t}(\mathbf{r}, \mathbf{p}) = t(\mathbf{r}, \mathbf{p}) + \bar{t}^{E}(\mathbf{r}, \mathbf{p}). \tag{12.57b}$$

Only the direct matrix element is to be computed in calculating $\mathcal{T}_{\mathrm{fi}}$. This is of course a great simplification and one that is reasonably accurate, as numerical comparisons have shown (Petrovich, pers. com., 1982; Petrovich *et al.*, 1983).

F. MULTIPOLE EXPANSION OF THE INTERACTION

In the preceding section a local interaction possessing central, tensor, and spin–orbit components was incorporated into the properly antisymmetrized distorted-wave approximation to the transition amplitude. An approximation was made in the evaluation of the exchange terms, which led to an effective contact interaction in these terms. Consequently, under this approximation the entire effective interaction, including the Pauli exchange, is to be evaluated as a direct amplitude only. The full effective interaction is represented by (12.57).

This is a greatly simplified version of the calculation. In particular, under this approximation the scattered particle plays the role of an external one-body operator acting on the target nucleons. In that case any inelastic nuclear transition can be written as a superposition of single-particle transitions from the orbital j to j' with amplitudes $S_J^{j'j}(J_{\mathrm{f}}, J_{\mathrm{i}})$, as derived in Chapter 9. In fact, the entire discussion of the way in which nuclear structure information is to be embedded into inelastic and charge-exchange reactions carries over, provided only that we perform a multipole expansion of the interaction.

We saw how to construct the multipole expansion of a general central interaction in Chapters 7 and 9. There, the Legendre expansion was employed as the means of effecting a separation of angular coordinates into a scalar product of spherical tensors acting separately, one on the target and one on the projectile. Here we shall employ the Fourier transformation (Petrovich, 1975) as an alternative means of effecting the separation, as described in Petrovich *et al.* (1983). In either case, the multipole expansion provides the transformation properties of the interaction as regards its structure in the projectile and target coordinates. As a consequence, it provides a way of stating the conservation of the angular momentum that is transferred between target and projectile, as discussed in Chapters 7 and 9.

The starting point of the derivation is the Fourier transform of the effective force in (12.55). The momentum components are given in (12.38) and are related through such Bessel transforms as (12.21) to the original interaction (12.10)–(12.15). Because we now want to separate the interaction into products of operators acting separately on particle 1 and particle 2, we must display these dependences. The isospin is not coupled to the rotation properties in space and can be left implicit. The spin dependences can be displayed explicitly, in the central force for example, by writing

$$t_{\mathrm{C}}^{D}(q) + t_{\mathrm{C}}^{E}(Q) = \tilde{t}_{\mathrm{C0}}(q, Q) + \tilde{t}_{\mathrm{C1}}(q, Q)\boldsymbol{\sigma}_1 \cdot \boldsymbol{\sigma}_2. \tag{12.58}$$

The tilde sign $\sim$ over the t's on the right signifies that the direct and exchange contributions are contained therein.

For brevity we introduce the factor

$$\tilde{t}_{\mathrm{LS}}(q, Q) = \tfrac{1}{2}[t_{\mathrm{LS}}(q) - (q^2/Q^2)t_{\mathrm{LS}}(Q)], \tag{12.59}$$

so that the spin–orbit force is

$$(-i/q^2)\tilde{t}_{\mathrm{LS}}(q, Q)\mathbf{q} \times \mathbf{p} \cdot \boldsymbol{\sigma}. \tag{12.60}$$

Thus the effective force (12.57), which is just the Fourier transform in (12.55), can be rewritten as

$$\begin{aligned} \bar{t}(\mathbf{r}, \mathbf{p}) = (2\pi)^{-3} \int d\mathbf{q} \exp[-i\mathbf{q} \cdot (\mathbf{r}_1 - \mathbf{r}_2)] [\tilde{t}_{\mathrm{C0}}(q, Q) + \tilde{t}_{\mathrm{C1}}(q, Q)\boldsymbol{\sigma}_1 \cdot \boldsymbol{\sigma}_2 \\ -(i/q^2)\tilde{t}_{\mathrm{LS}}(q, Q)\mathbf{q} \cdot \mathbf{p} \times \boldsymbol{\sigma} - t_{\mathrm{T}}^{D}(q)S_{12}(\hat{\mathbf{q}}) - t_{\mathrm{T}}^{E}(Q)S_{12}(\hat{\mathbf{Q}})]. \end{aligned} \tag{12.61}$$

There are two ways in which we could proceed. One is to write the exponential as the product of two plane-wave series and employ the re-coupling techniques of angular momentum. There is, however, a more direct route to a convenient representation, in terms of transverse and longitudinal components. For this purpose, note the two identities

$$\begin{aligned} \boldsymbol{\sigma}_1 \cdot \boldsymbol{\sigma}_2 &= (\boldsymbol{\sigma}_1 \cdot \hat{\mathbf{q}})(\boldsymbol{\sigma}_2 \cdot \hat{\mathbf{q}}) + \boldsymbol{\sigma}_1 \times \hat{\mathbf{q}} \cdot \boldsymbol{\sigma}_2 \times \hat{\mathbf{q}}, \\ S_{12}(\hat{\mathbf{q}}) &= 2(\boldsymbol{\sigma}_1 \cdot \hat{\mathbf{q}})(\boldsymbol{\sigma}_2 \cdot \hat{\mathbf{q}}) - \boldsymbol{\sigma}_1 \times \hat{\mathbf{q}} \cdot \boldsymbol{\sigma}_2 \times \hat{\mathbf{q}}, \end{aligned} \tag{12.62}$$

representing components that are parallel and perpendicular to $\hat{\mathbf{q}}$.

With these identities the *spin*-independent part of the central force and the *direct* part of the tensor can be combined as

$$\begin{aligned} &t_{\mathrm{C1}}\boldsymbol{\sigma}_1 \cdot \boldsymbol{\sigma}_2 - t_{\mathrm{T}}^{D}S_{12}(\hat{\mathbf{q}}) \\ &\quad = (t_{\mathrm{C1}} - 2t_{\mathrm{T}}^{D})\boldsymbol{\sigma}_1 \cdot \hat{\mathbf{q}}\, \boldsymbol{\sigma}_2 \cdot \hat{\mathbf{q}} + (\tilde{t}_{\mathrm{C1}} + t_{\mathrm{T}}^{D})\boldsymbol{\sigma}_1 \times \hat{\mathbf{q}} \cdot \boldsymbol{\sigma}_2 \times \hat{\mathbf{q}} \\ &\quad = (\tilde{t}_{\mathrm{C1}} - 2t_{\mathrm{T}}^{D})\boldsymbol{\sigma}_1 \cdot \hat{\mathbf{q}}\, \boldsymbol{\sigma}_2 \cdot \hat{\mathbf{q}} + (\tilde{t}_{\mathrm{C1}} + t_{\mathrm{T}}^{D})[\boldsymbol{\sigma}_1 \cdot \boldsymbol{\sigma}_2 - \boldsymbol{\sigma}_1 \cdot \hat{\mathbf{q}}\, \boldsymbol{\sigma}_2 \cdot \hat{\mathbf{q}}]. \end{aligned} \tag{12.63}$$

At the same time, note that

$$\mathbf{q} \exp[-i\mathbf{q} \cdot (\mathbf{r}_1 - \mathbf{r}_2)] = i\boldsymbol{\nabla}_1 \exp(-i\mathbf{q} \cdot \mathbf{r}) = -i\boldsymbol{\nabla}_2 \exp(-i\mathbf{q} \cdot \mathbf{r}). \tag{12.64}$$

Define

$$\tau_{LSJ}(q\mathbf{r}) = j_L(qr) T_{LSJ}(\hat{\mathbf{r}}, \boldsymbol{\sigma}), \tag{12.65}$$

where T_{LSJ} was defined in (7.84). Making the plane-wave expansions (2.8) of the exponentials, we can write

$$(2\pi)^{-3} \int \exp(-i\mathbf{q} \cdot \mathbf{r})\, t(q)\, d\mathbf{q} = (2/\pi) \sum_J \int q^2\, dq\, t(q) \tau_{J0J}(1) \cdot \tau_{J0J}(2). \tag{12.66}$$

Using these results in (12.61), we find

$$\begin{aligned}
\bar{t}(\mathbf{r}, \mathbf{p}) = \frac{2}{\pi} \sum_J \int q^2\, dq \Bigg\{ \Bigg[\tilde{t}_{C0} + \frac{1}{q^2} (\tilde{t}_{C1} - 2t_T^D) \boldsymbol{\sigma}_1 \cdot \boldsymbol{\nabla}_1 \boldsymbol{\sigma}_2 \cdot \boldsymbol{\nabla}_2 \\
+ (\tilde{t}_{C1} + t_T^D) \left(\boldsymbol{\sigma}_1 \cdot \boldsymbol{\sigma}_2 - \frac{1}{q^2} \boldsymbol{\sigma}_1 \cdot \boldsymbol{\nabla}_1 \boldsymbol{\sigma}_2 \cdot \boldsymbol{\nabla}_2 \right) - t_T^E(Q) S_{12}(\hat{\mathbf{Q}}) \Bigg] \tau_{J0J}(1) \cdot \tau_{J0J}(2) \\
+ \frac{1}{2q^2} \tilde{t}_{LS} [\boldsymbol{\sigma}_1 \times \boldsymbol{\nabla}_1 [\tau_{J0J}(1) \cdot \tau_{J0J}(2)] \cdot \mathbf{p}_1 + \boldsymbol{\sigma}_2 \times \boldsymbol{\nabla}_2 [\tau_{J0J}(1) \cdot \tau_{J0J}(2)] \cdot \mathbf{p}_2 \\
- \boldsymbol{\sigma}_1 \times \boldsymbol{\nabla}_1 [\tau_{J0J}(1) \cdot \tau_{J0J}(2)] \cdot p_2 - \boldsymbol{\sigma}_2 \times \boldsymbol{\nabla}_2 [\tau_{J0J}(1) \cdot \tau_{J0J}(2)] \cdot \mathbf{p}_1] \Bigg\}.
\end{aligned} \tag{12.67}$$

The product $\boldsymbol{\sigma}_1 \cdot \boldsymbol{\sigma}_2$ with the operators τ_{J0J} can be written according to (7.85) as

$$(\boldsymbol{\sigma}_1 \cdot \boldsymbol{\sigma}_2) \tau_{J0J}(1) \cdot \tau_{J0J}(2) = \sum_L (-)^{L+1+J} \tau_{L1J}(1) \cdot \tau_{L1J}(2). \tag{12.68}$$

The problem of carrying out the gradient operations remains. Here one can employ a number of useful results having to do with vector spherical harmonics and the curl of such objects. The reader is referred to Edmonds (1957, Sections 5.9 and 5.10) for this material. The vector spherical harmonic is defined as

$$\mathbf{Y}_{JL}^M(\hat{\mathbf{r}}) = [Y_L(\hat{\mathbf{r}})\mathbf{e}]_J^M = \sum_{M_L \mu} C_{M_L \mu M}^{L\ 1\ J} Y_{LM_L} \mathbf{e}_\mu, \tag{12.69}$$

where $\mathbf{e}_x$, $\mathbf{e}_y$, $\mathbf{e}_z$ are orthogonal unit vectors and $\mathbf{e}_\mu$ are its spherical components (see Appendix). Also define

$$\mathbf{J}_{JL}^M(q\mathbf{r}) = j_L(qr) \mathbf{Y}_{JL}^M(\hat{\mathbf{r}}). \tag{12.70}$$

Both of these expressions represent vectors, whose spherical components are given by the coefficient of $\mathbf{e}_\mu$. One scalar product of interest to us is

$$\boldsymbol{\sigma} \cdot \mathbf{J}_{JL}^M = j_L(qr) [Y_L(\hat{\mathbf{r}}) \sigma]_J^M = \tau_{L1J}^{\ M}. \tag{12.71}$$

The divergence of τ_{J0J} can be found from Edmonds [1957, Eq. (5.9.17)] and by using the properties of spherical harmonics, with the result that

$$\nabla \tau_{J0J}^{\ M} = q \sum_{L} x_{JL} \mathbf{J}_{JL}^{M}, \tag{12.72}$$

where the coefficients x_{JL} are given by

$$x_{J,J-1} = \left(\frac{J}{2J+1}\right)^{1/2}, \qquad x_{J,J+1} = \left(\frac{J+1}{2J+1}\right)^{1/2}. \tag{12.73}$$

With these we can now write

$$(\boldsymbol{\sigma}_1 \cdot \nabla_1)(\boldsymbol{\sigma}_2 \cdot \nabla_2)\tau_{J0J}(1) \cdot \tau_{J0J}(2) = q^2 \tau_J^{\sigma\|}(1) \cdot \tau_J^{\sigma\|}(2), \tag{12.74}$$

which is one of the terms in (12.67). The operator appearing here is a longitudinal spin operator,

$$\tau_{JM}^{\sigma\|} = \sum_{L} x_{JL} \tau_{L1J}^{\ M}, \qquad L = J \pm 1. \tag{12.75}$$

We also need the transverse spin operator,

$$\tau_{JM}^{\sigma\perp} = \sum_{L} y_{JL} \tau_{L1J}^{\ M}, \qquad L = J \pm 1, \tag{12.76}$$

where

$$y_{J,J-1} = \left(\frac{J+1}{2J+1}\right)^{1/2}, \qquad y_{J,J+1} = -\left(\frac{J}{2J+1}\right)^{1/2}. \tag{12.77}$$

The apparatus is now available for evaluating

$$\left(\boldsymbol{\sigma}_1 \cdot \boldsymbol{\sigma}_2 - \frac{1}{q^2} \boldsymbol{\sigma}_1 \cdot \nabla_1 \, \boldsymbol{\sigma}_2 \cdot \nabla_2\right) \tau_{J0J}(1) \cdot \tau_{J0J}(2) = \tau_J^{\sigma\perp}(1) \cdot \tau_J^{\sigma\perp}(2) - \tau_{J1J}(1) \cdot \tau_{J1J}(2). \tag{12.78}$$

The tensor exchange force is expressible in convenient form by defining the operator

$$Q_{L1J}^{\ M} = [\tau_{L0L}[Y_2(\hat{\mathbf{Q}})\boldsymbol{\sigma}_1]_1]_J^M \tag{12.79}$$

and its longitudinal and transverse components

$$Q_{JM}^{\|} = \sum_{L} x_{JL} Q_{L1J}^{\ M}, \tag{12.80}$$

$$Q_{JM}^{\perp} = \sum_{L} y_{JL} Q_{L1J}^{\ M}. \tag{12.81}$$

One then finds

$$S_{12}(\hat{\mathbf{Q}})\tau_{J0J}(1) \cdot \tau_{J0J}(2) = Q_{J1J}(1) \cdot \tau_{J1J}(2) - Q_J^{\|}(1) \cdot \tau_J^{\sigma\|} - Q_J^{\perp}(1) \cdot \tau_J^{\sigma\perp}. \tag{12.82}$$

The first spin–orbit term can be written as

$$\boldsymbol{\sigma}_1 \times \nabla_1[\tau_{JOJ}(1)\cdot\tau_{JOJ}(2)]\cdot\mathbf{p}_1 = q^2\tau_J^{\ell s}(1)\cdot\tau_{JOJ}(2), \tag{12.83}$$

where the superscript ℓs refers to the spin–orbit tensor,

$$\tau_{JM}^{\ell s}(1) = \frac{1}{q^2}\nabla_1\,\tau_{JOJ}^{\;M} \times \mathbf{p}_1\cdot\boldsymbol{\sigma}_1 \tag{12.84}$$

There remain the spin–orbit terms of the type,

$$\boldsymbol{\sigma}_1 \times \nabla_1[\tau_{JOJ}(1)\cdot\tau_{JOJ}(2)]\cdot\mathbf{p}_2 = \sum_{J\Lambda\mu}(-)^{J+1-\Lambda+\mu}\boldsymbol{\sigma}_1\cdot\nabla_1 \times \mathbf{J}_{\Lambda J}^{-\mu}(1)[\tau_{JOJ}(2)\mathbf{p}_2]_\Lambda^\mu. \tag{12.85}$$

The curl of $\mathbf{J}_{JL}$ can be evaluated from Edmonds [1957, Eqs. (5.9.18)–(5.9.20)] and by using properties of the spherical Bessel functions. The results are

$$\begin{aligned}
\boldsymbol{\sigma}\cdot\nabla \times \mathbf{J}_{JJ}^M &= iq\tau_J^{\sigma\perp},\\
\boldsymbol{\sigma}\cdot\nabla \times \mathbf{J}_{J,J-1}^M &= -iq\left(\frac{J+1}{2J+1}\right)^{1/2}\tau_{J1J}^{\;M},\\
\boldsymbol{\sigma}\cdot\nabla \times \mathbf{J}_{J,J+1}^M &= iq\left(\frac{J+1}{2J+1}\right)^{1/2}\tau_{J1J}^{\;M}.
\end{aligned} \tag{12.86}$$

Now, by defining the (transverse) current operators

$$\begin{aligned}
\tau_{JJM}^{\ell} &= \frac{1}{q}[\tau_{JOJ}\nabla]_J^M,\\
\tau_{JM}^{\ell} &= \frac{1}{q}\sum_L y_{JL}[\tau_{LOL}\nabla]_J^M,
\end{aligned} \tag{12.87}$$

we can write

$$\boldsymbol{\sigma}_1 \times \nabla_1[\tau_{JOJ}(1)\cdot\tau_{JOJ}(2)]\cdot\mathbf{p}_2 = -q^2[\tau_J^{\sigma\perp}(1)\cdot\tau_{JJ}^{\ell}(2) + \tau_{J1J}(1)\cdot\tau_J^{\ell}(2)]. \tag{12.88}$$

These are all that are needed to write the multipole decomposition of the force. It is useful to group the terms according to whether they lead to normal (N) or abnormal (A) parity transitions and according to whether or not spin is transferred to the projectile, indicated by 1 or 0. Thus

$$\begin{aligned}
\bar{t}^{\,\mathrm{N0}}(\mathbf{r},\mathbf{p}) &= (2/\pi)\sum_J\int q^2\,dq\,\{\tilde{t}_{\mathrm{C0}}(\tau_{JOJ}\cdot\tau_{JOJ})\\
&\quad+\tfrac{1}{2}\tilde{t}_{\mathrm{LS}}(\tau_{JOJ}\cdot\tau_J^{\ell s}) + \tfrac{1}{2}\tilde{t}_{\mathrm{LS}}(\tau_J^{\ell}\cdot\tau_{J1J})\}\\
\bar{t}^{\,\mathrm{N1}}(\mathbf{r},\mathbf{p}) &= (2/\pi)\sum_J\int q^2\,dq\,\{-[\tilde{t}_{\mathrm{C1}} + t_{\mathrm{T}}^{D}](\tau_{J1J}\cdot\tau_{J1J})\\
&\quad-t_{\mathrm{T}}^{E}(Q_{J1J}\cdot\tau_{J1J}) + \tfrac{1}{2}\tilde{t}_{\mathrm{LS}}(\tau_{J1J}\cdot\tau_J^{\ell}) + \tfrac{1}{2}\tilde{t}_{\mathrm{LS}}(\tau_J^{\ell s}\cdot\tau_{JOJ})\}
\end{aligned} \tag{12.89}$$

are the parts of the force that are responsible for exciting normal parity states in the nucleus. Such transitions were discussed in Chapter 9. The superscripts 1 and 0 denote whether spin is transferred to the projectile or not. We have streamlined the answer by omitting the arguments of the t's on the right side and by employing the convention in an operator product $\tau(1) \cdot \tau(2)$ that particle 1 always comes first. For the abnormal parity transitions,

$$\bar{t}^{A0}(\mathbf{r}, \mathbf{p}) = \frac{2}{\pi} \sum_J \int q^2\, dq\, \tfrac{1}{2}\tilde{t}_{LS}(\tau^{\ell}_{JJ} \cdot \tau^{\sigma\perp}_J),$$

$$\begin{aligned} t^{A1}(\mathbf{r}, \mathbf{p}) = \frac{2}{\pi} \sum_J \int q^2\, dq\, \{ & \tfrac{1}{2}\tilde{t}_{LS}(\tau^{\sigma\perp}_J \cdot \tau^{\ell}_{JJ}) \\ & + [\tilde{t}_{C1} - 2t^{D}_{T}](\tau^{\sigma\|}_J \cdot \tau^{\sigma\|}_J) + t^{E}_{T}(Q^{\|}_J \cdot \tau^{\sigma\|}_J) \\ & + [\tilde{t}_{C1} + t^{D}_{T}](\tau^{\sigma\perp}_J \cdot \tau^{\sigma\perp}_J) + t^{E}_{T}(Q^{\perp}_J \cdot \tau^{\sigma\perp}_J)\} \end{aligned} \tag{12.90}$$

The preceding formulas provide a desired multipole decomposition of the force into terms that consist of the scalar product of spherical tensor operators that act separately in the space of the target and the projectile. In Chapter 9 the advantages of having such a decomposition were discussed. Briefly, the multipolarity and structure of the operators denote the angular momentum and parity exchanged between the target and projectile.

The terms were arranged into four groups because in the plane-wave approximation there is no interference between them (Petrovich, pers. com., Petrovich *et al.*, 1983). This will be seen in the expressions for the plane-wave differential cross sections to be presented.

To further discuss the properties of this decomposition of the force, note that it has the general form

$$\bar{t}(\mathbf{r}, \mathbf{p}) = \sum_{\alpha\beta J} \int q^2\, dq\, t^{\beta\alpha}(q, Q) T^{\beta}_J(q\mathbf{r}_1) \cdot T^{\alpha}_J(q\mathbf{r}_2), \tag{12.91}$$

where $t^{\beta\alpha}$ denotes the Bessel transforms $\tilde{\tau}_{C0}, \tilde{\tau}_{LS}, \ldots$ appearing in (12.89) and (12.90) and T denotes the various operators.

The distorted-wave amplitude for inelastic scattering can be written, in view of (12.91), as

$$\mathscr{T}_{fi} = (2/\pi) \sum_{\alpha\beta JM} (-)^M \int q^2\, dq\, t^{\beta}(q, Q) \langle \chi_f | T^{\beta}_{J,-M} | \chi_i \rangle \langle \Phi^{M_f}_{J_f} | \sum_{i=1}^{A} T^{\alpha}_{J,M}(i) | \Phi^{M_i}_{J_i} \rangle. \tag{12.92}$$

Remember that $\bar{t}$ includes an approximation to the exchange integrals in the form of an effective operator that is to be computed as a direct integral. This has the great advantage that the scattered particle acts like a one-body operator in the space of the target nucleons, whereas the exchange integrals do not have this character. In such a case, as shown in Chapter 9, the nuclear matrix element for a general nuclear transition can be written as a linear combination of single-particle transitions. Thus, for the reduced matrix element in the target space, we may write, in general,

$$\langle \Phi_{J_f} \| \sum_{i=1}^{A} T_J(i) \| \Phi_{J_i} \rangle = \sum_{jj'} S_J^{j'j}(J_f, J_i) \langle j' \| T_J \| j \rangle. \tag{12.93}$$

The reduced matrix element on the right corresponds to the single-particle transition $j \to j'$. The amplitudes with which they are summed are given by (9.38)

$$S_J^{j'j}(J_f, J_i) = -\frac{1}{\sqrt{\hat{\mathbf{J}}}} \langle \Phi_{J_f} \| [\beta_{j'}^{\dagger} \tilde{\beta}_j]_J \| \Phi_{J_i} \rangle. \tag{12.94}$$

They can be computed, given model wave functions for the nuclear states in question. We presented numerous examples of such calculations in Chapter 9.

G. TRANSITION DENSITIES

In Chapter 9 we referred to the nuclear reduced matrix elements in the multipole expansion as form factors. Because of the way the decomposition was performed there (through the Legendre expansion), they still carried a dependence on the radial coordinate of the projectile. The present decomposition achieves a total separation of target and projectile coordinates. The reduced matrix elements will therefore be referred to as transition densities. We now discuss their properties. There are seven different operators in $\bar{t}$ that refer to the target space and three additional ones in the projectile space. They are listed in Tables I and II, together with the associated parity change, a notation for the corresponding transition density (or distortion function in the case of projectile), and a suitable name connoting the nuclear property that is probed. To understand the first entry in the table, note that the nucleon density at the point $\mathbf{r}$ can be found from

$$\rho(\mathbf{r}) = \int \Phi^*(\mathbf{r}_1, \mathbf{r}_2, \ldots \mathbf{r}_A) \sum_{i=1}^{A} \delta(\mathbf{r} - \mathbf{r}_i) \Phi(\mathbf{r}_1, \mathbf{r}_2, \ldots, \mathbf{r}_A)\, d\mathbf{r}_1 \cdots d\mathbf{r}_A, \tag{12.95}$$

so that we can call

$$\hat{\rho}(\mathbf{r}) = \sum_i \delta(\mathbf{r} - \mathbf{r}_i) \tag{12.96}$$

the nuclear density operator. In nondiagonal elements, as encountered in transitions, we can refer to (12.95) as the transition density, and we could

TABLE I

Operators Acting in the Target Space[a]

Operator	$\Delta\pi$	Density	Type of density
τ_{J0J}	$(-)^J$	ρ_J	Matter
τ_{J1J}	$(-)^J$	ρ^{σ}_{JJ}	spin
$\tau^{\ell s}_J$	$(-)^J$	$\rho^{\ell s}_J$	Spin-current
τ^{ℓ}_J	$(-)^J$	ρ^{ℓ}_J	Transverse current
$\tau^{\sigma\|\|}_J$	$(-)^{J+1}$	$\rho^{\sigma\|\|}_J$	Longitudinal spin
$\tau^{\sigma\perp}_J$	$(-)^{J+1}$	$\rho^{\sigma\perp}_J$	Transverse spin
τ^{ℓ}_{JJ}	$(-)^{J+1}$	τ^{ℓ}_{JJ}	Current

[a] This list includes their associated parity and transition density. The same operators and corresponding distortion functions are associated with the projectile.

place superscripts $\rho^{J_f J_i}(\mathbf{r})$ on it to denote the transition referred to, although we will not do so for reasons of notational convenience. Substitute into (12.95) the multipole expansion of the δ function,

$$\delta(\mathbf{r}-\mathbf{r}_i) = \frac{\delta(r-r_i)}{rr_i}\sum_J Y_J(\hat{\mathbf{r}})\cdot Y_J(\hat{\mathbf{r}}_i), \tag{12.97}$$

and use the Wigner–Eckart theorem to obtain

$$\rho(\mathbf{r}) = \sum_{JM}(-)^{J_f-M_i}\begin{pmatrix} J_f & J & J_i \\ -M_f & M & M_i \end{pmatrix}\rho_J(r)Y_J^{-M}(\hat{\mathbf{r}}), \tag{12.98}$$

where

$$\rho_J(r) = \langle\Phi_{J_f}\|\sum_{i=1}^{A}\frac{\delta(r-r_i)}{rr_i}Y_J(\hat{\mathbf{r}}_i)\|\Phi_{J_i}\rangle. \tag{12.99}$$

TABLE II

Operators Acting in Projectile Space[a]

Operator	$\Delta\pi$	Distortion function	Type of distortion
Q_{J1J}	$(-)^J$	D^Q_J	Tensor
$Q^{\|\|}_J$	$(-)^{J+1}$	$D^{Q\|\|}_J$	Longitudinal tensor
$Q^{\perp}_J$	$(-)^{J+1}$	$D^{Q\perp}_J$	Transverse tensor

[a] In the projectile space, all the operators in Table I act in addition to those listed here.

This is the reduced matrix element of the jth multipole of the density operator. We refer to it in brief as the matter transition density.

Now note that the Bessel transform of the transition density,

$$\rho_J(q) = \int j_J(qr)\rho_J(r)r^2\,dr = \langle \Phi_{J_f} \| \sum_{i=1}^{A} \tau_{J0J}(q\mathbf{r}_i) \| \Phi_i \rangle \tag{12.100}$$

is simply the reduced matrix element of the first of the operators listed in Table I, which occurs in the multipole decomposition of the force. This operator is therefore associated with the particle transition density.

Similarly, the operator

$$\hat{\rho}(\mathbf{r}, \sigma_\mu) = \sum_i \delta(\mathbf{r} - \mathbf{r}_i)\sigma_{i\mu} \tag{12.101}$$

is a spin density for the $\mu\,(= -1, 0, 1)$ spherical component of $\boldsymbol{\sigma}_i$. Again using (12.97) it can be written as

$$\hat{\rho}(\mathbf{r}, \sigma_\mu) = \sum_{JM} \left[\sum_i \frac{\delta(r - r_i)}{rr_i} T_{L1J}^{M}(\hat{\mathbf{r}}_i, \boldsymbol{\sigma}_i) \right] (-)^{M_L} C_{M_L \mu M}^{L\ 1\ J} Y_L^{-M_L}(\hat{\mathbf{r}}). \tag{12.102}$$

The quantity in brackets can be referred to as the LJ multipole of the spin-density operator and its reduced matrix element as the spin density,

$$\rho_{LJ}^{\sigma}(r) = \langle \Phi_{J_f} \| \sum_i \frac{\delta(r - r_i)}{rr_i} T_{L1J}(\hat{\mathbf{r}}_i, \boldsymbol{\sigma}_i) \| \Phi_{J_i} \rangle. \tag{12.103}$$

The Bessel transform in this case is

$$\rho_{LJ}^{\sigma}(q) = \int j_L(qr)\rho_{LJ}^{\sigma}(r)r^2\,dr = \langle \Phi_{J_f} \| \sum_{i=1}^{A} \tau_{L1J}(q\mathbf{r}_i) \| \Phi_i \rangle. \tag{12.104}$$

Not surprisingly, τ_{L1J} is associated with the spin density and, in the combinations (12.75) and (12.76), with the longitudinal and transverse components of the spin density. [The decomposition into those components was achieved through the identities (12.62).]

Operators (12.84) and (12.87) containing derivatives $\mathbf{p} = -i\boldsymbol{\nabla}$ are currents; and (12.84), because it also contains $\boldsymbol{\sigma}$, is associated with a spin current.

Having discussed the interpretation of the nuclear matrix elements that occur in the decomposition (12.92), we turn now to the projectile matrix elements. These we refer to as distortion functions,

$$D_{JM}^{\beta}(\theta, q) \equiv \langle \chi_f | T_{J,M}^{\beta} | \chi_i \rangle. \tag{12.105}$$

They depend on the scattering angle θ, on momentum q, and, of course, on the projectile energy. We can get a rough impression of the form of these functions by looking at the plane-wave approximation. Examine the one corresponding to particle density. It is

$$D_{JM}(\theta, q) \simeq \int \exp(-i\mathbf{k}_f \cdot \mathbf{r})\, \tau_{J0J}^{M}(q, r) \exp(i\mathbf{k}_i \cdot \mathbf{r})\, d\mathbf{r}$$

$$= 2\pi^2 i^{-J} Y_J^M(\hat{\mathbf{q}}') \frac{\delta(q' - q)}{qq'}, \tag{12.106}$$

$$\mathbf{q}' = \mathbf{k}_i - \mathbf{k}_f.$$

Therefore, we expect D to be peaked around the momentum transfer.

Let us now assemble the previously developed information into (12.92) to obtain

$$\mathscr{T}_{fi} = \frac{2}{\pi} \sum_{\alpha\beta JM} (-)^{J_f - M_i} \begin{pmatrix} J_f & J & J_i \\ -M_f & M & M_i \end{pmatrix} \int q^2\, dq\, D_{J-M}^{\beta}(\theta, q) t^{\beta\alpha}(q, Q) \rho_J^{\alpha}(q). \tag{12.107}$$

This has a very simple and instructive form. The plane-wave result for the distortion function suggests that it will peak near $q = |\mathbf{k}_i - \mathbf{k}_f|$. In any case, the integrand is a product of three factors, the other two being the Bessel transform of the force and the various transition densities. The momentum dependence of the force, at higher energy where the impulse approximation becomes valid, is reasonably well known from the free N–N data. This was discussed earlier in the chapter. The transition densities ρ_J^{α} can be calculated with the aid of (12.93), once model nuclear functions have been calculated. Alternatively, it is possible to deduce them in certain cases from electron scattering experiments, as will be mentioned again in connection with one of the examples.

H. CROSS SECTIONS

Finally, for completeness, we present the differential cross sections for $0^+ \to J^\pi$ normal and abnormal parity transitions in the plane-wave approximation (Petrovich, pers. com.; Petrovich *et al.*, 1983).

$$\begin{aligned} d\sigma^{N}/d\Omega = 4\pi (m_N/2\pi\hbar^2)^2 (2J+1)[&|A(q)\rho_J(q) + C(q)\rho_J^{\ell s}(q)|^2 \\ &+ |f(\theta)C(q)\rho_{JJ}^{\sigma}(q)|^2 + \tfrac{1}{2}|C(q)\rho_J^{\ell}(q) - F(q)\rho_{JJ}^{\sigma}(q)|^2 \\ &+ \tfrac{1}{2}|f(\theta)C(q)\rho_{JJ}^{\sigma}(q)|^2 + \tfrac{1}{2}|C(q)\rho_J^{\ell}(q) - B(q)\rho_{JJ}^{\sigma}(q)|^2], \end{aligned} \tag{12.108a}$$

$$\begin{aligned} d\sigma^{A}/d\Omega = 4\pi (m_N/2\pi\hbar^2)^2 (2J+1)[&|E(q)\rho_J^{\sigma\|}(q)|^2 + \tfrac{1}{2}|f(\theta)C(q)\rho_J^{\sigma\perp}(q)|^2 \\ &+ \tfrac{1}{2}|C(q)\rho_{JJ}^{\ell}(q) + B(q)\rho_J^{\sigma\perp}(q)|^2 + \tfrac{1}{2}|C(q)\rho_{JJ}^{\ell}(q) + F(q)\rho_J^{\sigma\perp}(q)|^2]. \end{aligned} \tag{12.108b}$$

Here m_N is the nucleon mass,

$$f(\theta) = (k^2/q^2)\sin\theta \tag{12.109}$$

and

$$A(q) = \hat{t}_{C0}(q, Q), \tag{12.110a}$$

$$C(q) = \tfrac{1}{2}\hat{t}_{LS}(q, Q), \tag{12.110b}$$

$$B(q) = \hat{t}_{C1}(q, Q) + t_T^D(q) + t_T^C(Q), \tag{12.110c}$$

$$E(q) = \hat{t}_{C1}(q, Q) - 2t_T^D(q) + t_T^C(Q), \tag{12.110d}$$

$$F(q) = \hat{t}_{C1}(q, Q) - t_T^D(q) - 2t_T^C(Q). \tag{12.110e}$$

These are just the results presented by Petrovich and Love (1981) and Love *et al.* (1983), written in terms of transverse and longitudinal densities with the tensor exchange included. These were first obtained without the target current terms by Kerman *et al.* (1959) whose derivation is more direct than that of Petrovich *et al.* (1983), which has been discussed here but is not amenable for extension to distorted-wave calculations.

A brief study of (12.108) bears out the lack of interference between the terms in the multipole decomposition that were separated in (12.89) and (12.90). Additional separations are also clear, such as the lack of interference between longitudinal and transverse terms, which one might have expected. Before going on to the discussion of some application of this formalism, we note that isospin dependence, which has been kept implicit in all of the preceding, is easily made explicit in (12.108) by replacing $A\rho_J$ by $\sum_T A_T\rho_J^T$, etc.; where the $T = 1$, interaction components are those that multiply $\bar{\tau}_i \cdot \bar{\tau}_2$, and the isospin transition densities contain the factors 1 and $\bar{\tau}_2$ for $T = 0$ and $T = 1$.

I. APPLICATIONS

The immediate goal of the theory of nuclear reactions is to formulate and then confirm the theory. The broader goal is, of course, to employ reactions as a probe of the structure of nuclei. The proceedings of several conferences will serve to illustrate some of the achievements in this latter respect. Here we draw on several examples that serve to illustrate and confirm some of the features of the effective interaction that were discussed earlier in this chapter.

We look first at the energy dependence of the central force, as illustrated in Fig. 12.1. Note that in the energy range from 50 to 200 MeV, t_0 and t_τ are decreasing rapidly, whereas $t_{\sigma\tau}$ is nearly constant. The energy dependence of the t_τ and $t_{\sigma\tau}$ components can be singled out for study by the (p, n) reaction. In particular, the so-called Gamow–Teller transitions, such as

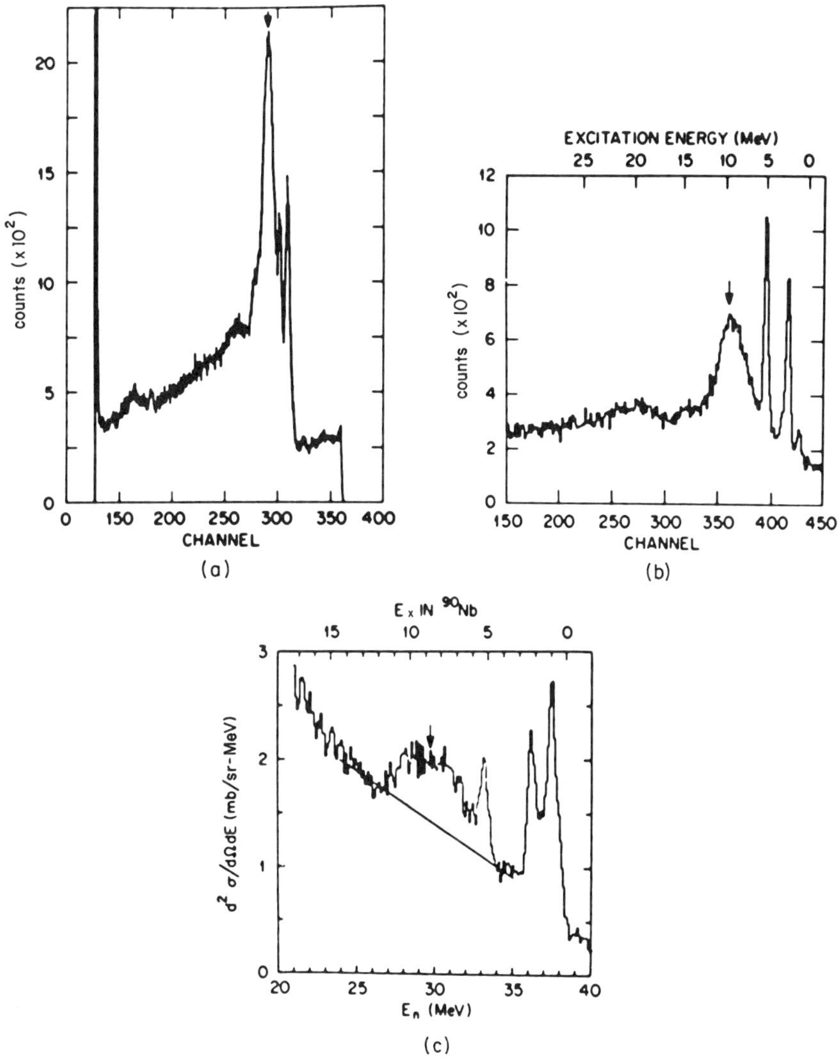

Fig. 12.4. The forward angle excitation functions for the reaction $^{90}Zr(p,n)^{90}Nb$ at three bombarding energies. Excitation energy of nuclear states runs from right to left. The Gamow–Teller transition is marked by an arrow in each figure. Its rapid rise above the background as the energy increases is attributed to the rapid decrease of t_0 and t_τ while $t_{\sigma\tau}$, which is responsible for the Gamow–Teller transition, remains roughly constant. (From Doering *et al.*, 1975; Bertsch *et al.*, 1977; Goodman, 1980; Love and Franey 1981; Petrovich and Love, 1981; Love *et al.*, 1983; Petrovich, 1983.)

$(0^+, T = 0) \rightarrow (1^+, T = 1)$, are caused by the $t_{\sigma\tau}$ part of the interaction. This can be understood as follows. The change in charge of the projectile obviously requires the τ part of the interaction. The fact that the parity is unchanged in the Gamow–Teller transition requires that the orbital angular momentum transfer L be even (recall Chapter 9), whereas the change in total angular momentum $0 \rightarrow 1$ in the preceding transition requires that $J = 1$. Consequently, $S = 1$ and $L = 0$ or 2. Therefore the operator L acting on the nuclear state that is needed to effect such a transition is $Y_L \sigma \tau$. Figure 12.4 shows the forward angle (p, n) cross section on ^{90}Zr at three different energies (see Goodman, 1980 and references cited therein.) At the low energy, according to Fig. 12.1, the t_τ and $t_{\sigma\tau}$ strengths are comparable, so that states excited by one or other of these components will have comparable cross sections. However, as the energy is raised, the t_τ strengths fall dramatically, whereas $t_{\sigma\tau}$ remains about the same. Corresponding to this, the 1^+, $T = 1$ state rises dramatically in cross section in comparison with neighboring states. In particular, the analog state, 0^+, $T = 0$ (5 MeV), which is prominent at the low bombarding energy, becomes relatively insignificant at higher energy. This is a single dramatic confirmation of the trend of the energy dependence of the interaction. It also shows how different types of excitations can be selected by the choice of experimental conditions. For a comprehensive study of Gamow–Teller transitions see the proceedings of the Telluride conferences (Goodman *et al.*, 1980; Petrovich *et al.*, 1983a).

The early work on inelastic scattering referred to in Chapter 9 employed a central force. It is therefore interesting to learn whether other components of the force play a role in the excitation of nuclear levels.

The importance of the spin–orbit interaction is confirmed in a study of natural parity states of ^{208}Pb (Adams *et al.*, 1980; Petrovich *et al.*, 1980a). In this study the nuclear transition densities ρ_J were inferred from electron scattering experiments and are shown in Fig. 12.5. The feature of particular importance is that the peak of the densities moves to larger momentum transfer as the multipolarity increases. When this fact is compared with the momentum dependence of the interaction (Fig. 15.3), one can infer that the low-spin states will be dominated by the central interaction, but the high states ($J > 8$) will be dominated by the spin–orbit interaction. Figure 12.6

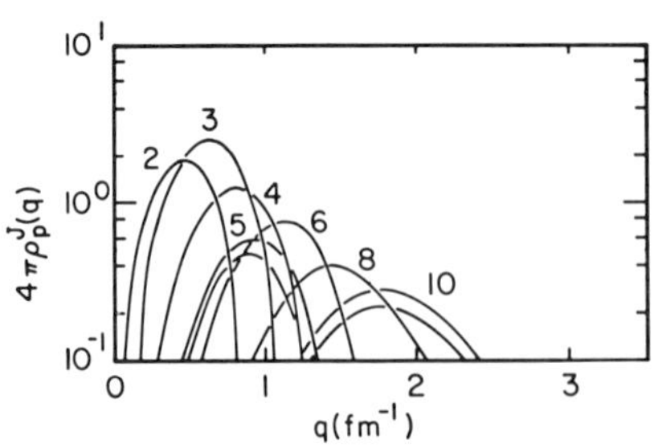

Fig. 12.5. Proton transition densities of various multipolarities, as inferred from electron scattering experiments (From Petrovich *et al.*, 1980a; Petrovich and Love, 1981 and Love *et al.* 1983.)

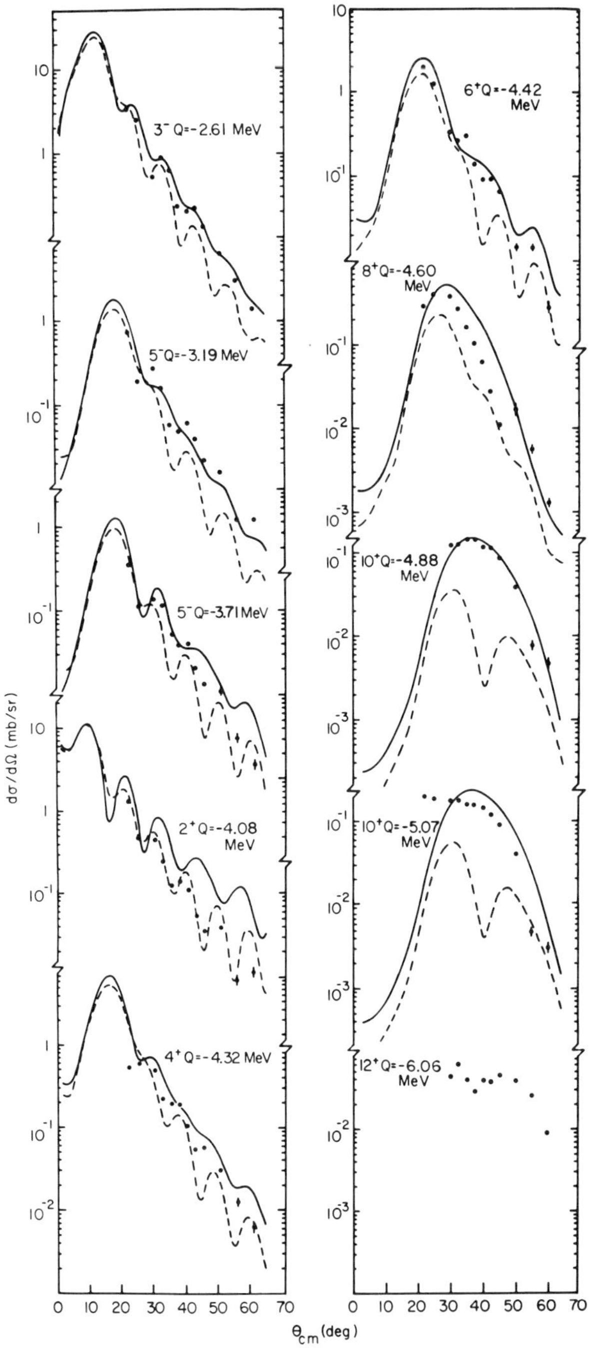

Fig. 12.6. Differential cross sections for proton inelastic scattering to various natural parity states in ^{208}Pb at 135.2 MeV. The solid curve (——) depicts the calculations corresponding to the central plus the spin–orbit parts of the interaction of Fig. 12.2. The dashes (– – –) correspond only to central. The shift in importance from central to spin–orbit interaction can be understood in light of Fig. 12.2 and the momentum dependence of the interaction [From Adams *et al.* (1980); Petrovich *et al.* (1980a); Petrovich and Love (1981); Love *et al.* (1983).]

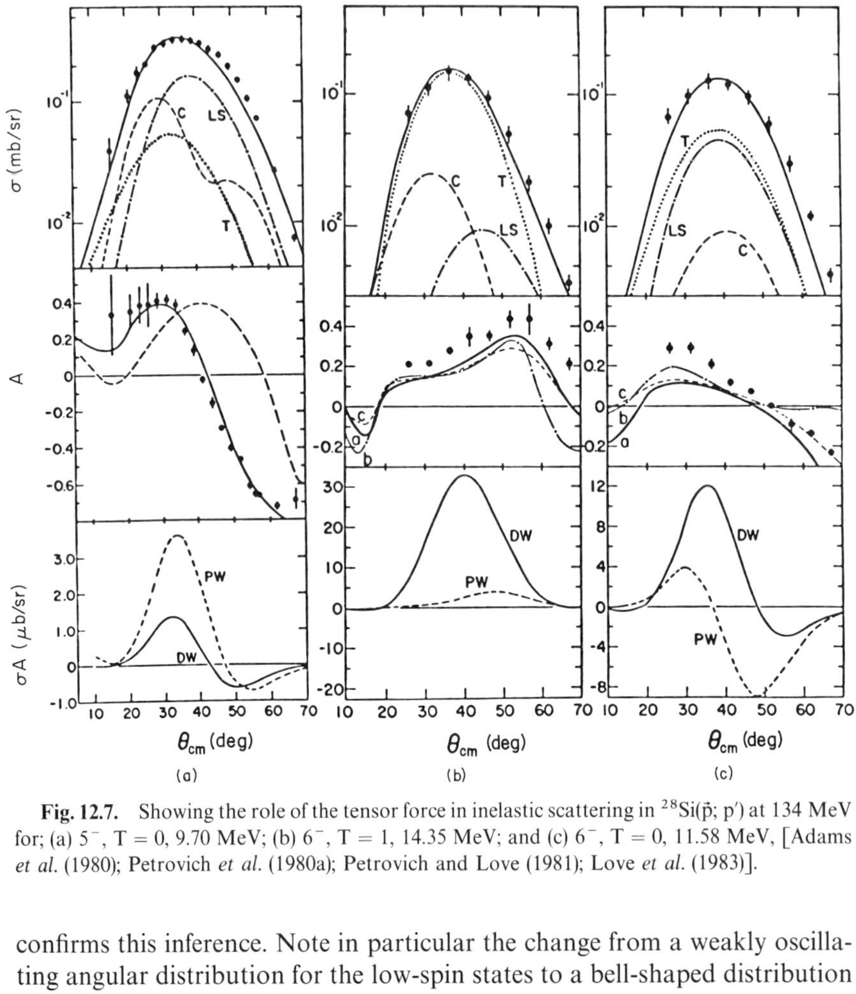

Fig. 12.7. Showing the role of the tensor force in inelastic scattering in $^{28}Si(\vec{p}; p')$ at 134 MeV for; (a) 5^-, T = 0, 9.70 MeV; (b) 6^-, T = 1, 14.35 MeV; and (c) 6^-, T = 0, 11.58 MeV, [Adams *et al.* (1980); Petrovich *et al.* (1980a); Petrovich and Love (1981); Love *et al.* (1983)].

confirms this inference. Note in particular the change from a weakly oscillating angular distribution for the low-spin states to a bell-shaped distribution for the high-spin states.

Although the role of the tensor force in inducing a *D*-state component into the wave function of the deuteron has been well known for many years, its relevance in heavier nuclei has generally been somewhat obscure. The recent studies of high-spin unnatural-parity states of stretched configuration have remedied this situation and provided definitive information about the high-momentum components of the tensor force (Adams *et al.*, 1977; Lindgren *et al.*, 1979; Petrovich *et al.*, 1980b; Yen *et al.*, 1981). Figure 12.7 supports the contention that the tensor force, as deduced from the free *t* matrix, gives a reasonable description of the cross sections for these levels.

As a final example concerned with the theory of the reaction mechanism, we look at very interesting evidence for the density dependence of the

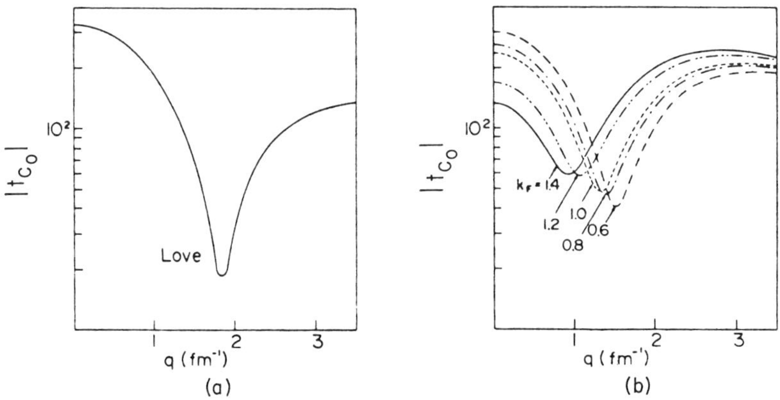

Fig. 12.8. For $E = 135$ MeV, the central part of the interaction appropriate to normal parity transitions is shown for the free t-matrix in part a (Bertsch *et al.*, 1977; Love and Franey, 1981) and for a local density approximation to the G-matrix in part b (Brieva and Rook, 1977, 1978). In the latter case, the local density is indicated by the value of k_F (From Kelley *et al.*, 1980).

effective interaction (Kelley *et al.*, 1980). Recall from the discussion of the G matrix that the intermediate states involved in its evaluation are usually approximated as the plane-wave states of nuclear matter that lie above the Fermi energy as defined by the local nuclear density. The latter is, of course, a function of the position in the nucleus. This is referred to in this context as the local density approximation and causes the G matrix, in applications to finite nuclei, to be a function of density or, equivalently, position. In Fig. 12.8 the central part of the interaction is shown as a function of momentum transfer, as extracted from the free t matrix, and as calculated by Brieva and Rook in the local density approximation to G. In the latter case it is labeled

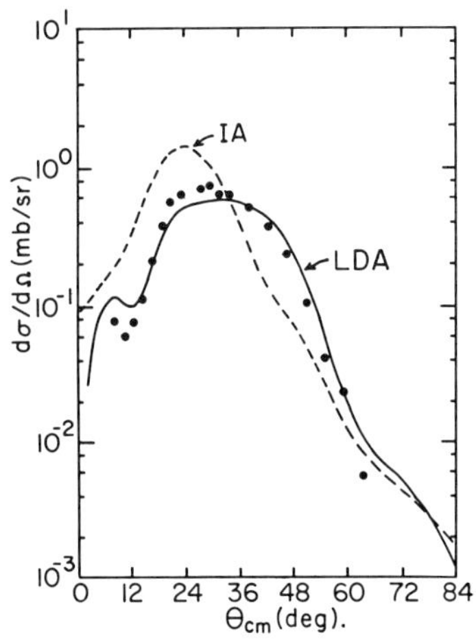

Fig. 12.9. The differential cross section at $E = 135$ MeV for the $^{16}O(p, p')$ reaction with 1^-, $T = 0$, $Q = -7.12$ MeV calculated using the free t matrix, denoted by IA (for Impulse Approximation), and using a G matrix calculated in the local density approximation, denoted by LDA (From Kelley *et al.*, 1980; Petrovich and Love, 1981; Love *et al.*, 1983).

by various values of k_F. Corresponding to these two interactions, Fig. 12.9 shows calculated cross sections for the 1^-, $T = 0$ state in ^{16}O. Only a qualitative agreement with the data is achieved when the interaction is approximated by the free t matrix. However, much better agreement is obtained with a density-dependent G matrix. The bombarding energy is $E = 135$ MeV, and we may infer that the neglect of the Pauli blocking, implicit when the t matrix is substituted for the G-matrix, is not fully justified at this low an energy.

In this kind of analysis we see the possibility of directly probing the properties of the G matrix through a careful analysis of the inelastic scattering data.

Chapter 13

Scattering from Deformed Rotational Nuclei

The scattering of alpha particles from deformed nuclei, although not directly related to microscopic theories of reactions, is important for several reasons. Historically it provided the first convincing evidence for higher-order components in the deformation of nuclei, and it also provided evidence that a few collective states by themselves contribute important components to the optical potential. The calculation of contributions to the optical potential arising from specific transitions has recently become an area of active research, so it is worth understanding the empirical basis that provided the initial motivation. Finally, the collective description of the inelastic transitions provides a very important basis for the understanding of transfer reactions between deformed nuclei (which is treated in Chapter 15).

A. WAVE FUNCTIONS OF DEFORMED NUCLEI

Strongly deformed nuclei are those that have an energy spectrum corresponding to a rigid rotor

$$E_J = E_0\, J(J+1). \tag{13.1}$$

This relation is obeyed with considerable accuracy by the first few states of many nuclei. In addition, the electromagnetic transition rates are enhanced over single-particle rates, and the relationship of the quadrupole transitions connecting consecutive members of a band are consistent with this interpretation. Of course, the nucleons are not frozen in place. Therefore, the rotor spectrum implies that the nucleon motion within the deformed nuclear potential is frequent enough, compared with the rotational frequency, so that the intrinsic structure is not disturbed by the rotation. It is able to readjust adiabatically to the rotation. This implies that there is a rotational band based on each of the various intrinsic states of nucleonic motion (Bohr and Mottelson, 1965 and 1975).

The wave functions for the rotor are the rotation matrices D^J_{MK} (denoted by Messiah by R), which are functions of the Euler angles defining the orientation of the rotor (Edmonds, 1957). They are eigenfunctions of angular momentum J^2 and the projection of J on the space-fixed Z axis and on the body-fixed 3 axis (denoted by 3 to distinguish from Z),

$$0 = (J^2 - \hbar^2 J(J+1))D^J_{MK} = (J_Z - \hbar M)D^J_{MK} = (J_3 - \hbar K)D^J_{MK}. \quad (13.2)$$

The ground state of even nuclei have $K = 0$, that is, the rotation occurs about an axis perpendicular to the nuclear symmetry axis. In this case the total nuclear wave function can be written as

$$\Phi_{JM} = (\hat{J}/8\pi^2)^{1/2}\, D^J_{MO}\, X_0(\mathbf{A}), \quad (13.3)$$

where $X_0(\mathbf{A})$ describes the internal nucleon motion. The factor in front normalizes the wave function.

B. MULTIPOLE EXPANSION OF THE INTERACTION

The nuclear shape is parametrized by

$$R(\theta', \phi') = R_0\left[1 + \sum_{\lambda\kappa} \alpha^*_{\lambda\kappa}\, Y_{\lambda\kappa}(\theta', \phi')\right] \equiv R_0 + \delta R, \quad (13.4)$$

where θ' and ϕ' are the angle coordinates corresponding to the body-fixed axis. Unlike the case for vibrations, the $\alpha_{\lambda\kappa}$ are constants, rather than dynamical variables. In order that R be real,

$$\alpha^*_{\lambda\kappa} = (-)^\kappa \alpha_{\lambda -\kappa}. \quad (13.5)$$

As already discussed, the interaction of a particle with the nucleus in which the intrinsic state of the nucleus is not changed can be represented by an optical potential. In the case of a deformed nucleus, the potential is expected to follow closely the same shape as the density. The nuclear part of the optical potential can be expanded about the spherical shape through Taylor's theorem as

$$U_{\text{nuc}}[r - R(\theta', \phi')] = U_{\text{nuc}}(r - R_0) + \sum_{n=1}^{\infty} \frac{(-\delta R)^n}{n!} \times \frac{\partial^n}{\partial r^n} [U_{\text{nuc}}(r - R_0)], \tag{13.6}$$

where U has, for example, the parametrization in terms of the Woods–Saxon form mentioned in Chapter 4.

By use of the addition theorem for spherical harmonics we are able to write

$$(\delta R/R_0)^n \equiv \left[\sum_{\lambda\kappa} \alpha_{\lambda\kappa} Y^*_{\lambda\kappa}(\theta'\phi')\right]^n = \sum_{LK} \delta^{(n)}_{LK} Y^*_{LK}(\theta'\phi'), \tag{13.7}$$

where

$$\delta^{(n)}_{LK} = \sum_{\lambda k \Lambda\kappa} \left(\frac{\hat{\lambda}\hat{\Lambda}\hat{L}}{4\pi}\right)^{1/2} \begin{pmatrix} \Lambda & \lambda & L \\ \kappa & k & -K \end{pmatrix} \times \begin{pmatrix} \Lambda & \lambda & L \\ 0 & 0 & 0 \end{pmatrix} \delta^{(n-1)}_{\Lambda\kappa} \alpha_{\lambda k}, \tag{13.8}$$

$$\delta^{(1)}_{\Lambda\kappa} = \alpha_{\Lambda\kappa}, \qquad \hat{\lambda} \equiv 2\lambda + 1.$$

Hence

$$U_{\text{nuc}}(r - R) = U_{\text{nuc}}(r - R_0) + \sum_{LK} N_{LK}(r) Y^*_{LK}(\theta', \phi')$$

$$N_{LK}(r) = \sum_{n=1}^{\infty} \frac{(-R_0)^n}{n!} \delta^{(n)}_{LK} \frac{\partial n}{\partial r^n} [U_{\text{nuc}}(r - R_0)]. \tag{13.9}$$

Because the deformation parameters are small ($\alpha \lesssim 0.3$) for typical nuclei, the convergence is rapid.

The Coulomb potential due to a charge distribution $\rho(\mathbf{r})$ is

$$V_{\text{coul}}(\mathbf{r}') = ZZ'e^2 \int \frac{\rho(r'')}{|r' - r''|} d\mathbf{r}'', \tag{13.10}$$

where the coordinates are referred to the body-fixed frame. Using the expression

$$\frac{1}{|\mathbf{r}' - \mathbf{r}''|} = \sum_{\lambda k} \frac{4\pi}{2\lambda + 1} \frac{r^\lambda_<}{r^{\lambda+1}_>} Y_{\lambda k}(\theta', \phi') Y^*_{\lambda k}(\theta'', \phi''), \tag{13.11}$$

we find, upon carrying out the integrations and further algebra, that

$$V_{\text{coul}}(\mathbf{r}') = V_{\text{coul}}(r) + \sum_{LK} C_{LK}(r) Y^*_{LK}(\theta', \phi'), \tag{13.12}$$

where

$$V_{\text{coul}}(r) = \begin{cases} \dfrac{ZZ'e^2}{r}, & r > R_2, \\[2ex] C_\delta \dfrac{3ZZ'e^2}{2R_c}\left[1 + \dfrac{\delta_0^{(2)}}{\sqrt{4\pi}} - \dfrac{1}{3}\left(\dfrac{r}{R_c}\right)^2\right], & r < R_1, \end{cases}$$

$$C_{LK}(r) = \frac{3ZZ'e^2}{2L+1} C_\delta \begin{cases} \dfrac{r^L}{R_c^{L+1}} I_{LK}, & r < R_2, \\[2ex] \dfrac{R_c^L}{r^{L+1}} O_{LK}, & r > R_1, \end{cases} \tag{13.13}$$

$$C_\delta = \frac{\rho_\delta}{\rho_0} = \left[1 + \frac{3\delta_0^{(2)} + \delta_0^{(3)}}{\sqrt{4\pi}}\right]^{-1},$$

$$I_{LK} = -\sum_{n=1}^{\infty} \frac{(-)^n}{n!} \frac{(L+n-3)!}{(L-2)!} \delta_{LK}^{(n)}, \qquad I_{00} = 0,$$

$$O_{LK} = \sum_{n=1}^{L+3} \frac{(L+2)!}{n!(L+3-n)!} \delta_{LK}^{(n)}, \qquad O_{00} = 0.$$

We have assumed a uniform charge distribution that, for the sphere, has radius R_c. The volumes of a sphere of radius R_c and the spheroid having a shape defined similar to (13.4) are different. The ratio of the corresponding densities defines C_δ, and R_1 and R_2 are the minimum and maximum values of $R(\theta, \phi)$.

The potentials are still expressed in the body-fixed coordinate system, whereas in the scattering problem they are needed in the laboratory fixed frame. The transformation is achieved through the rotation functions

$$Y_{LK}(\theta'\phi') = \sum_M D_{MK}^L Y_{LM}(\theta, \phi). \tag{13.14}$$

Adding together the nuclear and Coulomb parts of the potential we find

$$U[r - R(\theta', \phi')] = U(r - R_0) + \sum_{LM} Y_{LM}^*(\theta, \phi) \sum_{K \geq O} \times U_{LK}(r) \frac{D_{MK}^{L*} + D_{M-K}^{L*}}{1 + \delta_{KO}}, \tag{13.15}$$

where use was made of the easily proven fact that $U_{LK} = U_{L-K}$ and

$$U_{LK} = N_{LK} + C_{LK}, \qquad U = U_{\text{nuc}} + V_{\text{coul}}. \tag{13.16}$$

C. COUPLED EQUATIONS[‡]

The coupled equations that describe the scattering of an alpha particle from a deformed nucleus are a special case of those discussed in Chapters 8 and 9. In this case the truncated space consists of the lowest few levels of the ground-state rotational band. The excluded channels include the higher members of the ground band, as well as the intrinsic excitations and rotational states built on them. The effective interaction is supposed to carry the effects of this excluded space on the retained channels. Because, within the truncated space, all matrix elements are diagonal in the *intrinsic* coordinates, it is parametrized as a complex deformed optical potential, whose expansion in the deformation parameters is given in the preceding equations.

The channel functions, defined in analogy with Chapter 8, are

$$\begin{aligned} \phi_{\alpha\ell}^{m} &= [Y_{\ell}(\hat{\mathbf{r}})\Phi_0]_{\ell}^{m}, \qquad \text{ground state,} \\ \phi_{\alpha'\ell} &= [Y_{\ell'}(\hat{\mathbf{r}})\Phi_{J'}]_{\ell}^{m}, \qquad \text{excited state.} \end{aligned} \tag{13.17}$$

Note that because both alpha and the nucleus have zero spin in the entrance channel, the total channel angular momentum is $I = \ell$. The coupled equations take the explicit form

$$\left[\frac{\hbar^2}{2m}\left(\frac{d^2}{dr^2} - \frac{\ell'(\ell'+1)}{r^2}\right) - U_0(r) + E_{\alpha'}\right] u_{\alpha'}^{\alpha\ell}(r)$$
$$= \sum_{\alpha'' \neq \alpha'} \langle \phi_{\alpha'\ell}^{m} | U(r-R) | \phi_{\alpha''\ell}^{m} \rangle u_{\alpha''}^{\alpha\ell}(r). \tag{13.18}$$

In the nondiagonal matrix elements that appear on the right side, the spherical part does not contribute. On the left side, only the monopole ($L = 0$) part of (13.15) contributes to the diagonal elements, which we denote by U_0,

$$U_0(r) = U(r - R_0) + \frac{1}{\sqrt{4\pi}} U_{00}(r). \tag{13.19}$$

The matrix elements are a special case of those already evaluated,

$$\langle \phi_{\alpha'\ell}^{m} | U(r-R) | \phi_{\alpha''\ell}^{m} \rangle = \sum_{L} C_{L}^{\ell}(\alpha'\alpha'')\, \mathscr{F}_{L}^{\alpha'\alpha''}(r), \qquad \alpha' \neq \alpha'', \tag{13.20}$$

where

$$\mathscr{F}_{L}^{\alpha'\alpha''}(r) = U_{LK}(r) \langle J'K' \left\| \frac{D_{,K}^{L} + D_{,-K}^{L}}{1 + \delta_{K0}} \right\| J''K'' \rangle. \tag{13.21}$$

‡ See Glendenning (1969a) and Tamura (1965).

In our case, in which only the ground band is considered, and $K = 0$,

$$\mathscr{F}_L^{\alpha'\alpha''}(r) = U_{LO}(r)(\hat{J}'\hat{J}'')^{1/2}\begin{pmatrix} J' & L & J'' \\ 0 & 0 & 0 \end{pmatrix}. \tag{13.22}$$

This result follows from (13.21) according to the properties of the D functions (see Appendix). Finally, the geometric factor is

$$C_L^{\ell}(\alpha'\alpha'') = (-)^{\ell+J'+L}\left(\frac{\hat{\ell}'\hat{\ell}''\hat{L}}{4\pi}\right)^{1/2}\begin{pmatrix} \ell' & L & \ell'' \\ 0 & 0 & 0 \end{pmatrix}\begin{Bmatrix} \ell' & J' & \ell \\ J'' & \ell'' & L \end{Bmatrix}. \tag{13.23}$$

D. RELATION BETWEEN OPTICAL POTENTIAL OF SPHERICAL AND DEFORMED NUCLEI‡

The excitation of rotational states of the ground band of a deformed nucleus is exceptional in the sense that only the diagonal *intrinsic* matrix elements of the effective interaction are involved, in contrast to vibrations or other nuclear states where one state is distinguished from the other by differences in the nucleonic states. Of course, such excitations of a deformed nucleus also exist.

As discussed earlier, the elements of the effective interaction $\mathscr{V}$ that are diagonal in the intrinsic structure are very effectively and conveniently parametrized as an optical potential

$$\langle 0|\mathscr{V}|0\rangle = U_{\text{opt}}.$$

Referring to Chapter 8 for the theoretical structure of $\mathscr{V}$, we are able to write schematically, for the *elastic* optical potential,

$$U_{\text{opt}} = \langle 0|\mathscr{V}|0\rangle = \langle 0|V|0\rangle + \sum_Q \langle 0|V|Q\rangle \frac{1}{E - E_Q + i\varepsilon} \langle Q|V|0\rangle,$$

where Q is summed over all other channels. For spherical nuclei these include all the intrinsic excitations, whereas for deformed nuclei they include the rotations as well.

The elastic cross sections and corresponding optical-model parameters for alphas scattered from several spherical and deformed members of the samarium isotopes were shown in Chapter 4. It should be especially noted that the optical potential parameters of the deformed and spherical nucleus are markedly different. We shall now demonstrate that this difference arises because, in addition to the intrinsic excitations that all nuclei possess, deformed nuclei possess rotational states built on them. It is the strong transitions in the ground-state band that produce the difference in parameters. Extracting

‡ See Glendenning (1969a) and Glendenning *et al.*, (1968).

from the sum Q in U_{opt} those terms referring to the rotations of the ground band that are present if the nucleus is deformed, we obtain

$$U_{\text{opt}} = U_{\text{sph}} + \sum_R \langle 0|V|R\rangle \frac{1}{E - E_R + i\varepsilon} \langle R|V|0\rangle,$$

with

$$U_{\text{sph}} = \langle 0|V|0\rangle + \sum_{Q \neq 0 \text{ or } R} \langle 0|V|Q\rangle \frac{1}{E - E_Q + i\varepsilon} \langle Q|V|0\rangle$$

Clearly, this latter quantity U_{sph} is the optical potential that would be appropriate if there were no rotational states, that is, to a spherical nucleus, whereas the sum on R in U_{opt} is the additional contribution to the optical potential arising from the rotations in the case of deformed nuclei.

Now, the key point is that if the coupling to the rotational states is explicitly included in the scattering calculations in the manner of the previously described coupled equations, then the rotational terms R ought not to appear in the corresponding optical potential of the coupled equations. This leaves U_{sph}. In other words, the optical model parameters V, W, r_v, r_w, a_v, and a_w for a deformed and spherical nucleus ought to be essentially the same when the strong collective rotational excitations are explicitly coupled to the elastic channel. Indeed, this was found to be the case. A convenient region in which these ideas can be tested is one that spans a transition from spherical vibrational nuclei to statically deformed nuclei. The collectivity is only moderate in the former but strong in the latter. The samarium isotopes are a good example. Their spectra for the first several levels are shown in Fig. 13.1, illustrating the characteristic level spacing of vibrator and rotor at the extremes.

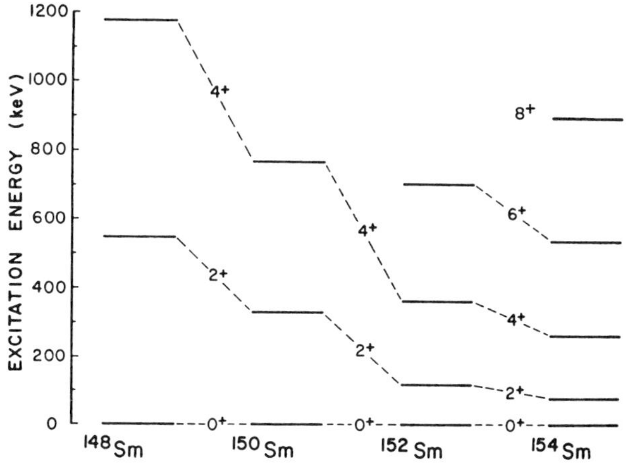

Fig. 13.1. Energy levels of samarium isotopes showing the vibrational spectrum in ^{148}Sm and the rotational spectrum fully developed in ^{152}Sm and ^{154}Sm.

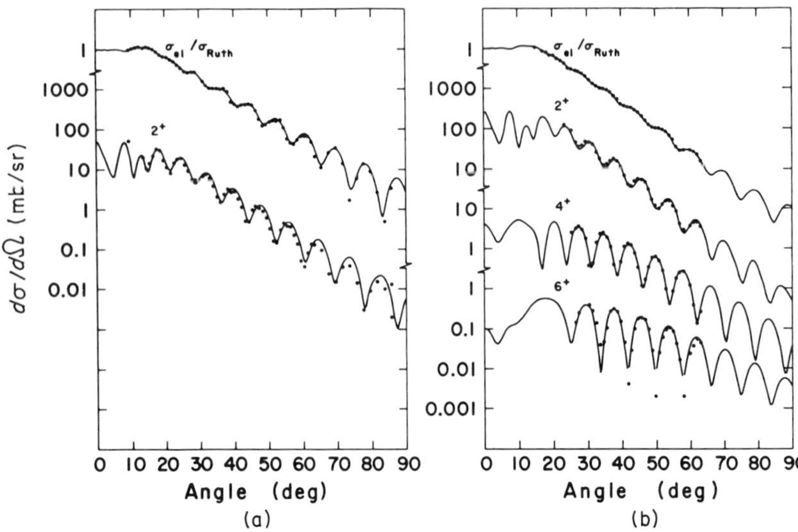

Fig. 13.2. Elastic and inelastic scattering of 50-MeV alpha particles by the spherical ^{148}Sm (a) and deformed ^{154}Sm (b) nucleus. Solid curves are coupled-channel calculations of cross sections based on a vibrational description of 148 and a rotational description of 154. In each case the *same* optical potential was used (Table 1) even though the elastic cross sections are different. Shape parameters β_λ for each nucleus are: for (a) $\beta_2 = 0.11$; for (b) $\beta_2 = 0.225$, $\beta_4 = 0.045$, $\beta_6 = -0.015$ (from Glendenning *et al.*, 1968).

The results of two calculations, one for the spherical ^{148}Sm nucleus and one for the deformed ^{154}Sm nucleus, are shown in Fig. 13.2, and the optical-model parameters are shown in Table I. The calculation was actually performed with an additional refinement. The collective vibrational 2^+ state of ^{148}Sm was also included explicitly, thus eliminating its contribution to the optical potential for this nucleus. It is found that the same potential can be used for both nuclei when the coupling to the collective states is included, even though the elastic cross sections, because of the coupling, are markedly different. When the elastic cross sections alone are calculated, thus requiring the optical parameters to account for the different cross sections, then as

TABLE I

Optical-Model Parameters for the Woods–Saxon Shape Used in the Coupled Channel Calculations.

V	W	$r_v = r_w$	$a_v = a_w$
−65.9	−27.3	1.440	0.637

seen in Table I of Chapter 4, the *elastic* optical parameters *are* markedly different in the deformed case but show little difference in the weaker vibrational case.

In this way we have seen that a few strongly collective levels can make significant contributions to the optical potential, which otherwise accounts implicitly for the enormous number of open channels in a typical situation. In the following chapter we shall discuss how the contribution of a specific collective level to the optical potential can be calculated.

E. ALPHA SCATTERING FROM DEFORMED RARE-EARTH NUCLEI‡

The deformed rare-earth nuclei are believed to have symmetry about their intrinsic 3 axis, so that the deformation parameters $\alpha_{\lambda k}$ exist only for $k = 0$. We use the conventional abbreviation

$$\alpha_{\lambda k} = \delta_{k0}\beta_{\lambda}.$$

Alpha particles are an excellent probe of the nuclear surface and hence the nuclear shape. This is so because they tend to be broken up if they enter the nucleus, and therefore, alpha particles that survive a collision must have interacted predominantly in the nuclear surface. Moreover, their wave number $k = (2\mu E/\hbar^2)^{1/2}$ is twice as large as for a nucleon of the same energy, so that it carries twice the angular momentum for a surface collision and has half the wave length. The higher angular momentum is useful in probing higher multipoles L in the interaction. The short wavelength gives radial sensitivity.

First, it is established by Fig. 13.3 that the DWBA is inadequate to account for the cross sections. In this approximation only the direct excitation from the ground state is calculated. The square of the deformation constant β_λ scales the magnitude of the cross section but has no effect whatsoever on the angular distribution in the distorted-wave Born approximation, as is seen explicitly in (7.63). This is so because the approximation is of first order in the transition interaction. The sign of β cannot be determined at this order. In contrast, when the coupling of the channels is taken into account, the cross section to any particular channel becomes a coherent sum of all amplitudes for processes that lead to it.§ Thus, for example, the 4^+ level

‡ See Glendenning (1969a) and Hendrie *et al.*, (1968).

§ We use the language of perturbation theory for convenience. The solution to the coupled equations automatically and implicitly sums all orders of the interaction.

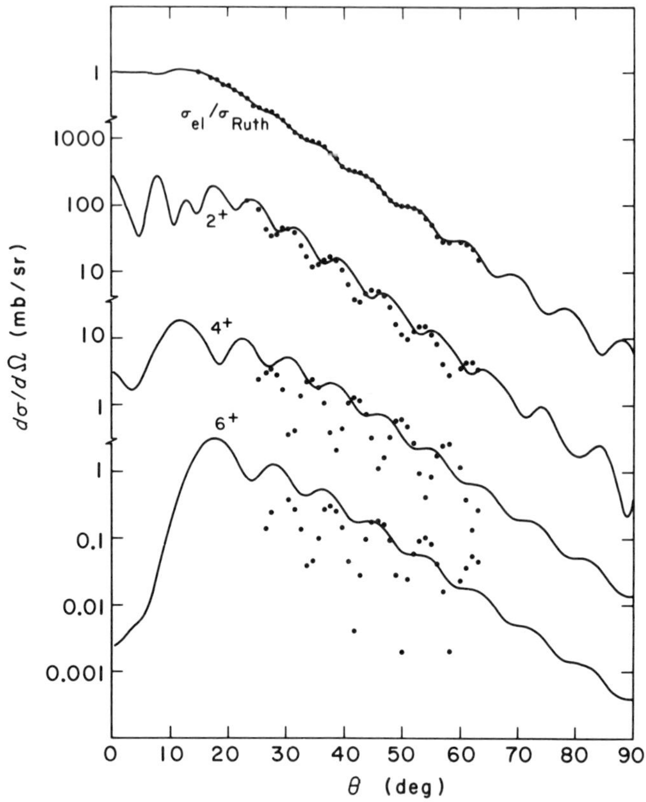

Fig. 13.3. Distorted-wave calculation with optical parameters that fit the elastic cross section as shown (listed in Chapter 4). $\beta_2 = 0.3$, $\beta_4 = 0.15$, $\beta_6 = 0.075$. DWBA, ^{154}Sm (from Glendenning, 1969a).

can be excited directly from the ground state through the term in the interaction proportional $\beta_4 Y_4$ [see Eqs. (13.6) and (13.7)] or indirectly by a cascade transition through the 2^+ state. The $0^+ \to 2^+$ transition involves β_2, whereas the $2^+ \to 4^+$ involves β_2, β_4, or β_6. Thus, this particular cascade transition involves β_2^2, $\beta_2\beta_4$, and $\beta_2\beta_6$ in lowest order. The coherent dependence on all amplitudes leading to a particular state can drastically alter the cross section of that state. This is indeed found to be so for rotational levels, for which the coupling is strong in this sense. Figure 13.4 shows the excellent agreement obtained when all possible couplings among the rotational levels are included (with all expansions computed to convergence). The deformation parameters that were adjusted to achieve the agreement are listed in the legend. The optical-model parameters *need not* be regarded as

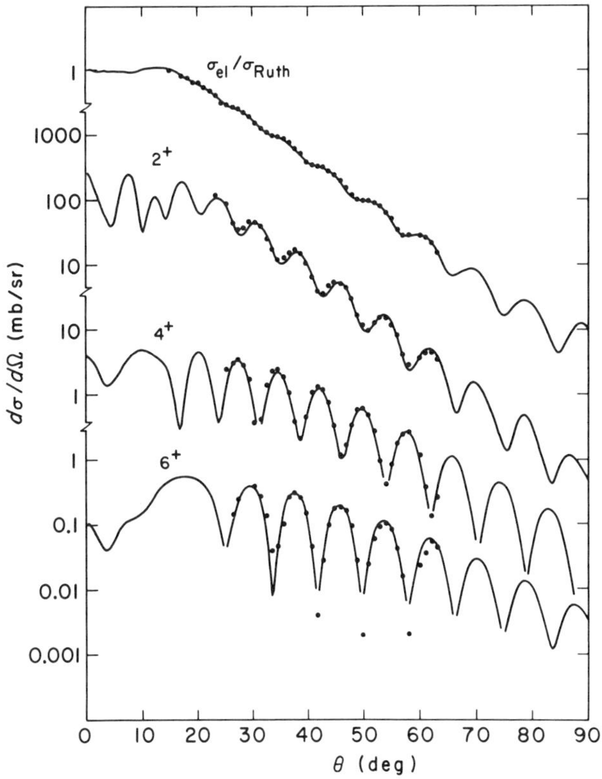

Fig. 13.4. Cross sections for 50-MeV alpha-excitation ground-state rotational band of ^{154}Sm. Curves are coupled-channel calculation as described in text. The data were taken at the Berkeley 88-in. Cyclotron. $\beta_2 = 0.225$, $\beta_4 = 0.05$, $\beta_6 = -0.015$ (from Harvey *et al.*, 1967; Hendrie *et al.* 1968; calculation by Glendenning 1969a).

adjusted parameters because of their determination as parameters of the nearby spherical nuclei, as discussed in the preceding section. The rationale of that section therefore provides very convincing evidence for the accuracy with which the deformation constants are determined.

The sensitivity of the calculation to the deformation parameters is illustrated in Fig. 13.5. It is clear that the wealth of precision data on several levels permits on accurate reading of the deformation parameters. The analysis of some other nuclei is shown in Fig. 13.6. In cases where the 8^+ cross section was measured, β_6 could be determined. (Truncation at $L + 2$ allows accurate determination of β_L.) These analyses provided the first evidence for the existence of a significant hexadecapole ($L = 4$) component in the nuclear shape. The β_4 values for a number of nuclei are shown in

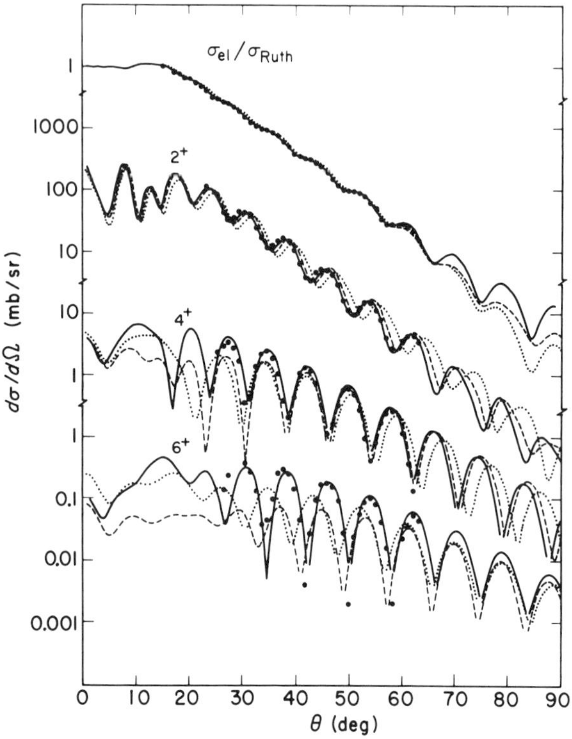

Fig. 13.5. Extreme sensitivity of 2^+, 4^+ and 6^+ to sign and value of β_4 is indicated in 50-MeV alpha scattering from ^{154}Sm. Values of β_4 are: 0.05 (solid curve); 0.0 (dashes); -0.05 (dots) (from Glendenning, 1969a; Hendrie *et al.*, 1968.)

Fig. 13.7. The nuclear shapes associated with both signs of β_4 are illustrated in Fig. 13.8.

These very precise data (Harvey *et al.*, 1967; Hendrie *et al.* 1968) and their analysis as detailed here permitted the first confirmation and determination of hexadecapole deformation in nuclear shapes. The $\lambda = 6$ deformation is less precisely determined, but its presence is also established.

Further corroboration of the physical interpretation just given was obtained through similar measurements made at other energies (Aponick *et al.*, 1970). At these energies, however, the data for the vibrational state in ^{148}Sm were not available. The optical potential parameters were determined for the elastic cross section and therefore carry the burden of implicitly describing the effect of the collective state on the elastic cross section. However, we saw in the case of the 50-MeV data that this did not have a large effect for the spherical nucleus (compare Table I with Table I of Chapter 4).

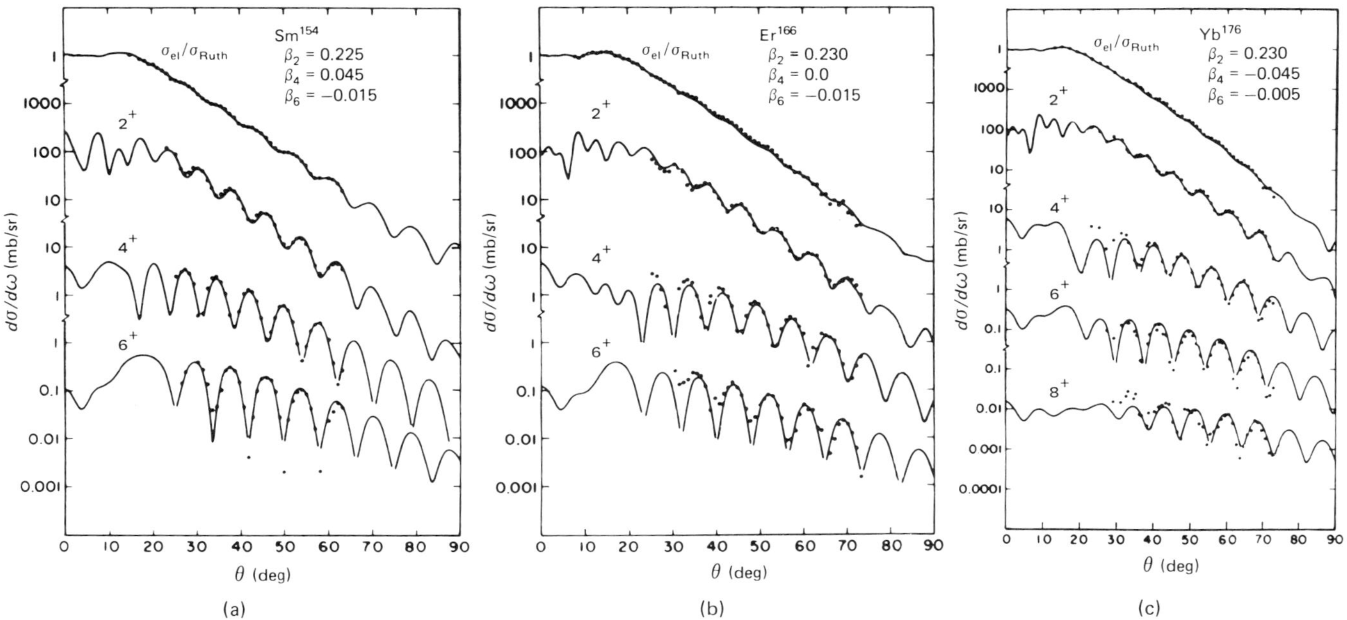

Fig. 13.6. Coupled-channel calculations of the differential cross sections for 50-MeV alpha particles of three nuclei, exhibiting, respectively: positive, zero, and negative values of β_4, are compared with the data. The shape parameters in each case are exhibited in the figure (From Hendrie *et al.*, 1968).

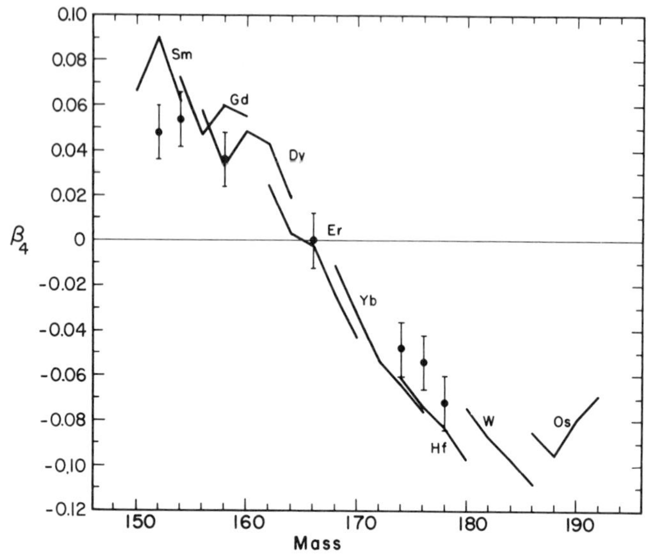

Fig. 13.7. Solid lines indicate the values of β_4 calculated in first-order perturbation theory based on the Nilsson scheme. The values of β_4, multiplied by $(r_0/1.2)$, that were obtained from an analysis of the scattering data are shown by solid dots. The error bars indicate our feeling of the precision with which the parameters can be extracted (from Hendrie *et al.*, 1968).

The elastic data and calculated cross sections are shown in Fig. 13.9 for three energies around 30 MeV and the optical potential parameters are listed in Table II.

The differential cross sections for one isotope, ^{154}Sm, are shown at two energies in Fig. 13.10, again with the optical potential determined by the analysis of data on the spherical nucleus ^{148}Sm. Here too the Coulomb excitation is important. However, separate parameters corresponding to the charge could not be extracted from an analysis of the data because any set consistent with the quadrupole moment was compatible with the data.

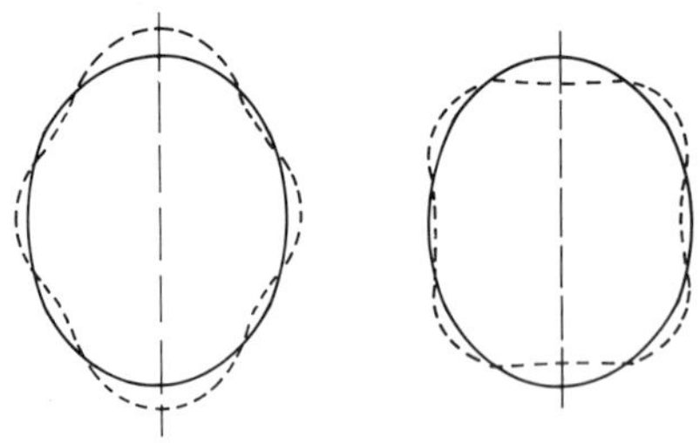

Fig. 13.8. Illustration of two nuclear shapes for $R = R_0[1 + \beta_2 Y_{20} + \beta_4 Y_{40} + \cdots]$. On the left $\beta_4 = 0.2$, on the right $\beta_4 = -.02 = -\beta_2$.

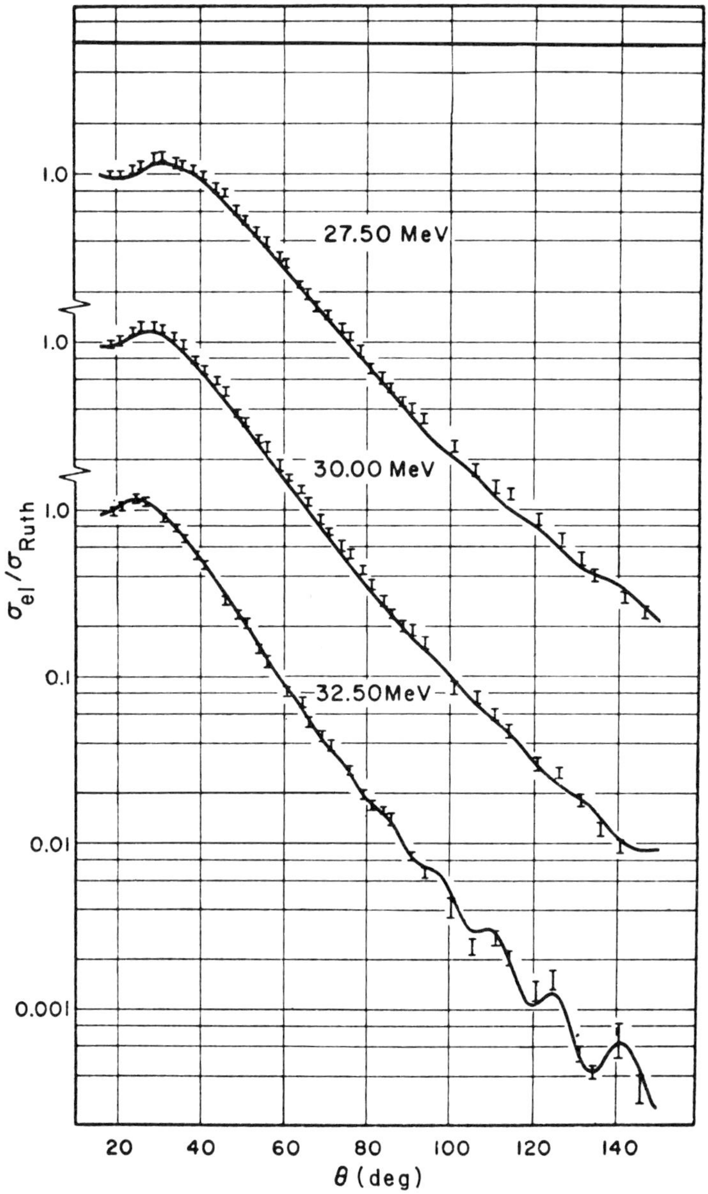

Fig. 13.9. Elastic cross sections for ^{148}Sm (α, α_0) at three energies and the optical-model cross sections (from Aponick *et al.*, 1970).

TABLE II

Optical-Model Parameters for Elastic α-Particle Scattering from ^{148}Sm

V	W	$r_v = r_w$	$a_v = a_w$	r_{coul}
50	11.2	1.492	0.6049	1.45

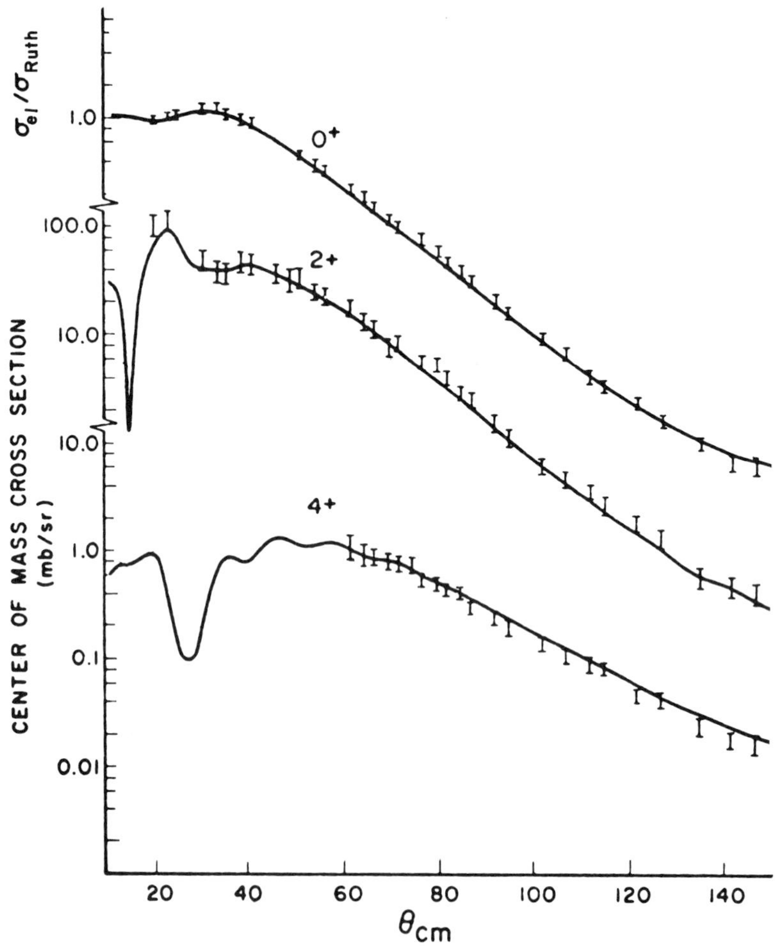

(a)

Fig. 13.10. Cross sections to the ground state rotational band of $^{1.54}$Sm (α, α) at (a) 27.50 MeV and (b) 32.50 MeV energies. [From Aponick *et al.*, 1970.]

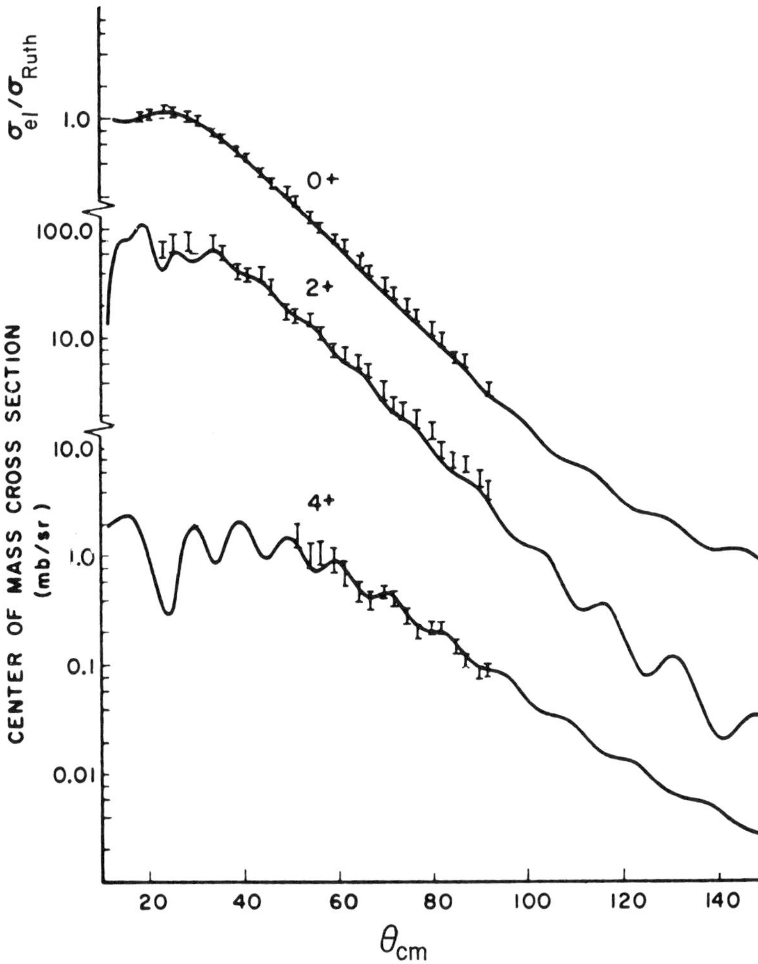

(b)

The deformation parameters determined at these lower energies are consistent within small errors with those corresponding to the 50-MeV data.

Thus, a very consistent picture has emerged. As discussed in Chapter 8, once strongly coupled excited states have been explicitly included in the coupled-channel description of the scattering, their effect on the optical potential is removed and the residual optical potential should be dominated by the high density of levels at high excitation. It should therefore be largely independent of the peculiarities of any particular nucleus and consequently should vary smoothly over large regions of mass. This has had the important consequence in the present situation of separating the determination of the nuclear shape from the determination of the optical potential, thus permitting the first measurement of higher-order components in the optical potential. In a broader context it is an important concept and one that will facilitate attempts to calculate the optical potential from first principles.

Notes to Chapter 13

Rotational nuclei are discussed in great detail in the two volume by Bohn and Mottelson, (1969, 1975).

Chapter 14

Calculation of Specific Components of the Optical Potential

A. INTRODUCTION

The optical potential plays a vital role in any discussion of direct reactions, because it was learned at an early stage that the scattering in the average field acting between target and projectile distorts the relative motion before and after the direct reaction. Although the optical potential was introduced originally as a phenomenological potential (which is complex to account for the partial absorption out of the elastic channel), its theoretical structure, as described in Chapter 8, was later to be appreciated. Moreover, it was even learned (see Chapter 13) that in those nuclei where there are very strong transitions, their contribution to optical potential can be very specific and identifiable (Glendenning *et al.*, 1968; Glendenning, 1969a). In this chapter we discuss recent work aimed at calculating those components in the optical potential that arise from a particular inelastic mode. The possible advantages lie in the fact that, for the most part, the optical potential receives small contributions from an enormous number of channels at typical collision energies. Because of the exponential growth of the density of nuclear levels with energy, the potential, for the most part, is dominated by the multitude of levels at high excitation. We expect that the higher excitations of neighboring nuclei are basically similar; that the spectra differ principally in the low-energy region. The conclusion is that if the optical potentials of neighboring nuclei differ very much from each other, they do so because of strong collective transitions that are different in the two cases (Glendenning *et al.*, 1968; Glendenning, 1969a). It is therefore interesting to isolate their contributions, because the remaining optical potential should be a smoothly varying function of mass and energy, for which systematic variations in the parameters can be identified.

B. POTENTIAL COMPONENTS OF A SINGLE LEVEL‡

Suppose we have truncated the scattering problem to two levels, the ground state and one excited state. In this case the effective interaction, and in particular its diagonal nuclear matrix element which is parametrized as an optical potential, carries the effect of all the eliminated channels on the two remaining levels. For simplicity of notation we consider the scattering of two spinless nuclei, say, an alpha with an even–even nucleus. The coupled equations have the form (Chapter 13)

$$(E_0 - T_\ell - U)u^\ell_{\ell 0}(r) = \sum_{\ell'} ([Y_{\ell'}(\hat{\mathbf{r}})\Phi_{J'}]_\ell | V | [Y_\ell(\hat{\mathbf{r}})\Phi_0]_\ell) u^\ell_{\ell' J'}(r), \tag{14.1}$$

$$(E_{J'} - T_{\ell'} - U)u^\ell_{\ell' J'}(r) = ([Y_\ell(\hat{\mathbf{r}})\Phi_0]_\ell | V | [Y_{\ell'}(\hat{\mathbf{r}})\Phi_{J'}]_\ell) u^\ell_{\ell 0}(r), \qquad \text{for each } \ell'. \tag{14.2}$$

The range of ℓ' is given by angular momentum and parity conservation to be those that satisfy

$$|J' - \ell| \le \ell' \le J' + \ell, \qquad (-)^{\ell'}\pi_{J'} = (-)^\ell \pi_0. \tag{14.3}$$

As always in this context, the matrix elements denote integration of the nuclear coordinates and the angle coordinate $\hat{\mathbf{r}}$ of $\mathbf{r}$, leaving a function of r. Also note that the matrix elements can be written quite generally as

$$([Y_{\ell'}\Phi_{J'}]_\ell | V | [Y_\ell \Phi_0]_\ell) = \sum_L C^\ell_L(J', 0)\mathscr{F}^{J'0}_L(r), \tag{14.4}$$

which follows from the general theory of Chapter 9 [Eq. (9.14)] or the development appropriate to our specialization here of Chapter 13. Here C is a purely geometrical factor and $\mathscr{F}$ the nuclear form factor. This form arises because we can always write (see Chapter 7, Section J)

$$V = \sum_{LSJ} T_{LSJ} \cdot \mathscr{T}_{LSJ}, \tag{14.5}$$

‡ See Rao *et al.* (1973); Coulter and Satchler (1977); and Baltz *et al.* (1978, 1979).

where $\mathscr{T}$ and T are spherical tensors acting on the nuclear coordinate and on the scattered particle, respectively.

Now, consistent with our purpose, we want to calculate the contribution to the *elastic* optical potential of the level J'. This can obviously be done by eliminating the channels referring to this level. In Chapter 5 we derived the Green's function corresponding to the homogeneous counter-part of Eq. (14.2), i.e., (5.8). It is

$$g_{\ell'}(r, r') = -\frac{2m}{\hbar^2}\frac{1}{k'} f_{\ell'}(r_<)h_{\ell'}^{(+)}(r_>). \tag{14.6}$$

Now the solution to Eq. (14.2) can be written as

$$\begin{aligned} u_{\ell' J'}(r) &= \int dr'\, g_{\ell'}(r, r') \sum_L C_L^{\ell}(0, J')\mathscr{F}_L^{0J'}(r')u_{\ell 0}^{\ell}(r') \\ &= -\frac{2m}{k'\hbar^2}\sum_L C_L(0, J')\int dr'\, f_{\ell'}(r_<)h_{\ell'}^{(+)}(r_>)\mathscr{F}_L^{0J'}(r')u_{\ell 0}^{\ell}(r'). \end{aligned} \tag{14.7}$$

This in turn is substituted in the right side of Eq. (14.1) to obtain

$$(E_0 - T_\ell - U)u_{\ell 0}^{\ell}(r) - \int U_\ell(r, r')u_{\ell 0}^{\ell}(r')\, dr' = 0, \tag{14.8}$$

where

$$U_\ell(r, r') = -\frac{2m}{k'\hbar^2}\sum_{\ell' LL'} C_L^{\ell}(J', 0)C_{L'}^{\ell}(0, J')\mathscr{F}_L^{J'0}(r)\mathscr{F}_{L'}^{0J'}(r')f_{\ell'}(r_<)h_{\ell'}^{(+)}(r_>).$$

From its definition (13.23),

$$C_L^{\ell}(0, J') = (-)^{\ell'}\left(\frac{\hat{\ell}'}{4\pi}\right)^{1/2}\begin{pmatrix} \ell & L & \ell' \\ 0 & 0 & 0 \end{pmatrix}\delta_{J'L} = C_L^{\ell}(J', 0). \tag{14.9}$$

Thus

$$U_\ell(r, r') = -\frac{2m}{k'\hbar^2}\mathscr{F}_{J'}^{J'0}(r)\mathscr{F}_{J'}^{0J'}(r')\sum_{\ell'}\frac{2\ell' + 1}{4\pi}\begin{pmatrix} \ell & J' & \ell' \\ 0 & 0 & 0 \end{pmatrix}^2 f_{\ell'}(r_<)h_{\ell'}^{(+)}(r_>). \tag{14.10}$$

Equations (14.8)–(14.10) embody the solution to the problem. Equation (14.8) is the one remaining equation for the elastic channel after all other channels have been eliminated. We can use a simpler notation in this case,

$$(E_0 - T_\ell - U)u_\ell(r) - \int U_\ell(r, r')u_\ell(r')\, dr' = 0, \tag{14.11}$$

where the u_ℓ are the radial functions of a partial-wave expansion.

The contribution to the optical potential U of the eliminated J' level is the nonlocal angular-momentum-dependent potential $U_\ell(r, r')$, which is also explicitly energy dependent through k'.

As a concrete example, consider the vibrational or rotational states. To first order in the deformation, we find from Chapter 13 that

$$\mathscr{F}_L^{0L}(r) = V_L(r) = -R_0\beta_L^N \frac{\partial U}{\partial r} + \frac{3ZZ'e^2}{2L+1}\frac{R_c^L}{r^{L+1}}\beta_L^c, \qquad r > R_c \quad (14.12)$$

for the nuclear and Coulomb contributions to the transition potential.

Sometimes the Coulomb part is expressed in terms of the $B(E\lambda\uparrow)$ of electromagnetic transitions, where our symbol stands for the $0^+ \to 2^+$ transition

$$B(E2\uparrow) = \frac{5}{16\pi} e^2 Q_0^2 = \left(\frac{3}{4\pi} Z'eR_c^2\beta_2^c\right)^2, \quad (14.13)$$

and where we have used

$$eQ_0 \equiv \sqrt{\frac{16\pi}{5}} \int_0^{R(\theta,\phi)} \rho_c r^2 Y_2^0 \, d\mathbf{r} = \sqrt{\frac{16\pi}{5}} \rho_c R_c^5 \beta_2^c + 0(\beta^2),$$

$$Z'e = \int_0^{R(\theta,\phi)} \rho_c \, d\mathbf{r} = \frac{4\pi}{3} R_c^3 \rho_c.$$

Then for quadrupole transitions, (14.12) can be written

$$V_2(r) = -\beta_2^N R_0 \frac{dU}{dr} + \sqrt{B(E2\uparrow)}\,\frac{4\pi}{5}\, Ze\, \frac{1}{r^3}. \quad (14.14)$$

C. TRIVIALLY EQUIVALENT LOCAL POTENTIAL

Although nonlocal equations can be solved numerically, it is sometimes useful to define the local equivalent potential

$$U_\ell(r) \equiv \frac{1}{u_\ell(r)} \int dr'\, U_\ell(r, r') u_\ell(r'). \quad (14.15)$$

Obviously the Schrödinger equation

$$(E_0 - T_\ell - U - U_\ell) u_\ell(r) = 0 \quad (14.16)$$

is equivalent to Eq. (14.11). Of course, to evaluate this potential U_ℓ exactly, one needs to know the solution u_ℓ itself. Equation (14.16) is circular, but it can be used as the basis for an iteration scheme to find solutions of the nonlocal problem. As the first step in the iteration, solve

$$(E_0 - T_\ell - U)u_\ell^{(0)} = 0$$

and evaluate

$$U_\ell^{(1)}(r) = \frac{1}{u_\ell^{(0)}(r)} \int dr' \, U_\ell(r, r') u_\ell^{(0)}(r').$$

Next solve

$$(E_0 - T_\ell - U - U_\ell^{(1)})u_\ell^{(1)} = 0$$

and evaluate

$$U_\ell^{(2)}(r) = \frac{1}{u_\ell^{(1)}(r)} \int dr' \, U_\ell(r, r') u_\ell^{(1)}(r')$$

Continue this iteration until consecutive solutions $u_\ell^{(n-1)}(r)$ and $u_\ell^{(n)}(r)$ are sufficiently close to each other. Assuming that convergence is obtained, one then has a numerical solution to the nonlocal Schrödinger equation, and one also has the trivially equivalent local potential $U_\ell(r)$.

D. LONG-RANGE ABSORPTION DUE TO COULOMB EXCITATION‡

Perhaps one of the most dramatic contributions to the optical potential is due to the Coulomb excitation of a collective state. In the first place, collective transitions are strong. In the second, the Coulomb multipole potentials have a long range, namely, $1/r^{L+1}$ [see Eq. (14.12)]. Thus particles can be removed from the elastic channel at long range through the Coulomb excitation of collective states. Noting Eq. (14.14), the nonlocal potential in the elastic channel arising from such a process involving a 2^+ excited state is

$$U_\ell(r, r') = -\frac{2m}{k'\hbar^2}\frac{4\pi}{25} B(E2\uparrow)(Ze)^2 \frac{1}{r^3}\frac{1}{r'^3} \times \sum_{\ell'} (2\ell' + 1)\begin{pmatrix} \ell & 2 & \ell' \\ 0 & 0 & 0 \end{pmatrix}^2 F_{\ell'}(r_<)H_{\ell'}^{(+)}(r_<). \qquad (14.17)$$

At this point we have replaced the radial functions f and h by their counterparts when the Coulomb force alone is present. The neglect of nuclear distortions in the propagator will be most valid for energies below the Coulomb

‡ See Baltz *et al.*, (1978, 1979).

barrier or for high partial waves whose classical turning point lies outside the range of the nuclear force.

Now we calculate, in the spirit of a perturbation theory, a local equivalent potential:

$$\begin{aligned}
U_\ell(r) &= \frac{1}{F_\ell(r)} \int dr'\, U_\ell(r, r') F_\ell(r') \\
&= -\frac{2m}{k'\hbar^2} \frac{4\pi}{25} (Ze)^2 B(E2\uparrow) \frac{1}{r^3} \sum_{\ell'} (2\ell' + 1) \begin{pmatrix} \ell & 2 & \ell' \\ 0 & 0 & 0 \end{pmatrix}^2 \\
&\times \left[i \frac{F_{\ell'}(r)}{F_\ell(r)} \int_0^\infty dr'\, F_{\ell'}(r') \frac{1}{r'^3} F_\ell(r') \right. \\
&+ \frac{F_{\ell'}(r)}{F_\ell(r)} \int_r^\infty dr'\, G_{\ell'}(r') \frac{1}{r'^3} F_\ell(r') \\
&+ \left. \frac{G_{\ell'}(r)}{F_\ell(r)} \int_0^r dr'\, F_{\ell'}(r') \frac{1}{r'^3} F_\ell(r') \right].
\end{aligned} \tag{14.18}$$

The last two terms, which are real, are small because G and F oscillate out of phase in the asymptotic region (4.12). Their insignificance compared with the first has been verified by computer calculation. Therefore they will be neglected. This leaves a pure imaginary potential. Writing out the sum on ℓ' and neglecting the excitation energy so that $k' = k$,

$$\begin{aligned}
U_\ell(r) = &-i \frac{2m}{k\hbar^2} \frac{4\pi}{25} (Ze)^2 B(E2\uparrow) \frac{1}{r^3} \\
&\times \left[(2\ell + 5) \begin{pmatrix} \ell & 2 & \ell + 2 \\ 0 & 0 & 0 \end{pmatrix}^2 \frac{F_{\ell+2}(r)}{F_\ell(r)} \int_0^\infty dr\, F_{\ell+2}(r) \frac{1}{r^3} F_\ell(r) \right. \\
&+ (2\ell + 1) \begin{pmatrix} \ell & 2 & \ell \\ 0 & 0 & 0 \end{pmatrix}^2 \int_0^\infty dr\, F_\ell(r) \frac{1}{r^3} F_\ell(r) \\
&+ \left. (2\ell - 3) \begin{pmatrix} \ell & 2 & \ell - 2 \\ 0 & 0 & 0 \end{pmatrix}^2 \frac{F_{\ell-2}(r)}{F_\ell(r)} \int_0^\infty dr\, F_{\ell-2}(r) \frac{1}{r^3} F_\ell(r) \right].
\end{aligned} \tag{14.19}$$

In the semiclassical approximation, $\ell + \frac{1}{2} \to \bar{\ell}$ and

$$\begin{aligned}
&(2\ell + 5) \begin{pmatrix} \ell & 2 & \ell + 2 \\ 0 & 0 & 0 \end{pmatrix}^2 = (2\ell - 3) \begin{pmatrix} \ell & 2 & \ell - 2 \\ 0 & 0 & 0 \end{pmatrix}^2 = \frac{3}{8}, \\
&(2\ell + 1) \begin{pmatrix} \ell & 2 & \ell \\ 0 & 0 & 0 \end{pmatrix}^2 = \frac{1}{4}.
\end{aligned} \tag{14.20}$$

There are closed forms for the integrals that in the same approximation, are (see Baltz *et al.*, 1979, for more detail)

$$\int_0^\infty dr\, F_{\ell+2}(r)\frac{1}{r^3}F_\ell(r) = \frac{k^2}{6}\frac{1}{\bar{\ell}^2+\eta^2},$$
$$\int_0^\infty dr\, F_{\ell-2}(r)\frac{1}{r^3}F_\ell(r) = \frac{k^2}{6}\frac{1}{\bar{\ell}^2+\eta^2}, \tag{14.21}$$
$$\int_0^\infty dr'\, F_\ell(r)\frac{1}{r^3}F_\ell(r) = \frac{k^2}{2\bar{\ell}^2}\left(1-\frac{\eta}{\bar{\ell}}\arctan\frac{\bar{\ell}}{\eta}\right).$$

Now we have

$$U_\ell(r) = -ik\frac{2m}{\hbar^2}\frac{4\pi}{25}(Ze)^2B(E2\uparrow)\frac{1}{r^3}\left[\frac{1}{8\bar{\ell}^2}\left(1-\frac{\eta}{\bar{\ell}}\arctan\frac{\bar{\ell}}{\eta}\right)\right.$$
$$\left.+\frac{1}{16(\bar{\ell}^2+\eta^2)}\left(\frac{F_{\ell+2}(r)}{F_\ell(r)}+\frac{F_{\ell-2}(r)}{F_\ell(r)}\right)\right]. \tag{14.22}$$

The recurrence relations for the Coulomb function can be used to obtain

$$\frac{F_{\ell+2}}{F_\ell}+\frac{F_{\ell-2}}{F_\ell} = -2+\frac{4}{\bar{\ell}^2+\eta^2}\left[\eta+\frac{\bar{\ell}}{kr}\right]. \tag{14.23}$$

This yields the remarkable result

$$U_\ell(r) = -i\frac{2m}{k\hbar^2}\frac{\pi}{50}(Ze)^2B(E2\uparrow)$$
$$\times\left[\left(\frac{\eta^2k^2(3\bar{\ell}^2+\eta^2)}{\bar{\ell}^2(\bar{\ell}^2+\eta^2)^2}-\frac{\eta k^2}{\bar{\ell}^3}\arctan\frac{\bar{\ell}}{\eta}\right)\frac{1}{r^3}\right.$$
$$\left.+\frac{4\eta k\bar{\ell}^2}{(\bar{\ell}^2+\eta^2)^2}\frac{1}{r^4}+\frac{2\bar{\ell}^4}{(\bar{\ell}^2+\eta^2)^2}\frac{1}{r^5}\right]. \tag{14.24}$$

Thus the locally equivalent contribution to the optical potential arising from the Coulomb excitation of a collective quadrupole state with $B(E2)$ is specifically angular-momentum-dependent and has a simple radial dependence with terms $1/r^3$, $1/r^4$, and $1/r^5$.

Love *et al.* (1977) have carried out the evaluation of the corresponding potential neglecting both the Coulomb distortion and the nuclear distortion. In this approximation the potential has a simple $1/r^5$ dependence and is independent of angular momentum. We have compared this approximation with our ℓ-dependent potential in Fig. 14.1. Interestingly, their approximation

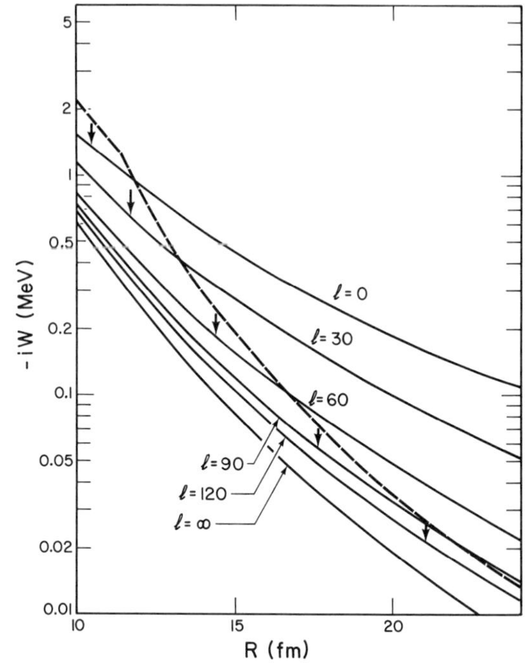

Fig. 14.1. The ℓ-dependent potential of $^{18}O + {}^{184}W$, at 90 MeV plotted for several values of ℓ (———). The Love-Terasawa-Satchler ℓ-independent potential (– – –) tracks it near the classical turning points indicated by the arrows (From Baltz *et al.* 1979.)

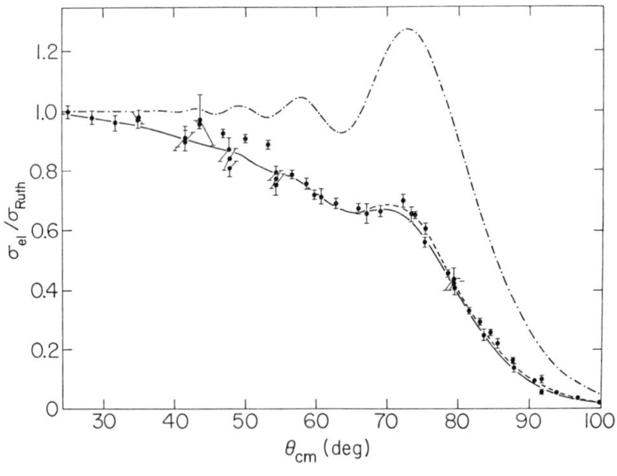

Fig. 14.2. The elastic cross section for $^{18}O + {}^{184}W$ at 90 MeV, where ℓ-dependent potential (———), LTS potential (– – –), and no long-range adsorption (–·–), compared with the theory. It is the long-range absorption that causes the fall off prior to the Fresnel peak, which is usually a fluctuation above 1. (Data is from Brookhaven, Calculations by Baltz *et al.*, 1979).

resembles the ℓ dependence of ours near the classical turning point of each ℓ-wave. In other words, it is the best possible approximation to an ℓ-dependent potential. It should be noted that the $1/r^5$ term in our potential dominates as $\ell \to \infty$, not the $1/r^3$ term. Thus the range is essentially $1/r^5$.

In semiclassical approximation it turns out to be possible to evaluate in closed form the contribution to the phase shifts of the potential $U_\ell(r)$. Employing Frahn's strong absorption formula for elastic scattering, a closed form can be obtained for the cross section (Baltz *et al.*, 1978, 1979).

Alternatively, $U_\ell(r)$ can be added to the usual optical potential and the corresponding scattering problem solved on a compter.

In Fig. 14.2 we compare the elastic cross section for the scattering of ^{18}O from ^{184}W, in which the long-range absorption due to the quadrupole excitation of ^{184}W was incorporated through the trivially equivalent local potential $U_\ell(r)$. The most remarkable aspect of the data, its strong deviation from unity before the Fresnel peak at 80° is reached, is very well reproduced.

Chapter 15

Two-Nucleon Transfer Reactions

A. CONTRAST BETWEEN ONE- AND TWO-NUCLEON TRANSFER REACTIONS

One-nucleon transfer reactions probe the single-particle structure of nuclear states. The angular distribution is sensitive to the orbital angular momentum of the state into which the nucleon is transferred. The magnitude of the cross section determines the spectroscopic factor, which measures the degree to which the nuclear state is a pure single-particle state. For two-nucleon transfer reactions, the angular distribution again is sensitive to the angular momentum transferred in the reaction. This must always be so when the direct single-step transition dominates a reaction, according to the general arguments given in Chapter 2. It is a reflection of the conservation of angular momentum. However, here the angular momentum is carried by a pair of nucleons, so that it does not directly reflect the angular momentum of the single-particle states into which the nucleons are transferred. Only to the extent that the two single-particle angular momenta must sum to the transferred angular momentum is there a constraint. Obviously, the angular momentum of the pair generally can be shared between them in many different ways, and nothing in the measurement of the transferred angular momentum distinguishes between these. Therefore, all such ways that the angular momentum can be shared, consistent with the structure of the nuclear states connected by the reaction, must contribute coherently to the reaction. This coherence can produce large cross sections in states for which it is constructive and very small ones in states for which it is destructive. The coherence depends on the correlation between the two nucleons—the degree to which it is similar in the two nuclear states between which they are transferred. Correlations in the motion of a pair of nucleons inside a nucleus depend on two factors. The angular momentum and parity of the nuclear state impose a certain minimum correlation because the motion of the nucleons must be consistent with these conserved quantities. We refer to this as a static correlation. In addition to this, the nucleon–nucleon interaction will induce spatial and spin correlations. We refer to this as dynamical correlation. In the language of the shell model, it is the interaction that is responsible for configuration mixing in the nuclear wave functions. Consequently, the two-nucleon transfer reactions provide a means of testing nuclear wave functions in details not accessible to single-particle transfer reactions. The role of correlations, an important feature of multi-nucleon transfer reactions, will be taken up again later in this chapter. Some of the early work on two-nucleon transfer can be found in the literature.[‡] Here we shall use the development by Glendenning (1963, 1965).

B. TRANSFER FROM A LIGHT-ION PROJECTILE

Nucleon transfer reactions in general are a valuable spectroscopic tool because the angular distribution is a rather direct signature of the angular momentum transferred in the reaction. In addition, two-nucleon transfer reactions have the special feature of being very sensitive to pair correlations as discussed above. These reactions were discovered by Cerny and Harvey to be highly selective, populating only a few nuclear states with appreciable cross section (Harvey and Cerny, 1960; Cerny *et al.*, 1962; Harvey *et al.*, 1962). The selectivity depends, of course, on the similarity of the pair correlation in the nuclear states that are connected by the reaction. Because one knows the nature of the correlation in light ions rather well, reactions that involve a light ion as one of the reaction partners are especially important, for in such cases one learns which of the states in the heavy nucleus possesses this correlation.

In this chapter we shall concentrate our discussion on those cases in which one of the reaction partners has mass number 4 or less. The general

[‡] See El Nadi (1957), Newns (1960), Glendenning (1962, 1963, 1965), Yoshida (1962), Henley and Yu (1964), and Lin and Yoshida (1964).

case is similar in most ways but is deferred to a later chapter, where it will be used as an example in handling a finite-range direct interaction.

We denote the reaction by

$$\underbrace{(\mathrm{b}+2)}_{\mathrm{a}}+\mathrm{A}\rightarrow \mathrm{b}+\underbrace{(\mathrm{A}+2)}_{\mathrm{B}}. \tag{15.1a}$$

With the restriction $\mathrm{a} \leq 4$, the nucleus a must be ^{4}He, ^{3}He, or ^{3}H, in which case b is a deuteron, neutron, or proton. The amplitude for the reaction (15.1) or its inverse are, of course, trivially related. We now calculate the amplitude for the transition

$$(J_{\mathrm{a}}M_{\mathrm{a}}T_{\mathrm{a}}\mu_{\mathrm{a}})+(J_{\mathrm{A}}M_{\mathrm{A}}T_{\mathrm{A}}\mu_{\mathrm{A}})\rightarrow(J_{\mathrm{b}}M_{\mathrm{b}}T_{\mathrm{b}}\mu_{\mathrm{b}})+(J_{\mathrm{B}}M_{\mathrm{B}}T_{\mathrm{B}}\mu_{\mathrm{B}}), \tag{15.1b}$$

where J denotes the total angular momentum and T the isospin with projections M and μ, respectively. According to Chapter 5, the distorted-wave Born approximation of the amplitude is

$$\mathscr{T}_{\beta\alpha}=W\int\chi_{\mathrm{b}}^{(-)*}(\mathbf{r}_\beta)\{\Phi_{\mathrm{B}}\Phi_{\mathrm{b}}|V|\Phi_{\mathrm{A}}\Phi_{\mathrm{a}}\}\chi_\alpha^{(+)}(\mathbf{r}_\alpha)\,d\mathbf{r}\,d\mathbf{R}\,d\boldsymbol{\rho}\,d\mathbf{b}\,d\mathbf{A}, \tag{15.2a}$$

$$W=\binom{\mathrm{A}+2}{2}^{1/2}\binom{\mathrm{a}}{2}^{1/2}, \tag{15.2b}$$

where the curly braces denote an integration over all spin and isospin coordinates. Recall that W counts the number of direct integrals that occur because of the total antisymmetrization of the wave functions. The integral in (15.2a) is a typical direct one, referring to a specific distribution of nucleon coordinates in the wave functions. We label the transferred pair 1 and 2. The integration is over the relative coordinates, which were discussed in Chapter 5 in connection with the multipole expansion of the amplitude. Recall that $\mathbf{b}$ and $\mathbf{A}$ denote the $3(b-1)$ and $3(A-1)$ intrinsic coordinates of b and A, r is the intrinsic coordinate of the transferred pair, and $\mathbf{R}$ and $\boldsymbol{\rho}$ are the convenient additional coordinates needed to make the complete set of $3(\mathrm{A}+\mathrm{a}-1)$ intrinsic coordinates. These coordinates are illustrated in Fig. 15.1. The direct

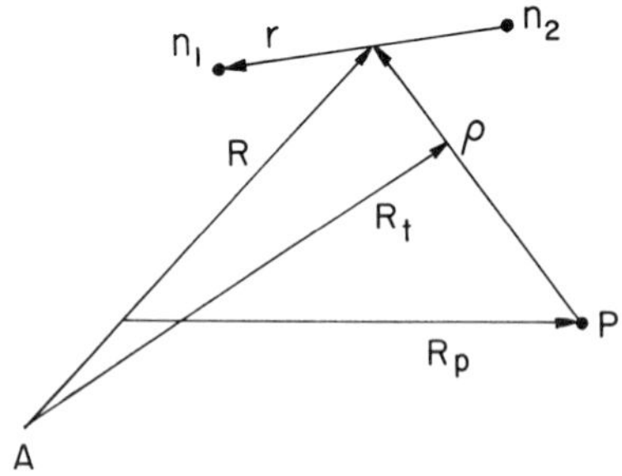

Fig. 15.1. Coordinates for the p, t reaction.

interaction in the amplitude is

$$\begin{aligned} V &= V_\beta - U_\beta = V_{\mathrm{bB}} - U_\beta = V_{1\mathrm{b}} + V_{2\mathrm{b}} + V_{\mathrm{Ab}} - U_\beta \\ &\approx V_{1\mathrm{b}} + V_{2\mathrm{b}} = \sum_{i \in \mathrm{b}} (V_{1i} + V_{2i}). \end{aligned} \tag{15.3}$$

To carry out the evaluation of the amplitude, we make a two-particle parentage expansion for the heaviest nucleus B:

$$\Phi_{\mathrm{B}}(\mathrm{A}+2) = \sum_{\substack{j_1 j_2 JT \\ J'_{\mathrm{A}} T'_{\mathrm{A}}}} \beta_{j_1 j_2 JT}(\mathrm{B}, \mathrm{A}') \mathscr{A}[\Phi_{J'_{\mathrm{A}} T'_{\mathrm{A}}}(\mathrm{A}) \Phi_{(j_1 j_2)JT}(1,2)]^{M_{\mathrm{B}}\mu_{\mathrm{B}}}_{J_{\mathrm{B}} T_{\mathrm{B}}}. \tag{15.4}$$

Here $\Phi_{(j_1 j_2)JT}$ denotes a two-particle shell-model state. Equation (15.4) is analogous to the parentage expansion that was used in the evaluation of the one-nucleon transfer amplitude. The $\mathscr{A}$ denotes the antisymmetrization operator. The parentage amplitudes β can be derived explicitly if shell-model wave functions for the nuclear states are available. In some cases they would be simply the configuration mixing amplitudes of a shell-model expansion for the two-particle configuration described by $\Phi_{(j_1 j_2)JT}$. Generally speaking, each nucleus is a special case for which the parentage amplitudes can be derived from an assumed or calculated nuclear wave function. Later in the chapter some examples will be discussed, and others can be found in the literature (Towner and Hardy, 1969; Asquitto and Glendenning, 1970b).

With the aid of the parentage expansion, the brace in (15.2) takes the form

$$\begin{aligned} F(\mathbf{R}, \boldsymbol{\rho}) &\equiv (\Phi_{\mathrm{B}}\Phi_{\mathrm{b}}|V|\Phi_{\mathrm{A}}\Phi_{\mathrm{a}}) \\ &= W \int d\mathbf{A}\, d\mathbf{b}\, d\mathbf{r}\, \{\Phi_{\mathrm{B}}\Phi_{\mathrm{b}}|V|\Phi_{\mathrm{A}}\Phi_{\mathrm{a}}\} \\ &= W \sum_{\substack{j_1 j_2 JT \\ J'_{\mathrm{A}} T'_{\mathrm{A}}}} \beta_{j_1 j_2 JT}(\mathrm{B}, \mathrm{A}') \sum_P \binom{\mathrm{A}+2}{2}^{-1/2} (-)^P \int d\mathbf{A}\, d\mathbf{b}\, d\mathbf{r} \\ &\quad \times \{P[\Phi_{J'_{\mathrm{A}} T'_{\mathrm{A}}}(\mathrm{A}) \Phi_{(j_1 j_2)JT}(1,2)]^{M_{\mathrm{B}}\mu_{\mathrm{B}}}_{J_{\mathrm{B}} T_{\mathrm{B}}} \Phi^{M_{\mathrm{b}}\mu_{\mathrm{b}}}_{J_{\mathrm{b}} T_{\mathrm{b}}}(\mathbf{b}) | V | \Phi^{M_{\mathrm{A}}\mu_{\mathrm{A}}}_{J_{\mathrm{A}} T_{\mathrm{A}}}(\mathrm{A}) \Phi^{M_{\mathrm{a}}\mu_{\mathrm{a}}}_{J_{\mathrm{a}} T_{\mathrm{a}}}(\mathbf{a})\} \\ &= \binom{\mathrm{a}}{2}^{1/2} \sum_{j_1 j_2 JT} \beta_{j_1 j_2 JT}(\mathrm{B}, \mathrm{A}) C^{J_{\mathrm{A}}\ J\ J_{\mathrm{B}}}_{M_{\mathrm{A}} M M_{\mathrm{B}}} C^{T_{\mathrm{A}} T T_{\mathrm{B}}}_{\mu_{\mathrm{A}} \mu \mu_{\mathrm{B}}} f(\mathbf{r}_\alpha, \mathbf{r}_\beta). \end{aligned} \tag{15.5}$$

The bracket { } denotes an integration over spin–isospin variables, as usual, and

$$\begin{aligned} f(\mathbf{R}, \boldsymbol{\rho}) &\equiv (\Phi_{(j_1 j_2)JT}^{\ M\mu}(1,2)\Phi_{\mathrm{b}}|V|\Phi_{\mathrm{a}}) \\ &= \int d\mathbf{b}\, d\mathbf{r}\, \{\Phi_{(j_1 j_2)JT}^{\ M\mu}(1,2)\Phi_{\mathrm{b}}|V|\Phi_{\mathrm{a}}\}. \end{aligned} \tag{15.6}$$

The integration over $\mathbf{b}$, the internal space coordinates of b, is over $\mathbf{r}_{34}$ for $a = 4$, and otherwise is absent because there is no internal space coordinate for the nucleon, here regarded as a fundamental particle. Notice that the

TABLE I

The rms Radius and Size Parameter for Light Nuclei[a]

Variable	^{4}He	^{3}He	^{3}H
$\langle r^2\rangle^{1/2}$	1.61	1.97	1.68
η	0.233	0.206	0.242

[a] The unit of length is fm $= 10^{-13}$ cm.

integration over **A** has selected out of the expansion (15.4) only those parent states in which the "core" nucleons A are in the same state before and after the reaction. This is because the interaction V does not contain any of those coordinates. This is a very important selectivity.

To evaluate (15.6) it is necessary to choose a particular description of the light ions a and b. The nucleons in the light ions He^4, He^3, H^3, and d are dominantly in the lowest $\ell = 0$ state (S-state) of relative motion.‡ The wave function is therefore symmetric in the spatial coordinates and consequently must be antisymmetric in the spin–isospin coordinates. It is a product,

$$\Phi_a = \Phi_a(\text{space})\Phi_a(\text{spin–isospin}). \tag{15.7}$$

For the space part we choose a normalized Gaussian wave function in the relative coordinates, which can be conveniently expressed as a product of harmonic oscillator functions (Glendenning, 1963, 1975a),

$$\begin{aligned}\Phi_a(\text{space}) &= \mathcal{N}\exp\left(-\eta^2 \textstyle\sum r_{ij}^2\right)\\ &= \begin{cases}\phi_{10}(\sqrt{8}\eta\rho)\phi_{10}(2\eta r_{12})\phi_{10}(2\eta r_{34}), & a = 4\\ \phi_{10}(2\eta\rho)\phi_{10}(\sqrt{3}\eta r_{12}), & a = 3.\end{cases}\end{aligned} \tag{15.8}$$

The parameter η is obviously related to the size of nucleus a and can be chosen so the rms radius implied by the wave function agrees with the value determined from electron scattering experiments, for example. Such values are shown in Table I. (See Appendix for properties of oscillator functions.)

The antisymmetric spin–isospin functions for three or four nucleons is

$$\Phi_a(\text{spin–isospin}) = \sum_{ST} A_{ST}[X_S(\sigma_1, \sigma_2)X_{J_b}(\sigma)]_{J_a}^{M_a}[X_T(\tau_1, \tau_2)X_{T_b}(\tau)]_{T_a}^{\mu_a}, \tag{15.9a}$$

‡ (Spectroscopic notation refers to $l = 0, 1, 2$ as S, P, D; the superscript refers to the value of the spin degeneracy $2S + 1$.)

where

$$A_{ST} = \frac{(-)^S}{\sqrt{2}} \frac{1-(-)^{S+T}}{2} \tag{15.9b}$$

$$X_{J_b}(\sigma) = \begin{cases} X_S(\sigma_3, \sigma_4)\, \delta_{J_b S}, & a = 4, \\ X_{1/2}(\sigma_3)\, \delta_{J_b 1/2} & a = 3; \end{cases} \tag{15.10}$$

X_{T_b} is similarly defined. One can verify that this spin–isospin function is antisymmetric under the exchange of any pair of spin–isospin coordinates and that it is normalized to unity.

The wave function for the light nucleus b is assumed to be

$$\Phi_b = \begin{cases} X_1^{M_b}(\sigma_3, \sigma_4) X_0^0(\tau_3, \tau_4) \phi_d(r_{34}) & a = 4 \\ X_{1/2}^{M_b}(\sigma_3) X_{1/2}^{\mu_b}(\tau_3), & a = 3, \end{cases} \tag{15.11}$$

The deuteron is known to have about 95% probability of being in the S state, with the remaining probability being the D state. So we assume with high accuracy, that ϕ_d is an S-state function.

We can now continue the evaluation of (15.6). In Chapter 16 we shall do this in the general case in which V is a spin-dependent and finite-range interaction. However, to bring out most clearly the main physical features of two-nucleon transfer reactions, we assume here that V has zero range. As it turns out, the angular distributions are affected very little by this assumption, although relative cross sections are moderately changed. For the present we make the simplifying assumption that

$$V = g\delta(\boldsymbol{\rho}). \tag{15.12}$$

This zero-range assumption has the effect that

$$\mathbf{r}_\alpha \to \mathbf{R}, \qquad \mathbf{r}_\beta \to \frac{A}{A+2}\mathbf{R}, \qquad \text{for} \quad \boldsymbol{\rho} \to 0. \tag{15.13}$$

Then (15.6) becomes

$$f(\mathbf{R}, \boldsymbol{\rho}) = \Omega_b g \delta(\boldsymbol{\rho}) \phi_{10}(\sqrt{4a - 8}\eta\boldsymbol{\rho}) \sum_{ST} A_{ST} C_{M_S M_b M_a}^{S\ J_b\ J_a} C_{\mu\ \mu_b\ \mu_a}^{T T_b T_a}$$
$$\times \int d\mathbf{r}\, \{\Phi_{(j_1 j_2)JT}^{M\mu}(r, R) | \phi_{10}(\sqrt{a}\eta\mathbf{r}) X_S^{M_S}(\sigma_1, \sigma_2) X_T^{M_T}(\tau_1, \tau_2)\}, \tag{15.14a}$$

where we have shown $\Phi_{(j_1 j_2)JT}$, a function of $\mathbf{r}_1$ and $\mathbf{r}_2$, as transformed to $\mathbf{r}$ and $\mathbf{R}$. We shall discuss the details shortly. Also, we defined

$$\Omega_b = \begin{cases} \int \phi_d^*(\mathbf{r}_{34}) \phi_{10}(2\eta\mathbf{r}_{34})\, d\mathbf{r}_{34}, & a = 4 \\ 1, & a = 3, \end{cases} \tag{15.14b}$$

which is an overlap of the spatial wave function of nucleus b with the corresponding part of the wave function of a.

TABLE II

Consequences of Spin–Isospin Antisymmetry

Reaction	Spin (S)	Isospin (T)
(α, d)	1	0
(t, p), (^{3}He, n)	0	1
(t, n), (^{3}He, p)	1	0
	0	1

C. SPIN–ISOSPIN SELECTION RULES

The selection rules on S and T follow readily from (15.14). The Clebsch–Gordan coefficient on isospin projections requires

$$\mu = \mu_{\mathrm{a}} - \mu_{\mathrm{b}}.$$

The antisymmetry of the spin-isospin function (15.9b) requires that $S + T =$ odd. The consequences are listed in Table II.

D. FORM FACTOR

The two-particle shell-model wave function $\Phi_{j_1 j_2}(1, 2)$ for a pure two-nucleon configuration is a function, of course, of the two space coordinates $\mathbf{r}_1$ and $\mathbf{r}_2$ of the neutrons (as well as their spin–isospin coordinates). A transformation can be carried out to the relative and c.m. coordinates $\mathbf{r}$ and $\mathbf{R}$ of the pair. When this is done, clearly the $\mathbf{r}$ integration in the bracket of Eq. (15.14a) can be performed. It remains a function of R, which we denote by

$$F_{(j_1 j_2)J}{}^{M}{}_{S}{}^{M_S}{}_{T}^{\mu}(\mathbf{R}) = \int \{\Phi_{(j_1 j_2)J}{}^{M}{}_{T}^{\mu} | \phi_{10}(\sqrt{3\eta^2}\mathbf{r}) X_S^{M_S} X_T^{\mu}\} \, d\mathbf{r}. \tag{15.15}$$

The DWBA integral is thus seen to be a certain sum of integrals of the form

$$\int \chi_\beta^{(-)*}\left(\frac{\mathrm{A}}{\mathrm{A}+2}\mathbf{R}\right) F_{(j_1 j_2)J}{}^{M}{}_{S}{}^{M_S}{}_{T}^{\mu}(\mathbf{R}) \chi_\alpha^{(+)}(\mathbf{R}) \, d\mathbf{R}. \tag{15.16}$$

We now calculate this $F(\mathbf{R})$ in detail. The meaning of $\Phi_{(j_1 j_2)}$ is

$$\Phi_{(j_1 j_2)J}{}^{M}{}_{T}^{\mu}(1, 2) = \frac{X_T^{\mu}(1, 2)}{[2(1 + \Delta_{j_1 j_2})]^{1/2}} \{\phi_{(j_1 j_2)J}^{M}(1, 2) + (-)^T \phi_{(j_1 j_2)J}^{M}(2, 1)\}, \tag{15.17}$$

$$\Delta_{j_1 j_2} = \delta_{n_1 n_2} \delta_{\ell_1 \ell_2} \delta_{j_1 j_2}. \tag{15.18}$$

This is a fully antisymmetric two-nucleon shell-model wave function in spin, isospin, and space. The ϕ on the right side of (15.17) is

$$\phi_{(j_1 j_2)J}^{\ M} = [\phi_{n_1 \ell_1 j_1}(1)\phi_{n_2 \ell_2 j_2}(2)]_J^M, \tag{15.19}$$

where $\phi_{n\ell j}$ is a shell model spin–orbit wave function [see (7.89)].

To prepare for the integration over spin coordinates in F, use can be made of an angular momentum transformation (see Appendix)

$$|(\ell_1 s_1)j_1, (\ell_2 s_2)j_2; J\rangle = \sum_{LS} \begin{bmatrix} \ell_1 & s_1 & j_1 \\ \ell_2 & s_2 & j_2 \\ L & S & J \end{bmatrix} |(\ell_1 \ell_2)L, (s_1 s_2)S; J\rangle, \tag{15.20}$$

where the square bracket, denoting a transformation coefficient, is related to the $9-j$ symbol. The bracket notation of (15.20) is a commonly used one when many angular-momenta couplings are involved. The wave function on the right is sometimes referred to as an L–S function, as contrasted with that on the left, called a j–j function. In more detail,

$$|(\ell_1 \ell_2)L, (s_1 s_2)S; JM\rangle \equiv [[\phi_{n_1\ell_1}\phi_{n_2\ell_2}]_L[X_{s_1}X_{s_2}]_S]_J^M.$$

At this point the Talmi–Moshinsky transformation from individual to relative and c.m. coordinates can be performed (Talmi, 1952; Moshinsky, 1959; Brody and Moshinsky, 1960). This can be done exactly if the single-particle states $\phi_{n\ell j}$ are harmonic-oscillator wave functions in their radial part (see Appendix). Otherwise it can be done approximately by expanding the wave function in a finite number of oscillator wave functions (Glendenning, 1965). The transformation reads

$$\begin{aligned}&[\phi_{n_1\ell_1}(\sqrt{v}\mathbf{r}_1)\phi_{n_2\ell_2}(\sqrt{v}\mathbf{r}_2)]_L^M \\ &\quad = \sum_{n\lambda N\Lambda} \langle n\lambda N\Lambda; L|n_1\ell_1 n_2\ell_2; L\rangle[\phi_{n\lambda}(\sqrt{v/2}\mathbf{r})\phi_{N\Lambda}(\sqrt{2v}\mathbf{R})]_L^M,\end{aligned} \tag{15.21}$$

$$\mathbf{r}_1 = \mathbf{r}_1 - \mathbf{r}_2, \qquad 2\mathbf{R} = \mathbf{r}_1 + \mathbf{r}_2.$$

This is a very useful and important transformation in nuclear physics because nuclear wave functions are typically expressed in terms of the coordinates of the nucleons, whereas potential interactions are a function of relative coordinates. The summation is restricted by a condition that follows from the fact that the energy associated with an oscillator state $\phi_{n\ell}(\sqrt{v}r)$ for a particle of mass m is

$$[2(n-1) + \ell + \tfrac{3}{2}]\hbar\omega, \qquad \omega = \hbar v/m, \qquad n \geq 1.$$

Therefore energy conservation implies that

$$2(n_1 + n_2) + \ell_1 + \ell_2 = 2(n + N) + \lambda + \Lambda, \tag{15.22a}$$

and parity (see A8) implies

$$(-)^{\ell_1+\ell_2} = (-)^{\lambda+\Lambda} \tag{15.22b}$$

In terms of these transformations, the two-particle wave function can now be written as

$$\begin{aligned}\Phi_{(j_1 j_2) J}{}^{M\mu}_{T}(1,2) = \left(\frac{2}{1+\Delta_{j_1 j_2}}\right)^{1/2} X_T^\mu(1,2) \sum_{LS} \sum_{n\lambda N\Lambda} \frac{1+(-)^{\lambda+S+T+1}}{2} \\ \times \begin{bmatrix} \ell_1 & \frac{1}{2} & j_1 \\ \ell_2 & \frac{1}{2} & j_2 \\ L & S & J \end{bmatrix} \langle n\lambda N\Lambda; L | n_1\ell_1 n_2\ell_2; L\rangle \\ \times [[\phi_{n\lambda}(\sqrt{v/2}\,\mathbf{r})\phi_{N\Lambda}(\sqrt{2v}\,\mathbf{R})]_L, X_S(1,2)]_J^M.\end{aligned} \tag{15.23}$$

Throughout, for spin or isospin two-particle wave functions we use the notation

$$X_S^M(1,2) \equiv [X_{1/2}(\sigma_1)X_{1/2}(\sigma_2)]_S^M = X_S^M(\sigma_1, \sigma_2). \tag{15.24}$$

The integration over $\mathbf{r}$, as well as over spin and isospin in the expression for F [Eq. (15.15)], can now be performed and the result expressed by

$$F_{(j_1 j_2) J}{}^{M M_S}_{S\ T}(\mathbf{R}) = \frac{1-(-)^{S+T}}{2} \sum_L C^{L\ \ S\ \ J}_{M_L M_S M} f^{j_1 j_2}_{LSJ}(R) Y_L^{M_L *}(\hat{\mathbf{R}}), \tag{15.25}$$

where the radial part of the form factor is

$$f^{j_1 j_2}_{LSJ}(R) = \sum_N G^{j_1 j_2}_{NLSJ} u_{NL}(\sqrt{2v}\,R), \tag{15.26}$$

and the coefficients G are referred to as structure amplitudes for two-particle transfer reactions. They have been tabulated for the entire range of quantum numbers that occur in nuclei (Glendenning 1975a) and from the preceding development are defined by

$$\begin{aligned} G^{j_1\ j_2}_{NLSL} &= \Omega_n \left(\frac{2}{1+\Delta_{j_1 j_2}}\right)^{1/2} \begin{bmatrix} \ell_1 & \frac{1}{2} & j_1 \\ \ell_2 & \frac{1}{2} & j_2 \\ L & S & J \end{bmatrix} \langle n0NL; L | n_1\ell_1 n_2\ell_2; L\rangle, \\ \Omega_n &= \int \phi^*_{n0}(\sqrt{v/2}\,\mathbf{r})\phi_{10}(\sqrt{a\eta^2}\,\mathbf{r})\,d\mathbf{r}. \end{aligned} \tag{15.27}$$

Note that the assumption (true to very good approximation for light nuclei, $a \le 4$) that the relative motion in the triton is S state and that the direct interaction is zero range in the coordinate between the separate fragments of the projectile (15.12), has projected the $\lambda = 0$ terms in (15.23). This causes the c.m. coordinate to carry all the orbital angular momentum $\Lambda = L$. The

Talmi-Moshinsky bracket enforces the parity selection rule that $(-)^L$ has to equal the parity carried by the transferred nucleons $(-)^{\ell_1+\ell_2}$, which is of course equal to the parity change between initial and final nuclear states. The $\lambda = 0$ projection also has the effect of allowing only those F with $S + T$ being odd to have finite values, as (15.25) shows.

E. INTERPRETATION OF THE FORM FACTOR

From (15.5), (15.14), (15.15), and the results of the last section, we learn that the amplitude contains the following linear combination,

$$\tilde{F}(\mathbf{R}) = \sum_{j_1 j_2} \beta_{j_1 j_2 JT}(\mathrm{B, A}) F_{(j_1 j_2)}{}^{M M_S}_{J\ S\ T}(\mathbf{R}). \tag{15.28}$$

The meaning of this function is clear. The right side of the bracket of the defining equation (15.15) for F contains the part of the wave function of nucleus a that defines the state of motion of the two transferred nucleons in that nucleus. The left side of the bracket, taken in linear combination with the parentage factors β, defines their state of motion in the nucleus A + 2. The overlap projects out of this state of motion that part corresponding to the motion they possess in the nucleus a. Therefore, we reach the important conclusion that the reaction proceeds to the extent that the correlation of the pair in the nucleus A + 2 is the same as it is in the nucleus a, and at the same time the remaining A nucleons in A + 2 are in the same state as they are in A. The last part of the conclusion was remarked on following Eqs. (15.5) and (15.6). Note specifically that the correlation is singlet S state with a range specified by the size parameter η. Also, $\tilde{F}$ is, so to speak, that part of the neutron-pair c.m. wave function when the preceding projection has been made.

We note further from Eq. (15.25) that the *radial* wave functions for the c.m. of the pair, when they possess this correlation, are given by

$$\begin{aligned} \tilde{\phi}_{JSJ}(R) &= \sum_{j_1 j_2} \beta_{j_1 j_2 JT} f^{j_1 j_2}_{LSJ}(R) \\ &= \sum_N \left(\sum_{j_1 j_2} \beta_{j_1 j_2 JT} G^{j_1 j_2}_{NSLJ} \right) u_{NL}(\sqrt{2v}R). \end{aligned} \tag{15.29}$$

F. TRANSITION AMPLITUDE AND CROSS SECTION

The bracket in the transition amplitude (15.2) was evaluated through the sequence of results (15.5), (15.6), (15.14), (15.15), (15.25), and (15.29). Assembling these, we obtain for the transition amplitude between magnetic substates (15.1b),

$$\mathscr{T}_{\beta\alpha} = \Omega_{\mathrm{b}}\left(\tfrac{\mathrm{a}}{2}\right)^{1/2} \sum_{j_1 j_2 J T S} \beta_{j_1 j_2 J T}(\mathrm{B}, \mathrm{A}) A_{ST} C^{J_{\mathrm{A}}\, J\, J_{\mathrm{B}}}_{M_{\mathrm{A}} M M_{\mathrm{B}}} C^{T_{\mathrm{A}} T T_{\mathrm{B}}}_{\mu_{\mathrm{A}} \mu\, \mu_{\mathrm{B}}} C^{S\ \ J_{\mathrm{b}}\, J_{\mathrm{a}}}_{M_S M_{\mathrm{b}} M_{\mathrm{a}}} C^{T T_{\mathrm{b}} T_{\mathrm{a}}}_{\mu\, \mu_{\mathrm{b}}\, \mu_{\mathrm{a}}}$$

$$\times \int \chi_{\mathrm{b}}^{(-)*}(r_\beta) g \delta(\rho) \phi_{10}(\sqrt{4\mathrm{a} - 8}\,\eta\rho) F_{(j_1 j_2) J\ S\ T}^{\ M M_S}(\mathbf{R}) \chi_{\mathrm{a}}^{(+)}(\mathbf{r}_\alpha)\, d\mathbf{R}\, d\rho$$

$$= \Omega_{\mathrm{b}} D_0 \left(\tfrac{\mathrm{a}}{2}\right)^{1/2} \sum_{LSJT} A_{ST} C^{J_{\mathrm{A}}\, J\, J_{\mathrm{B}}}_{M_{\mathrm{A}} M M_{\mathrm{B}}} C^{T_{\mathrm{A}} T T_{\mathrm{B}}}_{\mu_{\mathrm{A}} \mu\, \mu_{\mathrm{B}}}$$

$$\times C^{S\ \ J_{\mathrm{b}}\, J_{\mathrm{a}}}_{M_S M_{\mathrm{b}} M_{\mathrm{a}}} C^{T T_{\mathrm{b}} T_{\mathrm{a}}}_{\mu\, \mu_{\mathrm{b}}\, \mu_{\mathrm{a}}} C^{L\ \ S\ \ J}_{M_L M_S M} i^L \hat{L}^{1/2} B^{M_L}_{LSJ}(\mathbf{k}_\beta, \mathbf{k}_\alpha), \tag{15.30}$$

where we define

$$D_0 = g\phi_{10}(0) = [(1/\sqrt{\pi})(\sqrt{4\mathrm{a} - 8}\,\eta)^3]^{1/2} g \tag{15.31}$$

and the kinematic amplitude

$$\mathrm{B}^{M_L}_{LSJ}(\mathbf{k}_\beta, \mathbf{k}_\alpha) = i^{-L} \hat{L}^{-1/2} \int \chi_\beta^{(-)*}\left(\mathbf{k}_\beta, \frac{\mathrm{A}}{\mathrm{A}+2}\mathbf{R}\right) \tilde{\phi}_{LSJ}(R) Y_L^{M_L *}(\hat{\mathbf{R}})$$

$$\times \chi_\alpha^{(+)}(\mathbf{k}_\alpha, \mathbf{R})\, d\mathbf{R}. \tag{15.32}$$

The cross section corresponding to unaligned target, unpolarized projectile, and the observation of all magnetic substates produced is given by (3.39) and (3.40). The sum over magnetic substates can be accomplished as in Chapter 5 by using the properties of the Clebsch–Gordan coefficients. The answer can be put in a convenient form by defining

$$\mathrm{b}_{ST} = \hat{S}^{-1/2} \left(\tfrac{\mathrm{a}}{2}\right)^{1/2} A_{ST} C^{T T_{\mathrm{b}} T_{\mathrm{a}}}_{\mu\, \mu_{\mathrm{b}}\, \mu_{\mathrm{a}}}, \tag{15.33}$$

which is a property of the light ions, and

$$C_{ST} = \mathrm{b}_{ST} C^{T_{\mathrm{A}} T T_{\mathrm{B}}}_{\mu_{\mathrm{A}} \mu\, \mu_{\mathrm{B}}} \tag{15.34}$$

Then we can write

$$\frac{d\sigma}{d\Omega_{\alpha\to\beta}} = \frac{m_\alpha m_\beta}{(2\pi\hbar^2)^2} \frac{k_\beta}{k_\alpha} \frac{\hat{J}_{\mathrm{B}}}{\hat{J}_{\mathrm{A}}} (\Omega_{\mathrm{b}} D_0)^2 \sum_{LSJM_L} C^2_{ST} |B^{M_L}_{LSJ}|^2, \tag{15.35}$$

which is the stripping cross section. It is an incoherent sum over LSJ. Whereas in single-particle transfer, these quantum numbers belonged to the transferred nucleon, here they refer to the pair. The sum over the pair configurations that contribute to a given set of LSJ appears coherently in the form factor or projected wave function $\tilde{\phi}$ that appears in B_{LSJ}. This is in accord with the discussion of coherence in the introduction to this chapter.

The various light-ion reactions can be conveniently compared by tabulating the numbers b_{ST}, which give the weight with which spin singlet and triplet transfers contribute. This can be done for a more general case than the spin-independent interaction (15.12) that we have used so far. More

TABLE III

Values of $|b_{ST}|$ for Listed Reactions and Their Inverse

$\lvert b_{ST}\rvert$	$S = 0, T = 1$	$S = 1, T = 0$
t, p	$2a_0$	0
He^3, n	$2a_0$	0
t, n	$\sqrt{2}a_0$	$\sqrt{2}a_1$
He^3, p	$\sqrt{2}a_0$	$\sqrt{2}a_1$
α,d	0	$4a_1$

generally, let V have a zero range as before, but a spin dependence

$$V_{ij} \simeq g\delta(\rho) \sum_{ST} V_{ST} P_{ST}^{ij}, \tag{15.36}$$

where P_{ST}^{ij} is a projection operator for the total spin–isopin S and T of particles i and j. Of course, i refers to one of the transferred nucleons, 1 or 2, and j to any nucleon in b. The integration over spin and isospin is now more complicated than before. It can be facilitated by the use of recoupling coefficients. This is done, for example, in Chapter 16, where the force is also allowed to have a finite range. Here we simply state the result as follows. The quantity b_{ST} of (15.33) is replaced by

$$b_{ST} = (\tfrac{a}{2})^{1/2} \hat{S}^{-1/2} C^{T T_b T_a}_{\mu \mu_b \mu_a} \sum_{S'T'} A_{S'T'} (\hat{S}\hat{S}' \, \hat{T}\hat{T}')^{1/2}$$

$$\times \begin{Bmatrix} \frac{1}{2} & \frac{1}{2} & S' \\ \frac{1}{2} & J_a & S \end{Bmatrix} \begin{Bmatrix} \frac{1}{2} & \frac{1}{2} & T' \\ \frac{1}{2} & T_a & T \end{Bmatrix} V_{S'T'}. \tag{15.37}$$

Table III lists the values for various reactions where

$$a_0 = \tfrac{1}{4}V_{01} + \tfrac{3}{4}V_{10}, \tag{15.38}$$
$$a_1 = \tfrac{3}{4}V_{01} + \tfrac{1}{4}V_{10}.$$

G. CORRELATIONS‡

From the preceding development we learn that two features of nuclear structure govern the strength of a two-nucleon transfer reaction. Both involve a question of parentage. How much does the state of A + 2 in question look like the ground state of A plus two additional nucleons, and how much

‡ See Glendenning (1968b, 1975c).

is the motion of these two nucleons in A + 2 correlated in the same way that they are in the light nucleus in which they reside?

If one of the partners in the reaction is a light nucleus, such as the triton in a (t, p) reaction, then the correlation is in the angular momentum state $\ell = 0$. In the (t, p) reaction the neutrons have their spin coupled to $S = 0$ so as to be antisymmetric. The correlation is therefore referred to as singlet S, or ^1S in spectroscopic notation.

That the ground states of light nuclei are predominantly S-state, and in particular, that the like particles are in ^1S states is not an accident. Like nucleons can exist in only the singlet-even and triplet-odd states. The former is attractive and the latter is weak and possibly repulsive. Therefore, in the lowest energy state two neutrons will be correlated mostly in the lowest singlet-even state, namely, the ^1S.

A similar argument holds for any nucleus. However, it is important to note that the residual interaction is weak compared with the central field of the nucleus. (Two-particle matrix elements are of the order of 1 MeV, whereas nucleons are bound in a nucleus by the order of 10 MeV.) Therefore, like nucleons tend to be correlated in the singlet-even states to the extent consistent with the shell structure, the Pauli principle, and their interaction with the unlike nucleons. The interaction between unlike nucleons can take place in all four states of motion, of which the triplet-even is strongest, followed by singlet-even. The odd-state interactions are weak or repulsive. Therefore a neutron–proton pair tends to be correlated in the triplet-even states. The lowest-lying state of given spin and parity possesses more of these favored correlations than any other such states in the nucleus. That is why it lies lowest. This statement is locally true, that is, in any given neighborhood of levels falling within the energy interval of the order of the interaction energy, it must be true. It need not be true over a much larger energy interval, only because the interaction between nucleons is a small perturbation on the shell-model central interaction. If this were not so, nuclei would not possess a shell structure.

The form factor (15.28) is a measure of the degree to which the parentages of A + 2 and A are favorable and to which the pair is correlated as in the light nucleus from which they are removed. In particular, for the (t, p) reaction on an even–even target nucleus, it is

$$\tilde{F}_J(\mathbf{R}) = (\Phi_J(\mathrm{A}+2)|\Phi_0(\mathrm{A})\phi_{10}(\sqrt{3\eta^2}\mathbf{r})X_0(\sigma_1\sigma_2)), \qquad (15.39)$$

for which a detailed reduction was previously carried out [Eqs. (15.25)–(15.28)]. Corresponding to a shell-model description (True and Ford, 1958) of ^{208}Pb and ^{206}Pb, its radial part is shown in Fig. 15.2. (Glendenning, 1968b). Not only does this illustrate that the correlation is concentrated in the ground state but also that it is concentrated in the nuclear surface region.

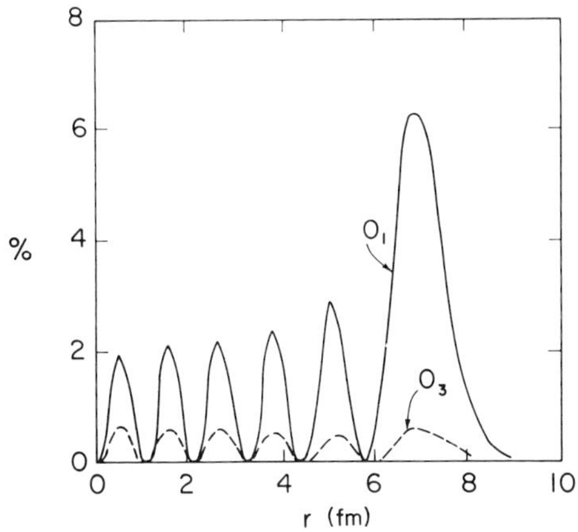

Fig. 15.2. The radial probability distribution $[r\tilde{\phi}(r)]^2$ of the ^1S correlation in the ground and O_3 state for $^{208}Pb(p, t)$. Note the concentration in the surface for the enhanced ground state transition, beside the fact that the total probability for it is much higher (Glendenning, 1968b).

Figure 15.3 shows the various 0^+ levels of ^{206}Pb formed on the left from the pure configurations at the energies prescribed by the single-particle energies of ^{207}Pb, and on the right after the residual interaction has mixed the configurations and changed the energies. The percentage with which the dominant configuration is represented in the wave function is also indicated. The percentage of ^1S correlation for pure and correlated states is shown also.

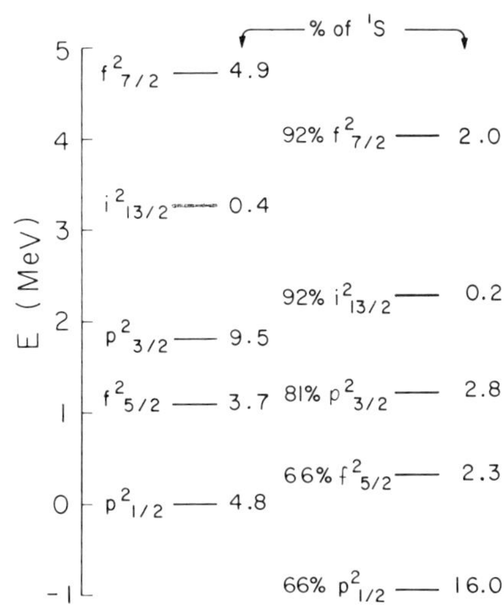

Fig. 15.3. For ^{206}Pb the pure and configuration-mixed spectra of O^+ states are shown on left and right, respectively. The percentage of ^1S correlation is indicated and here is concentrated in the lowest state (Glendenning, 1968b).

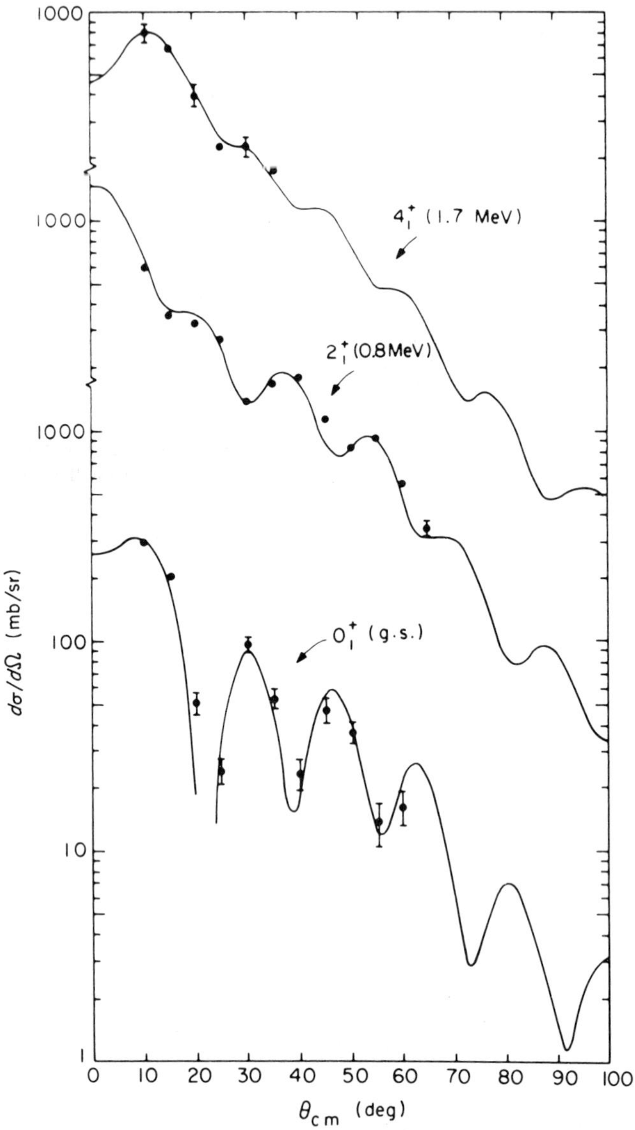

Fig. 15.4. Calculated angular distributions corresponding to 40-MeV protons in $^{208}Pb(p, t)$-^{206}Pb are compared with the Minnesota data of Reynolds *et al.* (1967), calculation by Glendenning (1967).

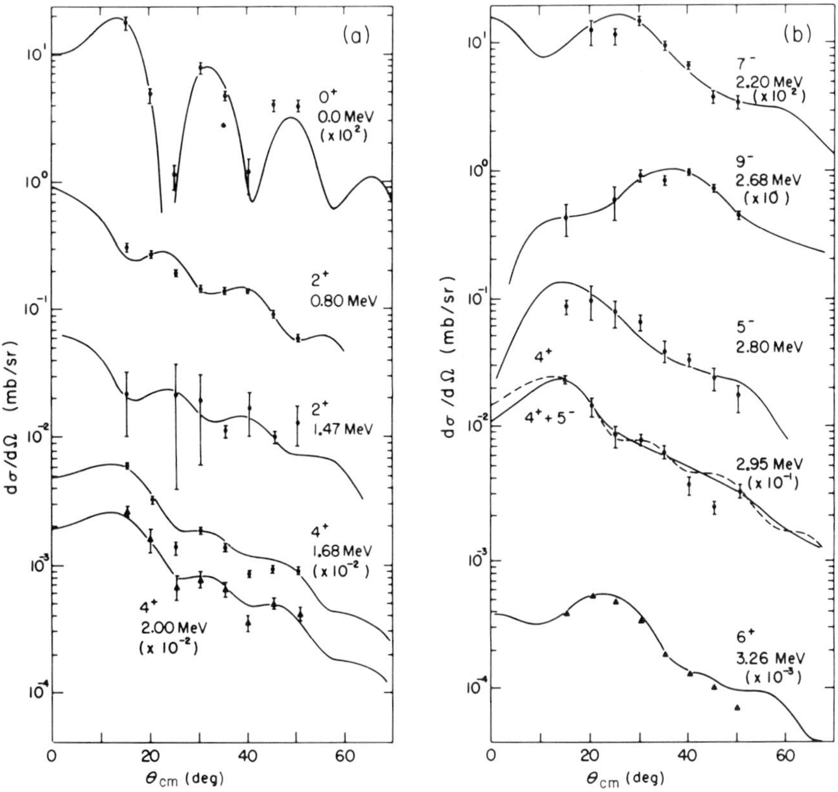

Fig. 15.5. Calculated angular distributions corresponding to 40-MeV protons in ^{208}Pb(p, t)-^{206}Pb to states ranging from spin 0 to 9 are compared with the MIT data of Smith *et al.* (1968).

The residual interaction has clearly redistributed the correlation so that the lowest 0^+ levels possess most of it. A similar situation holds for any other group of neighboring levels of the same spin and parity.

The cross sections, at two energies, were calculated (Glendenning, 1967, 1968b) in DWBA for the levels in ^{206}Pb using shell-model nuclear wave functions (True and Ford, 1958) and are compared with the data in Figs. 15.4–15.6. The calculation was normalized to the data. For the states shown in Fig. 15.6, the relative normalizations are shown in Fig. 15.7. Overall, the agreement between the theory and experiment is remarkably good, suggesting the appropriateness of the nuclear wave functions and the reaction mechanism.

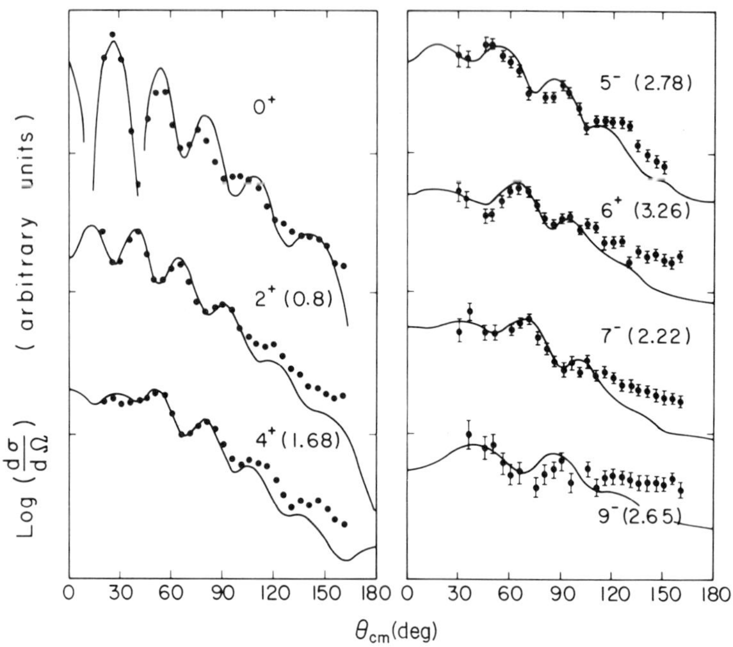

Fig. 15.6. Calculated angular distributions corresponding to 22-MeV protons for ^{208}Pb(p, t) reaction are compared with the Yale data of Bromley *et al.* (Calculation by Glendenning 1968b).

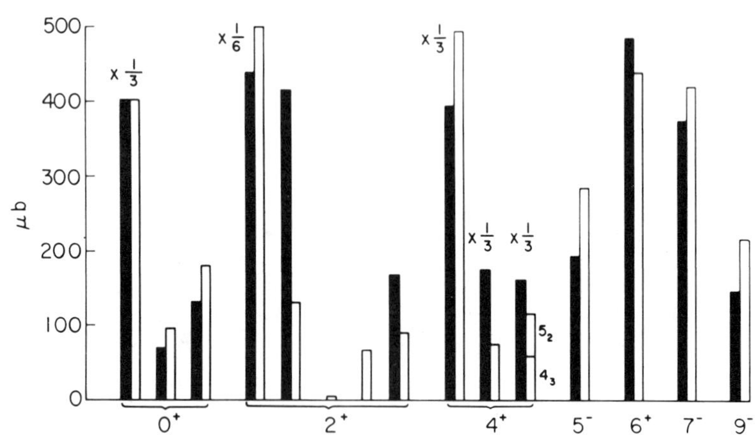

Fig. 15.7. Integrated cross sections from 20–90% for 22-MeV protons to a number of levels are shown. Black bars denote the experimental results of the Yale group, and open bars denote calculations based on True and Ford wave functions for ^{206}Pb (From Glendenning, 1968b).

H. PARENTAGE AMPLITUDE FOR TWO NUCLEONS

The discussion of parentage amplitudes for two nucleons is in many ways similar to that given in Chapter 7 for the one-nucleon transfer reactions. One major difference, however, is that the sum rules are of little use for two-nucleons transfer reactions. The reason for this difference is that the amplitude for one-nucleon transfer, B_ℓ^m of (7.17), involves the radial wave function of the orbital $n\ell j$. Once the angular momentum of the orbital is identified, which is easy to do, characterized as it is by the angular distribution, the transfer amplitude can be calculated with little uncertainty. The shell-model informs us of the likely values of the n and j quantum numbers, and the Q of the reactions tells us the separation energy which dictates the asymptotic behavior of the wave function. The solution of the radial Schrödinger equation with, say, a Woods–Saxon potential then provides the radial wave function. An analysis of the relevant elastic scattering provides the distorted waves. These are all the ingredients of B_ℓ^m. The spectroscopic amplitude can therefore be extracted as the only unknown quantity in a comparison with the data.

The analogous transfer amplitude for two nucleons (15.32) contain the function $\tilde{\phi}_{L0J}$, which is not a particle wave function nor the product of two of them, but a projection onto a pair wave function—a projection involving the relative motion in the light nucleus from which the pair is transferred. As previously discussed, depending on the two-particle correlations in the nuclear state, the radial distribution of $\tilde{\phi}_{L0J}(R)$ may be concentrated in the nuclear surface or not, and it may be large or small in comparison with a product of two single-particle wave functions, depending on the coherence in the nuclear state. A measurement of the angular distribution and Q value provides the multipolarity and the pair separation energy, but otherwise leaves $\tilde{\phi}$ completely unspecified. Therefore, although it would be possible (though more difficult) to derive sum rules, the experimental determination of the spectroscopic amplitudes entering the sum is not possible, in general. The value of two-nucleon transfer, as already stressed, is of a different nature. These reactions, being so sensitive to correlations, provide a rather severe test of nuclear wave functions that have been calculated from a nuclear model.

From Eq. (15.4) the parentage amplitude $\beta_{j_1 j_2 J}(\mathrm{B}, \mathrm{A})$ can be written, in terms of particle creation operators, as

$$\beta_{j_1 j_2 J}(\mathrm{B}, \mathrm{A}) = \sum_{M_\mathrm{A} M} (1 + \Delta_{12})^{-1/2} C^{J_\mathrm{A}\ J\ J_\mathrm{B}}_{M_\mathrm{A} M M_\mathrm{B}} \langle J_\mathrm{B} M_\mathrm{B} | [\mathrm{a}^\dagger_{j_1} \mathrm{a}^\dagger_{j_2}]^M_J J_\mathrm{A} M_\mathrm{A} \rangle, \quad (15.40)$$

where Δ_{12} is unity if the two orbitals $n_1\ell_1 j_1$ and $n_2\ell_1 j_2$ are the same and zero otherwise. This factor enters because otherwise the identical configuration would be counted twice in the wave function. As in (7.31), $\mathrm{a}^\dagger_{jm}$ creates a nucleon in the orbital $n\ell jm$, and

$$\tilde{a}_{jm} = (-)^{j-m} a_{j-m}$$

creates a hole. Both transform as an irreducible tensor of rank j and projection m. Some authors use the phase $(-)^{j+m}$. Because j is half odd integer, the operators so defined differ by a factor -1.

The Wigner–Eckart theorem (A.39) can be employed to show that β does not depend on magnetic qantum numberss. We find immediately that

$$\beta_{j_1 j_2 J}(\mathrm{B}, \mathrm{A}) = (-)^{2J}[\hat{J}_{\mathrm{B}}(1 + \Delta_{12})]^{-1/2}\langle J_{\mathrm{B}} \| [a^{\dagger}_{j_1} a^{\dagger}_{j_2}]_J \| J_{\mathrm{A}} \rangle \qquad (15.41)$$

This is a concise statement of how the parentage amplitude is to be calculated from the nuclear wave functions. The phase, of course, can be dropped because here it is always $+1$. We have left it in for comparison with the single-particle parentage amplitude of Chapter 7. There we needed the reduced matrix element of the particle creation operator and here of the normalized pair creation operator. To complete the picture, the amplitude for inelastic scattering was derived in Chapter 9 and is

$$S_{j_1 j_2 J}(\mathrm{A}', \mathrm{A}) = -(\hat{J})^{-1/2}\langle J' \| [a^{\dagger}_{j_1} \tilde{a}_{j_2}]_J \| J \rangle. \qquad (15.42)$$

Here we see a particle destruction operator $\tilde{a}$ (or equivalently, a hole creation operator) and a particle creation operator $a^{\dagger}$ corresponding to the promotion of a particle from the orbital j_2 to j_1 or the creation of a particle–hole pair $(j_1 j_2)$, depending on the language one prefers to use. These three amplitudes provide a clear and concise statement of what the nuclear-structure theorist must calculate to make his theory accessible to the experimental tests provided by one- and two-particle transfer and inelastic scattering.

Of course, in certain simple situations, such as several nucleons beyond closed shells, special and simple means are available for calculating the parentage amplitudes. Let us consider several examples. It is usual to calculate the structure of a nucleus in terms of a group of active or valence nucleons, which for the present we shall denote by the symbol V. The core nucleons are assumed to be in a degenerate state occupying the lowest shells. How does this affect our counting factors in connection with antisymmetrization, which should be correctly included in any case? From (15.4) we can write in abbreviated notation,

$$\beta_{j_1 j_2 JT}(\mathrm{B}, \mathrm{A}) = \binom{\mathrm{A} + 2}{2}^{1/2} \langle \Phi_{\mathrm{B}}(\mathrm{A} + 2) | [\Phi_{\mathrm{A}}(\mathrm{A})\phi_J]_{J_{\mathrm{B}}} \rangle, \qquad (15.43)$$

where the combinatorial factor comes from the normalization of the antisymmetrization factor in (15.4). We obtained (15.43) by taking the projection of (15.4) with a particular permutation of particles among Φ_A and Φ_J, say,

$$\Phi_A(1, 2, \ldots, A)\Phi_J(A + 1, A + 2).$$

[Note that, in the expression for the DWBA amplitude, the counting factor W for the direct terms contains precisely the same combinatorial factor (Chapter 5).] Suppose now that both A and B are described in terms of a core plus valence nucleons in the general form

$$\begin{aligned} \Phi_B(A + 2) &= \mathscr{A}_{CV}[\Phi_C(1, 2, \ldots, C)\Phi_V(C + 1, \ldots, A + 2)], \\ \Phi_A(A) &= \mathscr{A}_{CV'}[\Phi_C(1, 2, \ldots, C)\Phi_{V'}(C + 1, \ldots, A)], \end{aligned} \tag{15.44a}$$

where

$$\begin{aligned} \mathscr{A}_{CV} &= \binom{A+2}{C}^{-1/2} \sum (-)^{P_{CV}} P_{CV}, \\ \mathscr{A}_{CV'} &= \binom{A}{C}^{-1/2} \sum (-)^{P_{CV'}} P_{CV'} \end{aligned} \tag{15.44b}$$

are the normalized permutation operators. Consequently,

$$\begin{aligned} \beta_J(B, A) = {} & \binom{A+2}{2}^{1/2} \binom{A+2}{C}^{-1/2} \binom{A}{C}^{-1/2} \\ & \times \sum_{P_{CV}} \sum_{P_{CV'}} (-)^{P_{CV}} (-)^{P_{CV'}} \langle P_{CV}\Phi_C(1, 2, \ldots, C) \\ & \times \Phi_V(C + 1, \ldots, A+2)| \\ & \times P_{CV'}\Phi_C(1, 2, \ldots, C)[\Phi_{V'}(C + 1, \ldots, A)\Phi_J(A + 1, A + 2)]_{J_B}\rangle. \end{aligned}$$

Consider a particular permutation on the right side of the overlap. Through the orthonormality of the wave functions, it has a finite overlap with precisely one term in the sum over permutations of the particles in the left side of the overlap, namely, the one having precisely the same particles in Φ_C. However, there are $\binom{A}{C}$ permutations of the particles on the right, each with such a single finite and equal overlap. Therefore,

$$\beta_J(B, A) = \binom{V}{2}^{1/2} \langle \Phi_V(1, \ldots, V)|[\Phi_{V'}(1, \ldots, V - 2)\Phi_J(V - 1, V)]_{J_B}\rangle, \tag{15.45}$$

where $V = A + 2 - C$ and we have renamed the active nucleons. Equation (15.45) tells us how to count correctly when we deal with only V active nucleons.

Consider several simple shell-model examples now. Suppose A has closed shells and B is described as a configuration-admixed state

$$\Phi_B(A + 2) = \mathscr{A}[\Phi_0(A) \sum_{j_1 j_2} C_{(j_1 j_2)J}\Phi_{(j_1 j_2)J}(\mathbf{r}_1, \mathbf{r}_2)].$$

Then (15.45) gives us

$$\beta_{j_1 j_2 J} = C_{(j_1 j_2)J}.$$

Next, suppose that again we have closed shells of neutrons and protons plus n neutrons in the shell j in the nucleus B and $n - 2$ neutrons in the nucleus A. Then

$$\Phi_B = \Phi_0 |(j^n)0\rangle,$$
$$\Phi_A = \Phi_0 |(j^{n-2})vJ\rangle.$$

The symbol v denotes the seniority of the state and is needed because, for $n > 2$, J does not always uniquely specify a j^n configuration. Racah (1942, 1943) introduced the fractional parentage amplitudes precisely to deal with such wave functions. In terms of his fractional parentage coefficients we can write

$$|(j^n)0\rangle = \sum_{v,J} [(j^{n-2})vJ, (j^2)J; 0|\}(j^n)0] |(j^{n-2})vJ, (j^2)J; 0\rangle.$$

The coefficients are so constructed that the wave function remains totally antisymmetric, although the partition on the right is a definite one with respect to particle labels. Equation (15.45) yields, in the present example,

$$\beta_{j^2 J} = \binom{n}{2}^{1/2} [(j^{n-2})vJ, (j^2)J; 0|\}(j^n)0].$$

Of course, the usefulness of the fractional parentage expansion is that the coefficients can be written explicitly in many cases. Thus

$$[(j^{n-2})vJ, (j^2)J|\}(j^n)0] = \begin{cases} \left\{\dfrac{2(n-2)}{n-1} \dfrac{2J+1}{(2j-1)(2j+1)}\right\}^{1/2}, & v = 2, \quad J \neq 0, \\[2ex] \left\{\dfrac{2j+3-n}{(n-1)(2j+1)}\right\}^{1/2}, & v = 0 = J. \end{cases}$$

Many other explicit results could be cited. (See de-Shalit and Talmi, 1963.)

Let us turn now to the BCS pairing theory, which was briefly mentioned in Chapter 7. It is a theory that is relevant to single closed-shell nuclei, with a few nucleons of the other type beyond closed shells. The heavier nickel or tin isotopes are examples. For such a situation the pairing effects of the residual interaction can be approximately diagonalized by transforming to quasi-particles. The remaining interaction among quasi-particles is weak and can be treated through a diagonalization in a two quasi-particle basis for even nuclei. The transformation to quasi-particles, a creation operator for which is denoted by α^+, has the form

$$a^\dagger_{jm} = U_j \alpha^\dagger_{jm} - V_j \tilde{\alpha}_{jm}, \tag{15.46}$$

with $\tilde{\alpha}$ having a definition just like that of $\tilde{a}$. The ground state is the quasi-particle vacuum $|\omega\rangle$, and excited states are two (or four) quasi-particle states, such as

$$|JM\rangle = -\tfrac{1}{2} \sum_{j_1 j_2} \eta_{j_1 j_2} [\alpha^\dagger_{j_1} \alpha^\dagger_{j_2}]^M_J |\omega\rangle. \tag{15.47}$$

In terms of quasi-particles, the two particle operator (15.40) can be written as

$$\begin{aligned}[a^\dagger_1 a^\dagger_2]^M_J = {} & U_1 U_2 [\alpha^\dagger_1 \alpha^\dagger_2]^M_J + V_1 V_2 [\tilde{\alpha}_1 \tilde{\alpha}_2]^M_J \\ & - U_1 V_2 [\alpha^\dagger_1 \tilde{\alpha}_2]^M_J - V_1 U_2 [\tilde{\alpha}_1 \alpha^\dagger_2]^M_J \\ & + V_1 U_1 (\hat{j}_1)^{1/2} \delta_{12} \delta_{J0}. \end{aligned} \tag{15.48}$$

The calculation of the parentage amplitudes now involves use of the fermion anticommutation relations and angular momentum algebra. In some cases, this is easy and in others it is tedious but elementary. Many results of interest have been derived in the literature (Yoshida, 1962; Asquitto and Glendenning, 1970b).

We cite below results for the parentage amplitudes connecting vibrational states, which would be needed for two-nucleon stripping and pickup reactions.

$$\beta_{j_1 j_2 0}(0_A \leftrightarrow 0_{A+2}) \approx (j_1 + \tfrac{1}{2})^{1/2} V_{j_1} U_{j_1} \delta_{j_1 j_2}, \tag{15.49a}$$

$$\beta_{j_1 j_2 J}(0_{A+2} \leftrightarrow J_A) = \left(\frac{2J+1}{1 + \Delta_{j_1 j_2}} \right)^{1/2} \eta_{j_1 j_2 J} V_{j_1} V_{j_2}, \tag{15.49b}$$

$$\beta_{j_1 j_2 J}(0_A \leftrightarrow J_{A+2}) = -\left(\frac{1}{1 + \Delta_{j_1 j_2}} \right)^{1/2} \eta_{j_1 j_2 J} U_{j_1} U_{j_2}, \tag{15.49c}$$

$$\beta_{j_1 j_2 0}(J_A \leftrightarrow J_{A+2}) \approx \beta_{j_1 j_2 0}(0_A \rightarrow 0_{A+2}). \tag{15.49d}$$

Of particular interest is the opposite sign for the amplitudes connecting ground and excited states in the two reactions. This sign would be of no importance if the reactions were dominated by the direct transition. However, for reactions, just as for inelastic scattering (Chapters 8, 9, and 13), indirect transitions passing through intermediate states can sometimes compete in strength with the direct transition. We will take up this subject in Chapter 17. However, we can anticipate its importance and some consequences here by noting that the direct transition for stripping, $0_A \rightarrow 2_{A+2}$ and the indirect transitions $0_A \rightarrow 2_A \rightarrow 2_{A+2}$ and $0_A \rightarrow 0_{A+2} \rightarrow 2_{A+2}$ will interfere with opposite sign compared to the direct and indirect transitions for the pickup

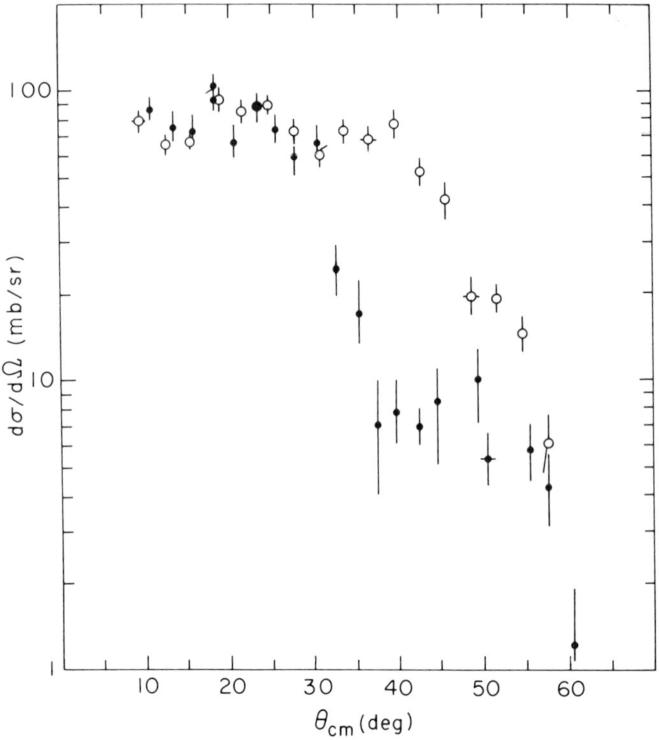

Fig. 15.8. Cross sections are compared for the pickup and stripping reaction leading to collective 2^+ states in neighboring nuclei. Solid dot is $^{74}Ge(^{18}O, {}^{16}O)^{76}Ge(2^+, 0.562)$ at 75 MeV and open dot $^{76}Ge(^{16}O, {}^{18}O)^{74}Ge^{2+}, 0.597)$ at 77.6 MeV (data from Bond *et al.* 1977).

reaction. If the direct transitions dominate, then the stripping and pickup cross sections to the collective 2^+ state in neighboring nuclei will be similar and proportional to the square of (15.49b) and (15.49c), whereas if the indirect transitions have comparable strength with the direct, the differential cross sections will be quite different because of the opposite sense of the interference. The theoretical prediction of this sensitivity (Ascuitto and Glendenning, 1973b) was subsequently observed in experiments (Scott et. al., 1975 and Bond et al., 1977), as illustrated in Fig. 8. The differences between the two cross sections is precisely of the nature anticipated (see Fig. 18.16).

Chapter 16

Finite-Range Interaction in Transfer Reactions

A. INTRODUCTION

The calculation of cross sections for one- or several-nucleon transfer reactions involves the evaluation of multidimensional integrals. This is complicated by the fact that the integrands consist of products of wave functions and a potential that depends on a number of vector coordinates. However, only several of them are the independent variables of integration. For light-ion reactions such as (d, p) reactions (Chapter 7) and (t, p) reactions (Chapter 15), the direct reaction causing the transition is often approximated by a zero-range potential. This is partially justified by the small size of the light nucleus (the deuteron or triton in the examples cited) in comparison with the size of the other nuclei. (Remember that direct reactions are dominantly surface reactions). In addition, the de Broglie wavelength associated with the relative motion is not small compared to the neglected distance (range of the potential). For a deuteron of say 10 MeV, it is

$$\lambdabar = k^{-1} = (2mE/\hbar^2)^{-1/2} = [2mc^2E/(\hbar c)^2]^{-1/2} \simeq 1 \quad \text{fm},$$

where m is the reduced mass of the deuteron and $\hbar c = 197$ MeV fm. As a consequence, the reaction does not have a great sensitivity to distances of this order. The zero-range assumption results in a great simplification; the arguments of the distorted waves become proportional to each other [see (15.13)], and the DWBA integrals reduce to an integration over a single vector coordinate, for which the angle integrals can be done in closed form. There then remains a one-dimensional integral (or rather a sum of such), which must be performed numerically because in general the distorted waves are not analytic functions but are obtained numerically as solutions to differential equations. The details of this reduction have been given in detail in the chapters previously referenced.

The assumption of a zero range for the interaction does not have any important consequences for the physics of the reaction mechanism in light-ion reactions. In particular, the discussion of the role of correlations in two-nucleon transfer reactions is unaffected. However, it is possible that the cross section and, to some extent, the angular distribution calculated under this assumption would be in some error. Therefore the magnitude of spectroscopic factors deduced by comparison with one-nucleon transfer data would have an error associated with this approximation. However, the angular momentum transfer would rarely be misassigned.

The situation is quite different for reactions between heavy nuclei (referred to usually, and somewhat inappropriately, as heavy ions because accelerators operate on ions of the nuclei). The large size of the nucleus from which the nucleon is transferred, together with the small wavelength associated with the relative motion, makes the zero-range approximation a poor one. The wavelength is small because the mass is large. Therefore, the reaction is sensitive to behavior of the wave functions on a short distance scale and renders approximations involving the vector coordinates less reliable than in the case of light-ion reactions. There is one approximation that has been frequently used, called the no-recoil approximation. It makes certain approximations to the vector arguments in the integrals, which simplifies their evaluation and at the same time retain's the finite range of the interaction (Buttle and Goldfarb, 1966, 1968; Schmittroth *et al.*, 1970; Ascuitto and Glendenning, 1972; Baltz and Kahana, 1972; Roberts, 1972). In some cases this is adequate for mapping out the phenomenology of reactions and their general behavior.

A number of approaches have been proposed and implemented for calculating the finite-range DWBA amplitude, and it would be difficult to say whether one is superior to another in general. One method uses the Monte Carlo technique to evaluate the six-dimensional integrals (Bayman, 1974). The other methods in one way or another achieve transformations of the functions in the integrands to others that depend on the independent coordinates of integration (Austern *et al.*, 1964; Sawaguri and Tobocman, 1967; Robson and Koskel, 1972; Charlton, 1973; McMahon and Tobocman, 1973; Glendenning and Nagarajan, 1974). When this is accomplished, the angular integrations can be performed in closed form, leaving only a double radial integration to be performed numerically.

In the following sections we shall outline the problem to be solved, introduce parentage expansions of the wave functions, and then specialize the general case to two particular cases that serve to illustrate the kinds of manipulations involved. These examples are the two-nucleon transfer (t, p) reaction and heavy ion reactions. We use two of these methods in achieving the goal of reducing the amplitude to two-dimensional integrals.

B. THE DWBA AMPLITUDE AND ITS COORDINATE DEPENDENCES

Denote the reaction by the atomic numbers of the participants,

$$\mathrm{A} + \underbrace{(\mathrm{b} + \mathrm{x})}_{\mathrm{a}} \rightarrow \underbrace{(\mathrm{A} + \mathrm{x})}_{\mathrm{B}} + \mathrm{b}, \tag{16.1}$$

where x denotes the transferred nucleon or group of nucleons. In specific examples we shall consider at most two-nucleon transfer, but at this stage the notation can be left more general with no loss of clarity. The DWBA amplitude, including the statistical factor (5.32) which counts the number of direct integrals, has the form

$$\begin{aligned} \mathcal{T} \simeq W \int \chi_\beta^{(-)*}(\mathbf{k}_\beta, \mathbf{r}_\beta)\{\Phi_\mathrm{B}\Phi_\mathrm{b}|(V_\beta - U_\beta)|\Phi_\mathrm{A}\Phi_\mathrm{a}\} \\ \times \chi_\alpha^{(+)}(\mathbf{k}_\alpha, \mathbf{r}_\alpha)\, d\xi_1\, d\xi_2 \cdots d\xi_{\mathrm{A}+\mathrm{a}-1}. \end{aligned} \tag{16.2}$$

At this stage, the approximation sign indicates only that the exchange integrals introduced by antisymmetrization are neglected for the reason discussed in Chapter 5. The expression in (16.2) is an integral over the position coordinates and a sum over spin and isospin coordinates that is denoted by { }. There are A + a nucleons, so there are 3(A + a) coordinates, three of which can be taken as the center of mass (c.m.) of the system in which we calculate. There remain 3(A + a) − 3 relative coordinates, which are denoted by ξ in the integral. A particular and convenient set of relative coordinates consists of

$$\mathbf{A}, \mathbf{x}, \mathbf{b}, \mathbf{R}, \boldsymbol{\rho}, \tag{16.3}$$

where **A**, **x**, and **b** denote 3(A − 1), 3(x − 1), and 3(b − 1) relative coordinates for the nuclei A, x, and b, respectively, and **R** and $\boldsymbol{\rho}$ denote relative vectors between A and x and x and b, as shown in Fig. 16.1. The channel coordinates appearing in the distorted waves, which connect A to a and B to b, are related

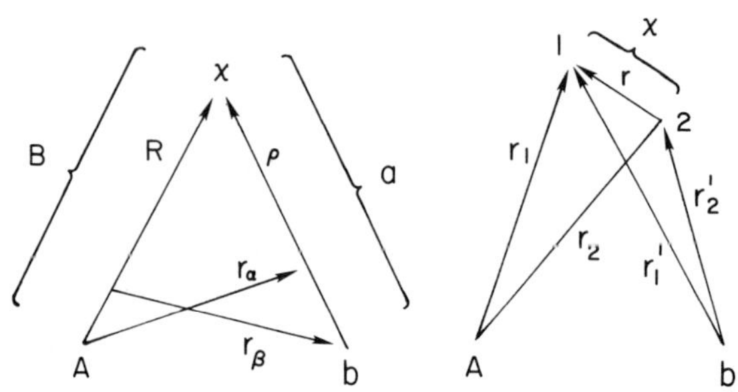

Fig. 16.1. On the left a convenient set of coordinates for describing the transfer of x from nucleus a to B. On the right, if x consists of two nucleons, the individual coordinates relative to A and b are shown.

to **R** and $\boldsymbol{\rho}$ by

$$\mathbf{r}_\alpha = \mathbf{R} - (\mathrm{b/a})\boldsymbol{\rho}, \qquad \mathbf{r}_\beta = (\mathrm{A/B})\mathbf{R} - \boldsymbol{\rho}. \tag{16.4}$$

Of course, it will not be convenient to regard the channel coordinates as dependent variables (appearing as they do in the distorted waves, which can be obtained only by numerical integration of the optical-potential scattering problem). Therefore we shall ultimately want to carry out the integration with these as variables of integration. The volume elements are related by a Jacobian

$$d\boldsymbol{\rho}\, d\mathbf{R} = J\, d\mathbf{r}_\alpha\, d\mathbf{r}_\beta, \tag{16.5a}$$

where J is a 6×6 determinant

$$J = \frac{\partial(\boldsymbol{\rho}, \mathbf{R})}{\partial(\mathbf{r}_\alpha, \mathbf{r}_\beta)} = \begin{vmatrix} \dfrac{\partial\boldsymbol{\rho}}{\partial\mathbf{r}_\alpha} & \dfrac{\partial\mathbf{R}}{\partial\mathbf{r}_\alpha} \\[2ex] \dfrac{\partial\boldsymbol{\rho}}{\partial\mathbf{r}_\beta} & \dfrac{\partial\mathbf{R}}{\partial\mathbf{r}_\beta} \end{vmatrix} = \left(\frac{\mathrm{a}}{x}\frac{\mathrm{B}}{\mathrm{A}+\mathrm{a}}\right)^3. \tag{16.5b}$$

We now write (16.2) as

$$\mathscr{T} = J \int \chi_\beta^{(-)*}(\mathbf{k}_\beta, \mathbf{r}_\beta)(\Phi_\mathrm{B}\Phi_\mathrm{b}|V_\beta - U_\beta|\Phi_\mathrm{A}\Phi_\mathrm{a})\chi_\alpha^{(+)}(\mathbf{k}_\alpha, \mathbf{r}_\alpha)\, d\mathbf{r}_\alpha\, d\mathbf{r}_\beta, \tag{16.6}$$

where the nuclear overlap function is,

$$(\Phi_\mathrm{B}\Phi_\mathrm{b}|V_\beta - U_\mathrm{B}|\Phi_\mathrm{A}\Phi_\mathrm{a}) = W \int d\mathbf{A}\, d\mathbf{b}\, d\mathbf{x}\, \{\Phi_\mathrm{B}\Phi_\mathrm{b}|V_\beta - U_\beta|\Phi_\mathrm{A}\Phi_\mathrm{a}\}. \tag{16.7}$$

Because the integrand is a function of $3(\mathrm{A} + \mathrm{a}) - 3$ coordinates but is integrated only over the internal coordinates shown, it remains a function of the two additional coordinates in the set (16.3). We do not mean to imply that these are the coordinates that naturally appear in the wave functions but only that such coordinates can be expressed in terms of these two, namely, **R** and $\boldsymbol{\rho}$, or through the inverse of (16.4) in terms of $\mathbf{r}_\alpha$ and $\mathbf{r}_\beta$. That is to say,

$$F(\mathbf{r}_\alpha, \mathbf{r}_\beta) \equiv (\Phi_B \Phi_b | V_\beta - U_B | \Phi_A \Phi_a). \tag{16.8}$$

The object is to obtain the explicit expression for the function F of the indicated coordinates. This function also depends on the angular momenta of the nuclei and their projections. We shall often suppress some or all of such labels on the left side of the equation.

C. PARENTAGE EXPANSIONS

We have already introduced the concept of the fractional parentage expansion. Recall that it is a way of writing a wave function for, say, n nucleons in terms of products of partitions of $n - z$ and z nucleons, for example. Fractional parentage coefficients were first used in shell-model calculations for atoms by Racah, where they expressed a partition of a pure configuration. An excellent exposition in the context of the nuclear shell-model is given by de-Shalit and Talmi (1963). A generalization of the parentage expansion is expressed by

$$\Phi_B(\mathbf{B}) \equiv \Phi_{J_B\,T_B}^{M_B \mu_B} = \sum_{J'_A T'_A J T} \beta_{JT}(\mathrm{B}, \mathrm{A}') \mathscr{A}_B [\Phi_{J'_A T'_A}(\mathbf{A}) \Phi_{JT}(\mathbf{r}_1, \mathbf{r}_2)]_{J_B\,T_B}^{M_B \mu_B}, \tag{16.9}$$

where, as usual, the square bracket denotes vector coupling and $\mathscr{A}_B$ denotes an antisymmetrization operator

$$\mathscr{A}_B = \binom{\mathrm{A} + 2}{2}^{-1/2} \sum_P (-)^P P. \tag{16.10}$$

Here P is an operator that permutes particles between the two Φ as discussed in Chapter 5, and the combinatorial factor normalizes the result to unity. The generalized parentage amplitude β should contain such labels as are necessary to completely specify all the states, and the sum includes all labels appearing only on the right side. The states of A are said to be parents of the state B. For concreteness we have specialized the discussion to two-nucleon transfer. The coordinates $\mathbf{r}_1$ and $\mathbf{r}_2$ are measured from the nucleus A and are appropriate for describing their motion when they are in the nucleus B (see Fig. 16.1). When they are in nucleus a, we shall denote the coordinate of the same nucleons measured from b as $\mathbf{r}'_1$ and $\mathbf{r}'_2$.

Recall from (15.17) that the function $\Phi_{JT}(\mathbf{r}_1, \mathbf{r}_2)$ represents an antisymmetric two-particle shell-model wave function, and that it can be transformed to relative and c.m. coordinates. From Fig. 16.1, the c.m. coordinates referring to A or b are, respectively, **R** and $\boldsymbol{\rho}$, whereas the relative coordinate is

$$\mathbf{r} = \mathbf{r}_1 - \mathbf{r}_2. \tag{16.11}$$

The transformation to these coordinates is shown in (15.23). If a is also a "heavy" nucleus, a similar parentage expansion can be made for it. The shell-model wave functions in this case will be centered on b, just as in the shell model description of B they are centered on A.

D. THE (t, p) REACTION

Light-ion reactions are a special case of the preceding general formulation in the sense that the wave functions of the "heavy" nuclei and the light ones in

$$\mathrm{t} + \mathrm{A} \rightarrow \mathrm{p} + \underbrace{(\mathrm{A} + 2)}_{\mathrm{B}}$$

are expressed in different ways. The triton wave function is better expressed directly in terms of intrinsic relative coordinates as in (15.8). On the other hand, the heavier nuclei, A and B, are usually represented by shell-model wave functions for which a parentage expansion like the preceding one can be used. The two transferred nucleons occupy shell-model states in B that are centered in the nucleus A. By contrast, in a "heavy ion" reaction all nuclei would be represented by shell-model wave functions, and two parentage expansions would be employed.

As in Chapter 15, the overlap function (16.8) corresponding to this reaction is evaluated first by introducing the parentage expansion for the nucleus B in terms of the parent states of A, and then integrating over the intrinsic coordinates A. The result can be written [see (15.5)] as

$$F(\mathbf{R}_\mathrm{p}, \mathbf{R}_\mathrm{t}) = \binom{\mathrm{a}}{2}^{1/2} \sum_{j_1 j_2 JT} \beta_{(j_1 j_2)JT}(\mathrm{B}, \mathrm{A}) C^{J_\mathrm{A} J\ J_\mathrm{B}}_{M_\mathrm{A} M M_\mathrm{B}} C^{T_\mathrm{A} T T_\mathrm{B}}_{\mu_\mathrm{A}\, \mu\, \mu_\mathrm{B}} f(\mathbf{R}_\mathrm{p}, \mathbf{R}_\mathrm{t}), \tag{16.12}$$

where

$$f(\mathbf{R}_\mathrm{p}, \mathbf{R}_\mathrm{t}) \equiv (\Phi^{M\mu}_{(j_1 j_2)JT} \Phi_\mathrm{p} | V | \Phi_\mathrm{t}) = \int d\mathbf{r}\, \{\Phi^{M\mu}_{(j_1 j_2)JT} \Phi_\mathrm{p} | V | \Phi_\mathrm{t}\}. \tag{16.13}$$

Notice that because the intrinsic coordinates **A** do not appear in V, the integration selects only the parent terms that leave the nucleons in A just as they were in the target nucleus. We have declared, by writing the functional dependence of f as $\mathbf{R}_\mathrm{p}$, $\mathbf{R}_\mathrm{t}$, that these are the coordinates in which we want to express the result of the integration, although the two unintegrated coordinates are $\mathbf{R}$ and $\boldsymbol{\rho}$. This corresponds, then, to the transition from (16.2) to (16.8) previously discussed.

The interaction in V, according to an argument similar to that expressed by (7.4), is the interaction between the particles in the light ion that are separated in the reaction

$$V = V(r_{1\mathrm{p}}) + V(r_{2\mathrm{p}}), \qquad \mathbf{r}_{1\mathrm{p}} = \mathbf{r}_1 - \mathbf{r}_\mathrm{p}. \tag{16.14}$$

We allow V to have both a finite range and a spin dependence, which we write in the form

$$V(r_{ij}) = \sum_{\bar{S}\bar{T}} V_{\bar{S}\bar{T}}(r_{ij}) P_{\bar{S}}^{ij} P_{\bar{T}}^{ij}, \tag{16.15}$$

where $P_{\bar{S}}^{ij}$ ($\bar{S} = 0, 1$) is the projection operator for spin singlet and triplet states of particles i, j. In terms of the Pauli spin matrices,

$$P_0^{ij} = (1 - \boldsymbol{\sigma}_i \cdot \boldsymbol{\sigma}_j)/4, \qquad P_1^{ij} = (3 + \boldsymbol{\sigma}_i \cdot \boldsymbol{\sigma}_j)/4, \tag{16.16}$$

which provides the connection with other representations of the interaction.

Recall from Chapter 15 the wave functions appearing in (16.13). Their symmetry under the interchange $1 \leftrightarrow 2$ can be exploited to simplify the calculation. We obtain

$$\begin{aligned} f(\mathbf{R}_\mathrm{p}, \mathbf{R}_\mathrm{t}) = \sum & A_{S'T'} B_{n\lambda N\Lambda LSJ\;T}^{\;M} C_{M_L M_S M}^{\;L\;\;S\;\;J} \\ & \times \langle X_T^{\mu}(\tau_1, \tau_2) X_{1/2}^{\mu_\mathrm{p}}(\tau_\mathrm{p}) | P_{\bar{T}}^{2\mathrm{p}} | [X_{T'}(\tau_1, \tau_2) X_{1/2}(\tau_\mathrm{p})]_{1/2}^{\mu_\mathrm{t}} \rangle \\ & \times \langle X_S^{M_S}(\sigma_1, \sigma_2) X_{1/2}^{m_\mathrm{p}}(\sigma_\mathrm{p}) | P_{\bar{S}}^{2\mathrm{p}} | [X_{S'}(\sigma_1, \sigma_2) X_{1/2}(\sigma_\mathrm{p})]_{1/2}^{m_\mathrm{t}} \rangle \\ & \times \{ [\phi_{n\lambda}(\sqrt{\tfrac{1}{2}v}\,\mathbf{r}) \phi_{N\Lambda}(\sqrt{2v}\,\mathbf{R})]_L^{M_L} | V(r_{2\mathrm{p}}) + (-)^\lambda V(r_{1\mathrm{p}}) | \Phi_\mathrm{t}(\text{space}) \} \end{aligned} \tag{16.17a}$$

The last bracket is an integral on $\mathbf{r}$. The coefficient $A_{S'T'}$ comes from (15.9) and is defined by

$$A_{S'T'} = \frac{(-)^{S'}}{\sqrt{2}} \frac{1 - (-)^{S'+T'}}{2}, \tag{16.17b}$$

whereas B is the coefficient in (15.23) obtained by rewriting the shell-model wave function of the two neutrons in nucleus B in relative and c.m. coordinates

$$\begin{aligned} B_{n\lambda N\Lambda LSJT}^{\;M} = & \left(\frac{2}{1 + \Delta_{j_1 j_2}} \right)^{1/2} \frac{1 + (-)^{\lambda + S + T + 1}}{2} \\ & \times \begin{bmatrix} \ell_1 & \tfrac{1}{2} & j_1 \\ \ell_2 & \tfrac{1}{2} & j_2 \\ L & S & J \end{bmatrix} \langle n\lambda N\Lambda; L | n_1 \ell_1 n_2 \ell_2; L \rangle. \end{aligned} \tag{16.17c}$$

We first show how to do the spin matrix elements. They are most easily evaluated by employing the recoupling algebra presented in the Appendix and discussed in greater detail by Edmonds (1957). Because the projection operator P acts on the pair 2, p, we want to recouple so that this pair is coupled together. Thereupon the action of P is immediate. We note that

$$
\begin{aligned}
&|[X_{S'}(\sigma_1, \sigma_2)X_{1/2}(\sigma_p)]^{m_t}_{1/2}\rangle \\
&\quad \equiv |(\tfrac{1}{2}\ \ \tfrac{1}{2})S', \tfrac{1}{2}; \tfrac{1}{2}m_t\rangle \\
&\quad = \sum_{K'} (\hat{S}'\hat{K}')^{1/2} \begin{Bmatrix} \frac{1}{2} & \frac{1}{2} & S' \\ \frac{1}{2} & \frac{1}{2} & K' \end{Bmatrix} |\tfrac{1}{2}, (\tfrac{1}{2}\ \ \tfrac{1}{2})K'; \tfrac{1}{2}m_t\rangle \\
&\quad = \sum_{K''} (\hat{S}'\hat{K}')^{1/2} \begin{Bmatrix} \frac{1}{2} & \frac{1}{2} & S' \\ \frac{1}{2} & \frac{1}{2} & K' \end{Bmatrix} |[X_{1/2}(\sigma_1)X_{K'}(\sigma_2, \sigma_p)]^{m_t}_{1/2}\rangle.
\end{aligned}
\tag{16.18}
$$

To accomplish the similar recoupling on the left side of the spin matrix element, we must first express the product in terms of states of total angular momentum. Thus

$$
\begin{aligned}
&|X^{M_S}_S(\sigma_1, \sigma_2)X^{m_p}_{1/2}(\sigma_p)\rangle \\
&\quad = \sum_{QM_Q} C^{S\ 1/2\ Q}_{M_S\, m_p\, M_Q} |(\tfrac{1}{2}\ \ \tfrac{1}{2})S, \tfrac{1}{2}; QM_Q\rangle \\
&\quad = \sum_{QM_Q} C^{S\ 1/2\ Q}_{M_S\, m_p\, M_Q} \sum_K (\hat{S}\hat{K})^{1/2} \begin{Bmatrix} \frac{1}{2} & \frac{1}{2} & S \\ \frac{1}{2} & Q & K \end{Bmatrix} |[X_{1/2}(\sigma_1)X_K(\sigma_2, \sigma_p)]^{M_Q}_Q\rangle.
\end{aligned}
\tag{16.19}
$$

Hence we obtain

$$
\begin{aligned}
&\langle X^{M_S}_S(\sigma_1, \sigma_2)X^{m_p}_{1/2}(\sigma_p)|P^{2p}_{\bar{S}}|[X_{S'}(\sigma_1, \sigma_2)X_{1/2}(\sigma_p)]^{m_t}_{1/2}\rangle \\
&\quad = C^{S\ 1/2\ 1/2}_{M_S\, m_p\ \ m_t}(\hat{S}'\hat{S})^{1/2}\hat{\bar{S}} \begin{Bmatrix} \frac{1}{2} & \frac{1}{2} & S' \\ \frac{1}{2} & \frac{1}{2} & \bar{S} \end{Bmatrix} \begin{Bmatrix} \frac{1}{2} & \frac{1}{2} & S \\ \frac{1}{2} & \frac{1}{2} & \bar{S} \end{Bmatrix}.
\end{aligned}
\tag{16.20}
$$

The isospin matrix element has the same structure and can be written immediately using this result.

We turn now to the radial matrix element in (16.17a). A possible triton wave function was given by (15.8) as.

$$\Phi_t(\text{space}) = \phi_{10}(2\eta\rho)\phi_{10}(\sqrt{3}\eta r). \tag{16.21}$$

The coordinates are shown in Fig. 16.1 where, in this case, b ≡ p. Notice that

$$
\begin{aligned}
r_{1p} &= |\boldsymbol{\rho} + \tfrac{1}{2}\mathbf{r}| = (\rho^2 + \tfrac{1}{4}r^2 + \rho r \cos\theta)^{1/2}, \\
r_{2p} &= |\boldsymbol{\rho} - \tfrac{1}{2}\mathbf{r}| = (\rho^2 + \tfrac{1}{4}r^2 - \rho r \cos\theta)^{1/2}.
\end{aligned}
\tag{16.22}
$$

Therefore, $V(r_{2p})$ is a function of the lengths ρ and r and the angle between the vectors $\boldsymbol{\rho}$ and $\mathbf{r}$,

$$\cos\theta = \mathbf{r}\cdot\boldsymbol{\rho}/r\rho. \tag{16.23}$$

Consequently, as in Austern *et al.* (1964), $V(r_{2p})$ can be expanded on a complete set of Legendre functions‡

$$V_{\bar{S}\bar{T}}(r_{2p}) = \sum_k \frac{2k+1}{4\pi} F_k^{\bar{S}\bar{T}}(\rho, r) P_k(\cos\theta), \tag{16.24}$$

where the coefficients can be evaluated by inverting this equation, using the orthogonality relation

$$\int_0^\pi P_\ell(\cos\theta) P_k(\cos\theta)\sin\theta\, d\theta = \frac{2}{2k+1}\delta_{\ell k}. \tag{16.25}$$

Thus

$$F_k^{\bar{S}\bar{T}}(\rho, r) = 2\pi \int_0^\pi P_k(\cos\theta) V_{\bar{S}\bar{T}}(r_{2p}) \sin\theta\, d\theta. \tag{16.26}$$

The advantage of this expansion is that $P_k(\cos\theta)$ can be written as a product of spherical harmonics depending on $\hat{\boldsymbol{\rho}}$ and $\hat{\mathbf{r}}$ [see (A9)]. Thus

$$V_{\bar{S}\bar{T}}(r_{2p}) = \sum_k F_k^{\bar{S}\bar{T}}(\rho, r) Y_k(\hat{\boldsymbol{\rho}})\cdot Y_k(\hat{\mathbf{r}}). \tag{16.27}$$

The integration on $\mathbf{r}$ in (16.17) can now be done with the result

$$\begin{aligned}\{[\phi_{n\lambda}(\sqrt{\tfrac{1}{2}}v\mathbf{r})\phi_{N\Lambda}(\sqrt{2}v\mathbf{R})]_L^{M_L}|V(r_{2p}) + (-)^\lambda V(r_{1p})|\Phi_t(\text{space})\}\\ = u_{N\Lambda}(\sqrt{2}vR) f_{n\lambda}^{\bar{S}\bar{T}}(\rho)[Y_\lambda(\hat{\boldsymbol{\rho}})Y_\Lambda(\hat{\mathbf{R}})]_L^{M_L*},\end{aligned} \tag{16.28}$$

where we have defined

$$f_{n\lambda}^{\bar{S}\bar{T}}(\rho) = \phi_{10}(2\eta\rho)\int r^2\, dr\, u_{n\lambda}(\sqrt{\tfrac{1}{2}}vr) F_\lambda^{\bar{S}\bar{T}}(\rho, r)\phi_{10}(\sqrt{3}\eta r). \tag{16.29}$$

In these equations $u_{N\Lambda}$ and $u_{n\lambda}$ are the radial parts of the wave functions of ϕ, as in (15.8). Gathering together the results so far, we find

$$\begin{aligned} f(\mathbf{R}_p, \mathbf{R}_t) = 2\sum & A_{S'T'} B^{M}_{n\lambda N\Lambda LSJT} C^{L\ S\ J}_{M_L M_S M} \\ &\times C^{S\ \ 1/2\ 1/2}_{M_S\ m_p\ m_t} C^{T\ 1/2\ 1/2}_{\mu\ \mu_p\ \mu_t} \hat{\bar{S}}\hat{\bar{T}}(\hat{S}\hat{S}'\hat{T}\hat{T}')^{1/2} \\ &\times \begin{Bmatrix} \frac12 & \frac12 & S \\ \frac12 & \frac12 & \bar{S} \end{Bmatrix} \begin{Bmatrix} \frac12 & \frac12 & S' \\ \frac12 & \frac12 & \bar{S} \end{Bmatrix} \begin{Bmatrix} \frac12 & \frac12 & T \\ \frac12 & \frac12 & \bar{T} \end{Bmatrix} \begin{Bmatrix} \frac12 & \frac12 & T' \\ \frac12 & \frac12 & \bar{T} \end{Bmatrix} \\ &\times u_{N\Lambda}(\sqrt{2}vR) f_{n\lambda}^{\bar{S}\bar{T}}(\rho)[Y_\lambda(\hat{\boldsymbol{\rho}})Y_\Lambda(\hat{\mathbf{R}})]_L^{M_L*}. \end{aligned} \tag{16.30}$$

‡ The same expansion holds for $V_{ST}(r_{1p})$ if we insert a factor $(-)^k$ on the right side [because of (16.22) and $P_k(x) = (-)^k P_k(-x)$].

At this point, all internal coordinates and spins have been integrated. The task that remains is to transform the function of $\mathbf{R}$ and $\boldsymbol{\rho}$ in the last line to a function of $\mathbf{R}_p$ and $\mathbf{R}_t$. There are several ways of doing this. One method is to employ an addition theorem (Moshinsky, 1959; Austern *et al.*, 1964) for solid spherical harmonics, which is a method that was first implemented for (d, p) reactions (Austern *et al.*, 1964) and is widely employed. Another method is to use an addition theorem for a product of a spherical Bessel function and a spherical harmonic of which the addition theorem for solid spherical harmonics is a special case (Glendenning and Nagarajan, 1974). The more general addition theorem is

$$j_\lambda(qr)Y_\lambda^m(\hat{\mathbf{r}}) = \sum_{\ell_1\ell_2} A_{\ell_1\ell_2\lambda}(-)^{\ell_1} j_{\ell_1}(\alpha q r_1) j_{\ell_2}(\beta q r_2)[Y_{\ell_1}(\hat{\mathbf{r}}_1)Y_{\ell_2}(\hat{\mathbf{r}}_2)]_\lambda^m, \quad (16.31)$$

where

$$A_{\ell_1\ell_2\lambda} = \left(\frac{4\pi\hat{\ell}_1\hat{\ell}_2}{\hat{\lambda}}\right)^{1/2} i^{\ell_1-\ell_2-\lambda} C_{0\,0\,0}^{\ell_1\ell_2\lambda}, \quad (16.32)$$

and

$$\mathbf{r} = \alpha\mathbf{r}_1 + \beta\mathbf{r}_2. \quad (16.33)$$

The special case for solid spherical harmonics is

$$r^\lambda Y_\lambda^m(\hat{\mathbf{r}}) = \sum_{\ell=0}^{\lambda} (4\pi/\hat{\ell})^{1/2} \binom{2\lambda+1}{2\ell}^{1/2} (\alpha r_1)^{\lambda-\ell}(\beta r_2)^\ell [Y_{\lambda-\ell}(\hat{\mathbf{r}}_1)Y_\ell(\hat{\mathbf{r}}_2)]_\lambda^m. \quad (16.34)$$

To use the first of these addition theorems we would have to expand $u_{N\Lambda}(\sqrt{2\nu}R)$ and $f_{n\lambda}^{\overline{ST}}(\rho)$ in a Fourier–Bessel series over a finite interval (Glendenning and Nagarajan, 1974), say, from 0 to R_{max}, where R_{max} is a radius outside which the contribution to the reaction has fallen to zero. To use the second addition theorem, we note that

$$\begin{aligned} \mathbf{R}_p &= [A/(A+2)]\mathbf{R} - \boldsymbol{\rho}, \\ \mathbf{R}_t &= \mathbf{R} - \tfrac{1}{3}\boldsymbol{\rho} \end{aligned} \quad (16.35)$$

can be inverted to give R and ρ in terms of R_p, and R_t, and $\cos\omega = \mathbf{R}_p \cdot \mathbf{R}_t/(R_pR_t)$, whence the Legendre expansion can again be made on

$$\begin{aligned} \rho^{-\lambda}R^{-\Lambda}f_{n\lambda}^{\overline{ST}}(\rho)u_{N\Lambda}(R) &= \sum_k v_k^{\overline{ST}}(R_p,R_t)Y_k(\hat{\mathbf{R}}_p)\cdot Y_k(\hat{\mathbf{R}}_t) \\ &= \sum_k \sqrt{2k+1}(-)^k v_k^{\overline{ST}}(R_p,R_t)[Y_k(\hat{\mathbf{R}}_p)Y_k(\hat{\mathbf{R}}_t)]_0^0 \end{aligned} \quad (16.36)$$

just as (16.24). In the last line we reexpressed the scalar product as a tensor product of rank zero (i.e., vector coupled the Y's).

Notice that in this application, the expansion (16.36) is valid for the function on the left because the left side is well behaved at the origin. This is so because, for small arguments,

$$f_{n\lambda}(\rho) \to \rho^{\lambda}, \qquad u_{N\Lambda}(R) \to R^{\Lambda}. \tag{16.37}$$

In some cases the inverse powers that one would have to introduce in using the solid spherical harmonic addition theorem would not be cancelled, and then the general addition theorem would be an appropriate vehicle for proceeding further.

Continuing the development, call the coordinate part of f, Eq. (16.30),

$$G_L^{M_L}(\mathbf{R}_p, \mathbf{R}_t) \equiv u_{N\Lambda}(\sqrt{2v}R) f_{n\lambda}^{\bar{S}\bar{T}}(\rho)[Y_\lambda(\hat{\rho})Y_\Lambda(\hat{\mathbf{R}})]_L^{M_L*}, \tag{16.38}$$

where we have suppressed mention on the left of some of the indices, as we have done before. It can now be written as

$$\begin{aligned} G_L^{M_L}&(\mathbf{R}_p, \mathbf{R}_t) \\ &= \sum_k (-)^k \sqrt{2k+1}\, v_k^{\bar{S}\bar{T}}(R_p, R_t) \sum_{\ell\ell'} \frac{4\pi}{(\hat{\ell}\hat{\ell}')^{1/2}} \binom{2\lambda+1}{2\ell}^{1/2} \binom{2\Lambda+1}{2\ell'}^{1/2} \\ &\quad \times (\alpha_1 R_p)^{\lambda-\ell}(\beta_1 R_t)^{\ell}(\alpha_2 R_p)^{\Lambda-\ell'}(\beta_2 R_t)^{\ell'} \\ &\quad \times [Y_k(\hat{\mathbf{R}}_p)Y_k(\hat{\mathbf{R}}_t)]_0^0 [[Y_{\lambda-\ell}(\hat{\mathbf{R}}_p)Y_\ell(\hat{\mathbf{R}}_t)]_\lambda [Y_{\Lambda-\ell'}(\hat{\mathbf{R}}_p)Y_{\ell'}(\hat{\mathbf{R}}_t)]_\Lambda]_L^{M_L*}, \end{aligned} \tag{16.39}$$

where $\alpha_1\beta_1$ and $\alpha_2\beta_2$ are defined in terms of the masses through the inverse of (16.35),

$$\begin{aligned} \boldsymbol{\rho} &= \alpha_1 \mathbf{R}_p + \beta_1 \mathbf{R}_t, \\ \mathbf{R} &= \alpha_2 \mathbf{R}_p + \beta_2 \mathbf{R}_t. \end{aligned} \tag{16.40}$$

At this point we recouple the spherical harmonics, so that those of the same argument stand together, and then we use (A11) in the form

$$\begin{aligned} [Y_{\ell_1}(\hat{\mathbf{r}})Y_{\ell_2}(\hat{\mathbf{r}})]_L^M &= (-)^L \left(\frac{\hat{\ell}_1\hat{\ell}_2}{4\pi}\right)^{1/2} \begin{pmatrix} \ell_1 & \ell_2 & L \\ 0 & 0 & 0 \end{pmatrix} Y_L^M(\hat{\mathbf{r}}) \\ &\equiv C_{\ell_1\ell_2 L} Y_L^M(\hat{\mathbf{r}}). \end{aligned} \tag{16.41}$$

The angle part of (16.39) can thus be rearranged to yield

$$\begin{aligned} \sum_{\lambda_p\lambda_t\Lambda_p\Lambda_t} &\begin{bmatrix} \lambda-\ell & \ell & \lambda \\ \Lambda-\ell' & \ell' & \Lambda \\ \lambda_p & \lambda_t & L \end{bmatrix} \begin{bmatrix} k & k & 0 \\ \lambda_p & \lambda_t & L \\ \Lambda_p & \Lambda_t & L \end{bmatrix} \\ &\times C_{(\lambda-\ell)(\Lambda-\ell')\lambda_p} C_{\ell\ell'\lambda_t} C_{k\lambda_p\Lambda_p} C_{k\lambda_t\Lambda_t} [Y_{\Lambda_p}(\hat{\mathbf{R}}_p)Y_{\Lambda_t}(\hat{\mathbf{R}}_t)]_L^{M_L*}. \end{aligned} \tag{16.42}$$

Hence,

$$G_L^{M_L}(\mathbf{R}_p, \mathbf{R}_t) = \sum_k (-)^k \sqrt{2k+1} \sum_{\ell\ell'} \frac{4\pi}{(\hat{\ell}\hat{\ell}')^{1/2}} \binom{2\lambda+1}{2\ell}^{1/2} \binom{2\Lambda+1}{2\ell'}^{1/2}$$

$$\times \sum_{\lambda_p\lambda_t\Lambda_p\Lambda_t} \begin{bmatrix} \lambda-\ell & \ell & \lambda \\ \Lambda-\ell' & \ell' & \Lambda \\ \lambda_p & \lambda_t & L \end{bmatrix} \begin{bmatrix} k & k & 0 \\ \lambda_p & \lambda_t & L \\ \Lambda_p & \Lambda_t & L \end{bmatrix}$$

$$\times C_{(\lambda-\ell)(\Lambda-\ell')\lambda_p} C_{\ell\ell'\lambda_t} C_{k\lambda_p\Lambda_p} C_{k\lambda_t\Lambda_t}$$

$$\times v_k^{ST}(R_p, R_t)(\alpha_1 R_p)^{\lambda-\ell}(\beta_1 R_t)^{\ell}(\alpha_2 R_p)^{\Lambda-\ell'}$$

$$\times (\beta_2 R_t)^{\ell'}[Y_{\Lambda_p}(\hat{\mathbf{R}}_p) Y_{\Lambda_t}(\hat{\mathbf{R}}_t)]_L^{M_L *}. \tag{16.43}$$

This is the required form. When assembled into $f(\mathbf{R}_p, \mathbf{R}_t)$, (16.30), and then into (16.12), which in turn goes under the integral (16.6), we then have an integral of the form

$$\mathrm{B}_L^{M_L}(\mathbf{k}_p, \mathbf{k}_t) = i^{-L}(2L+1)^{-1/2} \int \chi_p^{(-)*}(\mathbf{k}_p, \mathbf{R}_p) G_L^{M_L}(\mathbf{R}_p, \mathbf{R}_t) \chi_t^{(+)}(\mathbf{k}_t, \mathbf{R}_t)\, d\mathbf{R}_p\, d\mathbf{R}_t. \tag{16.44}$$

This can be reduced to a two-dimensional radial integral by expanding the distorted waves in partial wave series as in (5.14) and doing the angle integrals by a repetition of the preceding methods. In particular, there will occur a sum over a spherical harmonic from both entrance and exit channels that can be coupled together

$$Y_{\ell_p}^{m_p}(\hat{\mathbf{R}}_p) Y_{\ell_t}^{m_t}(\hat{\mathbf{R}}_t) = \sum_{\mathscr{L}\mathscr{M}} C_{m_p m_t \mathscr{M}}^{\ell_p \ell_t \mathscr{L}} [Y_{\ell_p}(\hat{\mathbf{R}}_p) Y_{\ell_t}(\hat{\mathbf{R}}_t)]_{\mathscr{L}}^{\mathscr{M}}. \tag{16.45}$$

Now the angle integrations in (16.44) can be easily done. We obtain from the orthonormality of the spherical harmonics a set of delta functions

$$\delta_{\mathscr{L}L}\delta_{\mathscr{M}M}\delta_{\ell_p\Lambda_p}\delta_{\ell_t\Lambda_t}. \tag{16.46}$$

There remains only a two-dimensional integral, which was our goal. It is as far as we can go by analytic means.

The differential cross section is obtained by inserting the transition amplitude $\mathscr{T}$ into (5.50), with the result that the cross section will be an incoherent sum over the amplitudes (16.44)

This calculation was carried through for the (t, p) reaction. However, it is trivial to obtain the amplitude for (t, n), (He^3, p), or (He^3, n) from these results. It involves merely assigning the appropriate isospin properties in the Clebsch–Gordan coefficient in (16.30). As an example for the (He^3, p) reactions, the He^3 has isospin projection $-\frac{1}{2}$ compared with $+\frac{1}{2}$ for t. Hence the Clebsch–Gordan coefficient $C^{T\ \ 1/2\ \ 1/2}_{\mu\ -1/2\ -1/2}$ should replace the one in (16.30). This demands that $\mu = 0$ and hence that T can be either 0 or 1.

At this point we can contrast the relationships between angular momenta in the zero-range approximation with the finite-range results obtained here. Additional values of transfer angular momenta are allowed by the finiteness of the range of interaction and its spin dependence. In particular, for the (p, t) reaction the isospin T of the pair of nucleons transferred to nucleus B must have $T = 1$, as is required by the Clebsch—Gordan coefficient in (16.30). This is true for both zero and finite-range interactions. However, because of the spin—isospin dependence of the the finite-range interaction, the value of isospin T' associated with this pair in the triton need not equal T. Referring to (16.30), if $f^{\bar{S}\bar{T}}_{n\lambda}$ did not depend on $\bar{S}$, $\bar{T}$, then the sum of these spins could be performed, and the orthonormality of the $6-j$ coefficients would yield $\delta_{TT'}\delta_{SS'}$. In the spin-independent case, $T = 1$ would thus require $S = 0$ for antisymmetry of the triton wave function. In case the interaction has spin dependence, then $S = 1$ is also allowed. In addition to this, the angular momentum L is constrained by a parity selection rule in the zero-range approximation but not for the finite-range interaction. This can be inferred from (16.43). To see that this is so, note that in zero-range approximation the directions $\hat{\mathbf{R}}_p$ and $\hat{\mathbf{R}}_t$ become the same. In this case, Eq. (16.43) would contain

$$[Y_{\Lambda_p}(\hat{\mathbf{R}}_p)Y_{\Lambda_t}(\hat{\mathbf{R}}_p)]_L^{M_L*},$$

which, according to the $3-j$ symbol of (16.41), will require that

$$(-)^{\Lambda_p+\Lambda_t} = (-)^L \qquad \text{(zero-range)} \tag{16.47}$$

The other four C coefficients in (16.43) can be used to show that

$$(-)^{\lambda+\Lambda} = (-)^{\lambda_p+\lambda_t} = (-)^{\Lambda_p+\Lambda_t}. \tag{16.48}$$

Recall now that through (15.21) the parity carried by the transferred particles is related to $\lambda + \Lambda$, namely,

$$(-)^{\ell_1+\ell_2} = (-)^{\lambda+\Lambda}, \tag{16.49}$$

whence

$$(-)^{\ell_1+\ell_2} = (-)^L, \qquad \text{(zero-range)} \tag{16.50}$$

which agrees with the assertion that in zero range $(-)^L$ must agree with the parity change. From (16.47) and (16.48), it also follows that in zero range

$$(-)^{\ell_p+\ell_t} = (-)^L. \qquad \text{(zero-range)} \tag{16.51}$$

On the other hand, for a finite-range interaction, L is not connected as in these last two equations (16.50, 16.51), and the loss of this connection proves to be very important in heavy-ion reactions.

For light-ion reactions, however, and in particular the (p, t) reaction, the terms that dominate (16.43) are the ones that obey the zero-range selection rules. This can be understood as follows: Recall from our previous discussion that, for a finite-range spin-dependent force, $S = 0$ and $S = 1$ are both allowed. Now recall also that the coefficient B in (16.30) is given by (16.17c), and it requires that

$$\lambda + S + T + 1 = \text{even}. \tag{16.52}$$

For the (t, p) reactions, the isospin projection on the Clebsch–Gordan coefficient with T in (16.30) requires $T = 1$. Therefore

$$\lambda + S = \text{even}. \tag{16.53}$$

This is an expression of the Pauli principle for the transferred neutrons. There are two physical effects that favor small values of λ (the smallest being $\lambda = 0$). First is the dominance of the S state in the triton, which in zero-range approximation would project on $\lambda = 0$ alone. Second is the small size of the triton, which through (16.29) would favor small λ. Third, the strong absorption of composite particles like the triton favors surface reactions. This selects states of large angular momentum Λ for the center of mass of the transferred pair and consequently, through (15.22a), small relative angular momentum λ. Therefore we expect the largest contribution to come from $\lambda = 0 = S$. When $\lambda = 0$, note that it follows automatically from the Moshinsky bracket in the factor B of (16.30) that it has the connection to the parity quoted in (16.50) and (16.51).

For these reasons it is not surprising to find that calculated cross sections that retain the finite range of the interaction are virtually identical in their angular distributions to those obtained with the zero-range approximation. Figure 16.2 shows such comparisons (Takimasa *et al.*, 1979) for the same reaction that was also considered in the last chapter. An incorrect assignment of spin and parity would not be made on account of the zero-range approximation. The calculations are, however, independently normalized to the data, and differences of up to 30% occur between the finite-range and zero-range calculations regarding the magnitude of the cross section. This is, of course, significant for testing the correlation properties of nuclear wave functions. Similar calculations of Takimasa *et al.* (1979) are shown for the Zr reaction with similar conclusions (see Fig. 16.3). Nagarajan and collaborators (Werby *et al.*, 1980; Feng *et al.*, 1981; Nagarajan, 1982) have been especially careful in their treatment of the triton wave function, allowing

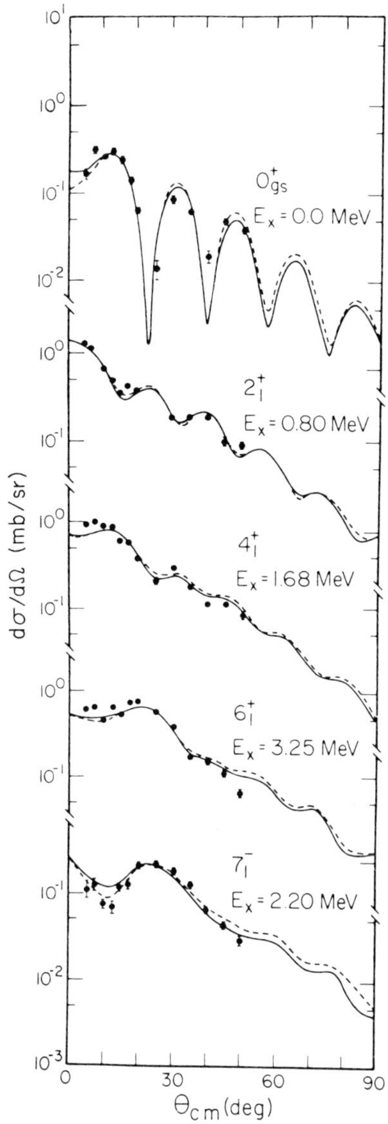

Fig. 16.2. Angular distributions (independently normalized to the data) for normal parity states calculated in zero-range approximation (---) and compared with the finite-range (———) calculation for ^{206}Pb reaction at 40 MeV. [Finite-range calculation from Takimasa *et al.* (1979); data from S. M. Smith, *et al.* (1968); zero-range calculation by Smith *et al.* (1970)].

for mixed symmetry and retaining self-consistency with the interactions employed. Their results generally confirm the situation previously described. Of course, the reward for the careful treatment is excellent agreement with the data in a calculation free of adjustable parameters. Their work therefore confirms the general theory of the reaction mechanism.

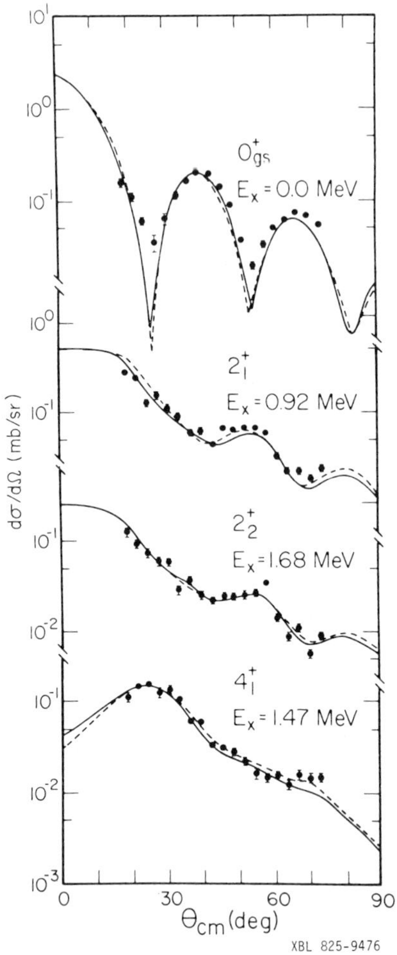

Fig. 16.3. Comparisons of zero-range (---) and finite-range (———) calculations for ^{94}Zr reaction at 20 MeV from Takimasa *et al.* (1979), (Data from Flynn, 1974.)

E. TWO-NUCLEON TRANSFER BETWEEN HEAVY IONS

As another example, we specialize to two-nucleon transfer between heavy ions. Some of the features of reactions that are peculiar to heavy ions will be taken up in a later chapter. We have already alluded to the fact that for these reactions the zero-range approximation would be poor, and that is why it is in this chapter that these reactions first appear as an example. The general development of scattering theory in the preceding chapters, of course, applies to these reactions as well as to light-ion reactions.

Let us look now at the direct interaction $V_\beta - U_\beta$. Separating out the two nucleons that are transferred and giving them the labels 1 and 2, we can write

$$V_\beta = \sum_{\substack{i\in \mathrm{b} \\ j\in \mathrm{B}}} V_{ij} = \sum_{i\in \mathrm{b}} \left(V_{i1} + V_{i2} + \sum_{j\in \mathrm{A}} V_{ij} \right). \tag{16.54}$$

As in the case of the (d, p) reaction, we assume that the optical potential U_β on the average cancels the last term. Then

$$V_\beta - U_\beta \simeq \sum_{i\in \mathrm{b}} V_{i1} + V_{i2}, \tag{16.55}$$

where the right side is the sum of interactions of the nucleons in b with 1 and 2, the two nucleons that are transferred in the reaction.

Let us look ahead now. We plan to use the parentage expansions such as (15.4) in (16.7). There will then occur the factor

$$\int d\mathbf{b}\, \phi_{j_\mathrm{b}\, t_\mathrm{b}}^{m_\mathrm{b}\mu_\mathrm{b}*}(\mathbf{b}) \sum_{i\in \mathrm{b}} (V_{i1} + V_{i2}) \phi_{j'_\mathrm{b}\, t'_\mathrm{b}}^{m'_\mathrm{b}\mu'_\mathrm{b}}(\mathbf{b}), \tag{16.56}$$

which is embedded with the other wave functions and integrated over all coordinates. The primes on the wave function on the right occur through the parentage expansion of Φ_a. However, because V_{i1} and V_{i2} are one-body operators as concerns nucleus b, the two states of b can differ by at most one nucleon and, in such an event, the interaction of the transferred nucleons 1 and 2 *must* be with that nucleon. However, for the diagonal elements, 1 and 2 (can interact coherently with each nucleon in b. For this reason we approximate (16.56) by its diagonal element and write

$$\int d\mathbf{b}\, \phi^*(\mathbf{b}) \sum_{i\in \mathrm{b}} (V_{i1} + V_{i2}) \phi'(\mathbf{b}) = [v(r'_1) + v(r'_2)]\delta_{j_\mathrm{b} j'_\mathrm{b}} \cdots. \tag{16.57}$$

We use the prime coordinates because, according to Fig. 16.1, these refer to b as a reference point. From this definition of $v(r'_1)$ it is a potential experienced by particle 1 in the field of the particles of the nucleus b. This would be typically represented by the shell-model potential of nucleus a with b as core.

Because none of the coordinates of A occur in the potentials, the integration on **A** will select out the $J'_\mathrm{A} = J_\mathrm{A}$ term in the parentage expansion for Φ_B when it is used in (16.7). Thus we will obtain

$$F(\mathbf{r}_\alpha, \mathbf{r}_\beta) = \sum_{JTJ'T'} \beta_{JT}(\mathrm{B}, \mathrm{A}) \beta_{J'T'}(\mathrm{a}, \mathrm{b}) \times \sum_{MM'\mu\mu'} C_{M_\mathrm{A} M M_\mathrm{B}}^{J_\mathrm{A}\, J\, J_\mathrm{B}} C_{\mu_\mathrm{A}\, \mu\, \mu_\mathrm{B}}^{T_\mathrm{A} T T_\mathrm{B}} C_{m_\mathrm{b} M' m_\mathrm{a}}^{j_\mathrm{b}\; J' j_\mathrm{a}} C_{\mu_\mathrm{b} \mu' \mu_\mathrm{a}}^{t_\mathrm{b}\; T' t_\mathrm{a}} f(\mathbf{r}_\alpha, \mathbf{r}_\beta), \tag{16.58}$$

where

$$f(\mathbf{r}_\alpha, \mathbf{r}_\beta) = \int d\mathbf{r}\, \phi_{JT}^{M\mu*}(\mathbf{r}_1, \mathbf{r}_2)[v(\mathbf{r}'_1) + v(\mathbf{r}'_2)] \phi_{J'T'}^{M'\mu'}(\mathbf{r}'_1, \mathbf{r}'_2). \tag{16.59}$$

(Recall that in the specialization to two-nucleon transfer, the integration $d\mathbf{x}$ becomes $d\mathbf{r}$.) By a succession of manipulations like those described for the (t, p) reaction, this can eventually be expressed in the form of a sum over such forms as

$$[Y_{\lambda_\alpha}(\hat{\mathbf{r}}_\alpha)Y_{\lambda_\beta}(\hat{\mathbf{r}}_\beta)]_L^{M_L}, \tag{16.60}$$

embedded in a function analogous to (16.43). Again, upon making a partial wave expansion of the distorted waves one can do the angle integrals in closed form. All integrals can thus be performed analytically except for a two-dimensional radial integral that has to be performed numerically. Because the manipulations are rather similar, we do not repeat them.

Whereas in light-ion reactions like (p, t) the zero-range approximation is rather good, for heavy-ion reactions it is very poor. Because of the large number of partial waves typically needed in heavy-ion calculations, approximation methods that at least retained the finite size were sought.

From the vector relations (16.4), we can derive

$$\mathbf{r}_\beta = \frac{\mathrm{A}}{\mathrm{B}}\mathbf{r}_\alpha - \frac{x}{\mathrm{a}}\frac{\mathrm{A}+\mathrm{a}}{\mathrm{A}+x}\boldsymbol{\rho}. \tag{16.61}$$

By neglecting the last term, the arguments of the distorted waves become proportional and their directions identical. The simplification in (16.43) is immediate and similar to the zero-range approximation yielding the same selection rules, which we refer to as "normal selection rules." Such approximations go under the name of no-recoil approximations (Buttle and Goldfarb,1966, 1968; Baltz and Kahana, 1972; Roberts, 1972).

However, the neglect of the last term in (16.61) is a poor approximation if the de Broglie wavelength of the relative motion between the nuclei is small compared with the neglected length (with $\rho \simeq$ radius of a). Dodd and Greider (1969) particularly emphasized the considerable effect on angular distributions that recoil approximations have. As already mentioned, the no-recoil approximation restricts the angular momentum transfer to the "normal" values. Because the kinematics of heavy-ion reactions generally favor large angular momentum transfer, if the "normal" values are smaller than those allowed in the general theory, then the no-recoil approximation can be especially poor (Kovar *et al.* 1973).

Nevertheless, because so many partial waves are needed in heavy-ion reaction calculations, approximation schemes were invented that treated the neglected quantity in first or second order (Nagarajan, 1972, 1973; Baltz and Kahana, 1974), thus achieving considerable economy. However, complete calculations (Bock and Yoshida, 1972; Devries, 1973; Devries and Kubo, 1973) were also carried out, and they demonstrated the very dramatic contributions of the nonnormal angular momenta (Devries, 1973; Devries and

Kubo, 1974). Figure 16.4 shows how, for the $^{12}C(^{14}N, ^{13}C)^{13}N$ reaction, the angular distribution corresponding to only the normal selection rules (in this case $\ell = 0$) is strongly oscillating, which is in disagreement with the data. However, the additional angular momentum ($\ell = 1$) admitted by the full calculation yields an angular distribution that is out of phase with the normal, and together they yield the featureless cross section observed.

The recoil effects are not always so dramatic, and many qualitative insights into the heavy-ion reaction mechanism can be gained by ignoring them. It is clear, however, that for accurate microscopic analysis they cannot be ignored. The subject of heavy-ion reactions will be taken up again in greater detail in Chapter 18.

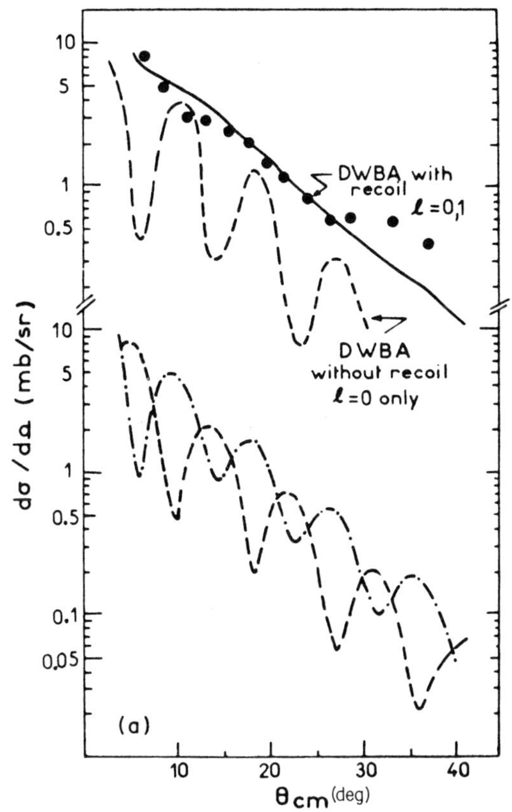

Fig. 16.4. The recoil effects in a heavy ion reaction for $^{12}C(^{14}N, ^{13}C)^{13}N$ reaction at 78 MeV. The top half shows DWBA with recoil (———) with $\sum \ell = 0, 1$ and DWBA without recoil (— — —) with $\ell = 0$. Lower half shows components of DWBA $\ell = 0$ (— · —) and $\ell = 1$ (— — —) from Devries and Kubo (1973).

Chapter 17

Higher-Order Processes in Particle Transfer Reactions

A. BEYOND THE DWBA

The three principal assumptions underlying the DWBA were discussed in Chapter 5. They are as follows:

(1) The transfer is assumed to take place directly from the initial to final state. All particles except those actually transferred are assumed to remain in their original states.
(2) The wave function for the relative motion between the partners in the reaction is assumed to be correctly described by the optical potential model, even in the interior region, which enters most crucially for reactions.
(3) The reaction is assumed to be sufficiently weak that it can be treated in lowest order.

Clearly there are conceivable nuclear states for which the cross section would vanish under the first assumption. Figure 17.1 illustrates, in an idealization, just such a situation. The first two states of the nucleus A + 1 can be reached by transferring a particle into the single-particle states j and j', respectively, whereas all other nucleons remain in their original state. The third state, however, cannot be produced in this way. Instead, it can be

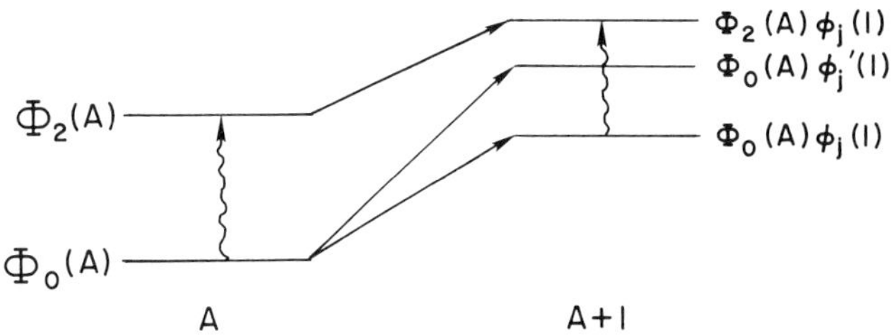

Fig. 17.1. Lower two states in A + 1 can be reached by stripping a particle into the *j* and *j'* orbital. The third state can be reached only by a second or higher order process involving excitation of the core as well. The two squiggly arrows show the strong inelastic transitions.

produced only if the core nucleons A are excited either before or after the transfer. More generally, if its structure is $a\Phi_2(A)\Phi_j(1) + b\Phi_0(A)\Phi_j$, then it can be reached both directly through the component in the wave function with amplitude b and indirectly by the two paths indicated in the figure through the component in the wave function with amplitude a. Because there is no way of distinguishing between the various paths by which a given level is produced, the total amplitude will be a superposition of all of them. Depending on the phases and kinematic factors of each amplitude, the interference of the subamplitudes can be of any degree from constructive to destructive. The cross section, therefore, can be larger or smaller than that for any one of the transitions or even larger than the sum of all individual transitions, and the angular distribution also can be radically altered from that of an individual path. Of course, as remarked, the cross sections of individual paths cannot be measured experimentally, but they can be isolated in our calculations, and this frequently helps in understanding the process. The interference among several transitions is a new element in our discussion of transfer reactions. Although it does make the analysis more difficult, a successful analysis of such a situation would yield more information about the nuclear wave function through such an interference between several amplitudes in the wave function.

In this chapter the theory will be developed to take account of possible indirect transitions. At the same time, as will become clear, the second approximation concerning the wave function for the relative motion will be improved. The method that relaxes the first two assumptions is known as the coupled-channel Born approximation. Finally, under those rarer circumstances for which the transfer amplitudes are large, we must relax the third assumption. The method for doing so is known as the coupled-reaction channel approximation.

Figure 17.1 makes it clear that, in general, transfer reactions involve not only the actual transfer of particles that are indicated by the diagonal arrows but also inelastic transitions of both initial and final nuclei. Obviously, transition paths having an inelastic transition will be more important in cases where such a transition is itself collective. In Chapters 8, 9, and 13, the coupled equations for scattering among channels within a given partition were derived and analyzed. In Chapter 5, the amplitude for a reaction leading from one partition to another was derived. A generalization is now needed in which scattering among channels of the initial and final partitions and reactions connecting the different partitions are simultaneously treated.

B. COUPLED-CHANNEL BORN APPROXIMATION (GENERALIZED DWBA)

The desired generalization is analogous to Eq. (5.17). In the derivation of (5.17), we assumed that the potentials U_α and U_β are optical potentials depending only on the relative coordinates r_α and r_β in the two partitions. In such a case they can cause neither excitation within a partition nor reactions from one partition to another. However, if, instead, we allow U to depend also on the nuclear coordinates, then such transitions are possible. Denote these more general optical potentials as $\hat{U}$. More particularly, suppose that $\hat{U}_\alpha$ is defined in such a way that transitions only within the partition α are possible, and similarly for $\hat{U}_\beta$. An example is the vibrational or deformed optical potential of Chapters 7 and 13. These are very appropriate examples because we anticipate that the higher-order transitions can compete with the first order only if there is an enchancement involved in the former, a suppression in the latter, or both.

The generalization of the Gell-Mann–Goldberger relation derived in Chapter 6 provides the basis of the coupled-channel Born approximation (CCBA). Recall that (6.39, 6.46) were derived without the aid of any particular assumption about the nature of the auxiliary potential U, other than that it is different from V. Thus those results certainly apply now to our generalized optical potentials $\hat{U}_\alpha$ and $\hat{U}_\beta$. We can restate the result in terms of $\hat{U}_\alpha$ and $\hat{U}_\beta$ as

$$\mathscr{T}_{\beta\alpha} = \langle \Psi_\beta^{(-)} | V_\alpha - \hat{U}_\alpha | \hat{\chi}_\alpha^{(+)} \rangle = \langle \hat{\chi}_\beta^{(-)} | V_\beta - \hat{U}_\beta | \Psi_\alpha^{(+)} \rangle. \tag{17.1}$$

This is still an exact result.‡ We have denoted the solution to the Schrödinger equation containing $\hat{U}_\beta^\dagger$ by $\chi_\beta^{(-)}$.§ It satisfies,

$$(E - H_\beta - T_\beta - \hat{U}_\beta^\dagger)\hat{\chi}_\beta^{(-)}(\mathbf{r}_\beta, \mathbf{B}, \mathbf{b}) = 0. \tag{17.2a}$$

‡ Since $\hat{U}$ causes no rearrangement collision by assumption, the first term of 6.39 and 6.46 vanishes.

§ Recall the convention of Chapter 6. If U is complex, then $\chi^{(-)'}$ satisfies the Schrödinger equation containing $U^\dagger$, whereas that for $\chi^{(+)}$ contains U.

This can be expanded in terms of the nuclear states that $\hat{U}_\beta$ can excite

$$\hat{\chi}_\beta^{(-)}(\mathbf{r}_\beta, \mathbf{B}, \mathbf{b}) = \sum_{\beta'} \chi_{\beta'}^{\beta(-)}(\mathbf{r}_\beta)\Phi_{\beta'}(\mathbf{B}, \mathbf{b}), \tag{17.2b}$$

where the $\chi_{\beta'}^{\beta}$ obey coupled equations as in Chapter 8. Two indices are needed on $\chi_{\beta'}^{\beta}$; the superscript denotes the original state of the system as being the channel β of the partition B + b, and the subscript denotes a possible other channel β' in the same partition that can be reached because of the interaction $\hat{U}_\beta$. The amplitude for the elastic or inelastic transition induced by $\hat{U}_\beta$ is given, according to the general result (3.33, 6.32), by

$$\mathscr{T}_{\beta'\beta}^{\hat{U}_\beta} = \langle \exp(i\mathbf{k}'_\beta \cdot \mathbf{r}_\beta)\Phi_{\beta'}|\hat{U}_\beta|\hat{\chi}_\beta^{(+)}\rangle = \langle \hat{\chi}_{\beta'}^{(-)}|\hat{U}_\beta| \exp(i\mathbf{k}_\beta \cdot \mathbf{r}_\beta)\Phi_\beta\rangle, \tag{17.2c}$$

where in the case of a nonhermitian potential $\hat{U}_\beta$, the convention described in Chapter 6 for the incoming wave solution $\hat{\chi}_{\beta'}^{(-)}$ is observed (c.f. 6.44).

For the channel α and potential $\hat{U}_\alpha$ we have a similar set of equations:

$$(E - H_\alpha - T_\alpha - \hat{U}_\alpha)\hat{\chi}_\alpha^{(+)}(\mathbf{r}_\alpha, \mathbf{A}, \mathbf{a}) = 0, \tag{17.3a}$$

$$\hat{\chi}_\alpha^{(+)}(\mathbf{r}_\alpha, \mathrm{A}, \mathrm{a}) = \sum_{\alpha'} \chi_{\alpha'}^{\alpha(+)}(\mathbf{r}_\alpha)\Phi_{\alpha'}(\mathbf{A}, \mathbf{a}), \tag{17.3b}$$

$$\mathscr{T}_{\alpha'\alpha}^{\hat{U}_\alpha} = \langle \exp(i\mathbf{k}_{\alpha'} \cdot \mathbf{r}_\alpha)\Phi_{\alpha'}|\hat{U}_\alpha|\hat{\chi}_\alpha^{(+)}\rangle = \langle \hat{\chi}_{\alpha'}^{(-)}|\hat{U}_\alpha| \exp(i\mathbf{k}_\alpha \cdot \mathbf{r}_\alpha)\Phi_\alpha\rangle, \tag{17.3c}$$

which describes the elastic and inelastic scattering in the partition α. These equations [(17.2), (17.3)] are more general than the optical-model equations employed in Chapter 5. There, U was a one-channel optical potential depending only on the relative channel coordinate and therefore could describe only elastic scattering.

Recall Eq. (6.38), which allows us to write

$$\Psi_\alpha^{(+)} = \hat{\chi}_\alpha^{(+)} + \frac{1}{E - H + i\varepsilon}(V_\alpha - \hat{U}_\alpha)\hat{\chi}_\alpha^{(+)}. \tag{17.4}$$

We now approximate the exact result (17.1) by using the first term only for $\Psi_\alpha^{(+)}$, obtaining,

$$\mathscr{T}_{\beta\alpha(\text{post})}^{\text{CCBA}} = \langle \hat{\chi}_\beta^{(-)}|V_\beta - \hat{U}_\beta|\hat{\chi}_\alpha^{(+)}\rangle = \sum_{\alpha'\beta'} \langle \chi_{\beta'}^{\beta(-)}\Phi_{\beta'}|V_\beta - \hat{U}_\beta|\chi_{\alpha'}^{\alpha(+)}\Phi_{\alpha'}\rangle. \tag{17.5a}$$

This is referred to as the post form of the amplitude, the interaction $V_\beta - \hat{U}_\beta$ referring to the final partition. Likewise, from Chapter 6, (6.47),

$$\Psi_\beta^{(-)} = \hat{\chi}_\beta^{(-)} + \frac{1}{E - H - i\varepsilon}(V_\beta - \hat{U}_\beta^\dagger)\hat{\chi}_\beta^{(-)}. \tag{17.6}$$

Approximating this by its first term in (17.1) gives

$$\mathscr{T}_{\beta\alpha(\text{prior})}^{\text{CCBA}} = \langle \hat{\chi}_\beta^{(-)}|V_\alpha - \hat{U}_\alpha|\hat{\chi}_\alpha^{(+)}\rangle = \sum_{\alpha'\beta'} \langle \chi_{\beta'}^{\beta(-)}\Phi_{\beta'}|V_\alpha - \hat{U}_\alpha|\chi_{\alpha'}^{\alpha(+)}\Phi_{\alpha'}\rangle. \tag{17.5b}$$

These two approximate results are equal, for if we take their difference, we find[‡]

$$\begin{aligned}\langle\hat{\chi}_\beta^{(-)}|V_\beta - V_\alpha - \hat{U}_\beta + \hat{U}_\alpha|\hat{\chi}_\beta^{(+)}\rangle \\ = \langle\hat{\chi}_\beta^{(-)}|(H_\alpha + T_\alpha + \hat{U}_\alpha) - (H_\beta + T_\beta + \hat{U}_\beta)|\hat{\chi}_\alpha^{(+)}\rangle = 0.\end{aligned} \qquad (17.7)$$

The difference vanishes "on the energy shell" according to (17.2a) and (17.3a). Of course, any subsequent approximation can violate the equality.

The approximation (17.5) is known as the coupled-channel Born approximation (CCBA). The $\alpha' = \alpha$ and $\beta' = \beta$ in (17.5) correspond to the usual DWBA, although they differ in one aspect that can be significant. The functions χ_α of the usual DWBA obey a single optical-model Schrödinger equation, whereas the $\chi_{\alpha'}^{\alpha}$ obey a set of coupled equations. Although, as remarked in connection with the approximations involved in the DWBA (Chapter 5), the optical potential can, in general, be adjusted to yield the phase shifts that correctly describe the elastic scattering, this only means that the corresponding wave function for the relative motion is asymptotically correct. It does not ensure that the wave function is correct in the region of the interaction, where it is needed in the DWBA matrix element. To the extent that the coupled functions $\chi_{\alpha'}^{\alpha}$ describe the scattering in several channels, they lie closer to the truth. Thus the generalization allows indirect transitions, which the original theory did not, *and* it has improved the description of the relative motion wave functions. It still treats the transition caused by V_β in first order.

The first calculations based on (17.5) were carried out by Penny and Satchler (1964). However, it remained for a later development to take place to provide the stimulus for fully calculating such an amplitude as (17.5). Early work allowed for inelastic transitions $\alpha \to \alpha'$ in the incident channel or $\beta \to \beta'$ in the final channel, but not both (Iano and Austern, 1966; Kozlowsky and deShalit, 1966; Levin, 1966). Evidently a complete calculation is much more time-consuming than DWBA. The reason is that in either case $\chi_{\alpha'}^{\alpha}(\mathbf{r})$ is actually found by making a partial-wave expansion, as in Chapter 8. Convergence is achieved only by carrying the expansion to $\ell_{\max} \lesssim k_\alpha R$, where R is the range of the interaction (several nuclear radii). Thus in DWBA there are $\ell_{\max}^2$ integrals of the form

$$\int (u_\ell Y_\ell)^* V(u_{\ell'} Y_{\ell'})\, d\mathbf{r}_\alpha\, d\mathbf{r}_\beta \qquad (17.8)$$

to be evaluated. For the generalized result (17.5), if there are N_α and N_β coupled equations in channels α and β, then there are $N_\alpha N_\beta \ell_{\max}^2$ such integrals.

‡ Recall Chapter 6 Section K for a rigorous proof of hermiticity as used in (17.7).

C. SOURCE-TERM METHOD

An alternative to evaluating the generalized DWBA integrals is provided by the so-called source-term method (Ascuitto and Glendenning, 1969). It is mathematically equivalent to the generalized DWBA (Mackintosh and Glendenning, 1960). It was by this method that the full complexity of the couplings were first taken into account. From a programming point of view, it does have an advantage, which is possibly why a full calculation was done with this method before Eq. (17.5) was fully calculated. From a computational point of view it is possibly also more efficient whenever the cross section to more than one final state is desired.

We note that the amplitude (17.5) requires the solution to two sets of coupled equations arising from use of the expansion (17.2) in (17.1) for the partition α, and similarly for β. They are

$$(E_{\alpha'} - T_{\alpha'} - U_{\alpha'})\chi^{\alpha}_{\alpha'} = \sum_{\alpha''} U_{\alpha'\alpha''}\chi^{\alpha}_{\alpha''}, \qquad \text{for each} \quad \alpha', \tag{17.9}$$

$$(E_{\beta'} - T_{\beta'} - U_{\beta'})\chi^{\beta}_{\beta'} = \sum_{\beta''} U_{\beta'\beta''}\chi^{\beta}_{\beta''}, \qquad \text{for each} \quad \beta', \tag{17.10}$$

where

$$U_{\alpha'\alpha''} = \langle \Phi_{\alpha'} | \hat{U}_{\alpha} | \Phi_{\alpha''} \rangle, \tag{17.11}$$

and

$$U_{\alpha} = U_{\alpha\alpha} \simeq U_{\alpha'\alpha'} = U_{\alpha'}. \tag{17.12}$$

Remember the earlier results and discussion (Chapter 8) that truncation to a finite basis implies the use of an effective interaction, and its diagonal elements are approximately the same, and generally taken to be a common optical potential in closely related channels α, α', α'', ...

The terms on the right sides of the differential equations (17.9, 17.10) are sources in the mathematical sense, so that, even if U were real, the flux in a given channel would not be conserved. That is to say, the terms on the right side shift flux among the channels. They represent, for channel β', the transitions indicated in Fig. 17.2 by wavy lines that touch β' with an arrow. In the sense of Fig. 17.2, the coupled equation (17.10) for β' is incomplete. What is needed are terms that represent the reactions that lead to β', such as the diagonal arrows in Fig. 17.2. Thus consider instead the equation

$$(E_{\beta'} - T_{\beta'} - U_{\beta'})\chi^{\alpha}_{\beta'} = \sum_{\beta''} U_{\beta'\beta''}\chi^{\alpha}_{\beta''} + \sum_{\alpha'} \rho^{\alpha}_{\beta'\alpha'}, \qquad \text{for each} \quad \beta', \tag{17.13}$$

where α' is summed over α, α', α'', ... This is an inhomogeneous equation, the homogeneous equation to which is (17.10). It should not be surprising,

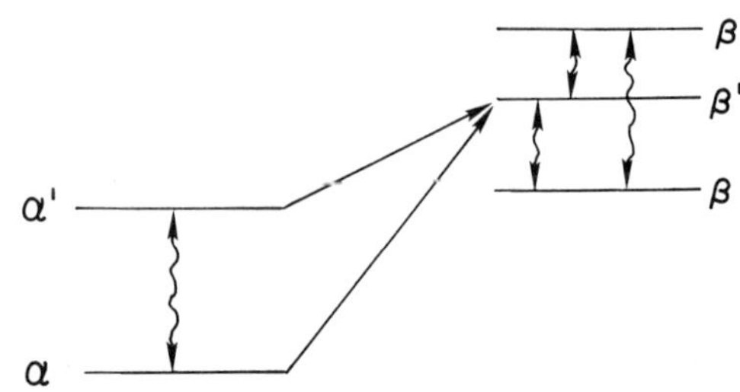

Fig. 17.2. Arrows directed to the level β' indicate the various ways it can be reached both directly and indirectly. Each such arrow is represented by a term in the coupled equations.

therefore, that if ρ is chosen to have the intuitively pleasing form

$$\rho^{\alpha}_{\beta'\alpha'}(\mathbf{r}_\beta) = (\Phi_{\beta'}|V_\beta - \hat{U}_\beta|\chi^{\alpha(+)}_{\alpha'}\Phi_{\alpha'}), \tag{17.14}$$

Then the amplitude of the outgoing wave part of $\hat{\chi}$ in channel β as $r_\beta \to \infty$ is given by (17.5) (Mackintosh and Glendenning, 1971). The superscript α in (17.13) denotes the entrance channel.

However, the aim is not to calculate the cross section by calculating (17.5). This would involve solving (17.9) and (17.10) and then computing the integrals in (17.5), subject to the indicated boundary conditions. Rather, the aim is to solve (17.9), and (17.13) and to find the S matrix simply and directly through the asymptotic conditions on the partial-wave solutions, thereby avoiding the computation of (17.5) altogether. These conditions are that the entrance channel α has incoming and outgoing spherical waves, consistent with there being a plane wave, representing the incident beam in this channel, and a scattered wave. All other channels $\alpha' \neq \alpha$ in partition α and all channels $\beta, \beta', \ldots$ in partition β have only outgoing waves. Specifically,

$$\chi^{\alpha(+)}_{\alpha'}(\mathbf{r}_\alpha) \xrightarrow[r_\alpha \to \infty]{} \delta_{\alpha'\alpha} \exp(i\mathbf{k}_\alpha \cdot \mathbf{r}_\alpha) + \frac{\exp(ik'_\alpha r_\alpha)}{r_\alpha} f_{\alpha'\alpha}(\theta), \tag{17.15}$$

$$\chi^{\alpha(+)}_{\beta}(\mathbf{r}_\beta) \xrightarrow[r_\beta \to \infty]{} \frac{\exp(ik_\beta r_\beta)}{r_\beta} f_{\beta\alpha}(\theta). \tag{17.16}$$

In a specific example, such as the (d, p) reaction, where we make the same assumptions as in Chapter 7,

$$\begin{aligned} \langle V_\beta - \hat{U}_\beta \rangle &\simeq V_{\mathrm{np}}, \\ V_{\mathrm{np}}\phi_{\mathrm{d}} &= D_0 \delta(\mathbf{r}_{\mathrm{n}} - \mathbf{r}_{\mathrm{p}}), \end{aligned} \tag{17.17}$$

then the source term takes the simpler form

$$\rho_{\mathrm{p'd'}}(\mathbf{r}_{\mathrm{p}}) = (\Phi_{\mathrm{B}}(\mathrm{A}, \mathbf{r}_{\mathrm{p}})|\Phi_{\mathrm{A}}(\mathrm{A}))\chi^{\mathrm{d}(+)}_{\mathrm{d'}}(\mathbf{r}_{\mathrm{p}}). \tag{17.18}$$

In practice, the three-dimensional Schrödinger equations are separated into radial and angular parts by performing a partial-wave expansion. This is done later in this chapter and in the literature (Ascuitto and Glendenning, 1969).

D. DERIVATION OF THE CCBA

The coupled-channel Born approximation amplitude (17.5) and the source-term equations, (17.9), (17.13), and (17.14), whose solutions yield precisely the same amplitude (Mackintosh and Glendenning, 1971), can be derived in another way, which will pave the way for the derivation of the coupled-reaction channel equations. The source-term equations were motivated by the discussion of the physical situation depicted in Fig. 17.2. These equations can be derived as an approximation to the scattering problem in the following way: The physical system is governed by the Schrödinger equation

$$(E - H)\Psi_{\alpha}^{(+)} = 0, \tag{17.19}$$

where, as in (3.13), H can be written in terms of any partition of the system. Recall the customary notation, according to which the subscript α denotes the particular solution of (17.19) that has the incident channel α and the superscript $(+)$ indicates that the solution is the one that possesses outgoing scattered waves in all open channels. Because $\Psi_{\alpha}^{(+)}$ is supposed to be a solution to the Schrödinger equation for the total system, it embodies all information on all channels. It is formally possible to represent Ψ_{α} on any complete set of basis functions (provided all continuum states are included). Two such representations are

$$\Psi_{\alpha}^{(+)} = \sum_{\alpha'} \chi_{\alpha'}^{\alpha(+)}(\mathbf{r}_{\alpha})\Phi_{\alpha'}, \tag{17.20a}$$

and

$$\Psi_{\alpha}^{(+)} = \sum_{\beta} \chi_{\beta}^{\alpha(+)}(\mathbf{r}_{\beta})\Phi_{\beta}, \tag{17.20b}$$

constructed of functions in the partitions α and β, respectively.

However, the accuracy with which this can be done is clearly limited by the number of coupled equations for the χ's that can be solved in practice. Therefore it is clear that if we wish to examine Ψ concerning the channels α, we will do better to employ (17.20a), whereas when we wish to examine it concerning the channels β, we will do better to seek its representation by (17.20b). Because our goal is to derive the cross section for the transfer reaction from $\alpha \to \beta$, we wish to examine Ψ_α as concerns its behavior in the β channels. Therefore, we write the Schrödinger equation (17.19) in the form [recall (3.13)]

$$(E - H_\beta - T_\beta - \hat{U}_\beta)\Psi_\alpha^{(+)} = (V_\beta - \hat{U}_\beta)\Psi_\alpha^{(+)}. \tag{17.21}$$

The generalized optical potential $\hat{U}$ that can give rise to some inelastic scattering, say among collective states, has been subtracted from both sides in (17.21). In accordance with the remark concerning (17.20) we should choose the representation of Ψ_α appropriate to the partition under examination. Therefore introduce (17.20b) in the left side. Then multiplying from the left by a typical Φ_β and taking nondiagonal terms to the right, we obtain

$$(E_\beta - T_\beta - U_\beta)\chi_\beta^{\alpha(+)} = \sum_{\beta'} U_{\beta\beta'}\chi_{\beta'}^{\alpha(+)} + (\Phi_\beta|V_\beta - \hat{U}_\beta|\Psi_\alpha^{(+)}), \tag{17.22}$$

where $U_{\beta\beta'}$ and U_β are given by (17.11) and (17.12). The first term on the right gives rise to inelastic transitions within the partition β. The second term represents nucleon transfer transitions from all open channels, but particularly from channels in the entrance partition α. These latter transitions will generally be most important because the ground-state channel is occupied to start, so that they are of lowest order. Therefore, we represent $\Psi_\alpha^{(+)}$ in the last term of (17.22) by (17.20a) to represent most accurately the α channels. Thus for each channel β we have

$$(E_\beta - T_\beta - U_\beta)\chi_\beta^{\alpha(+)} = \sum_{\beta'} U_{\beta\beta'}\chi_{\beta'}^{\alpha(+)} + \sum_{\alpha'} (\Phi_\beta|V_\beta - \hat{U}_\beta|\chi_{\alpha'}^{\alpha(+)}\Phi_{\alpha'}). \tag{17.23}$$

This is the source-term equation (17.13), with the last term being the source. The function $\chi_{\alpha'}^{\alpha(+)}$ that appears in the transfer source of (17.23) satisfies an equation analogous to (17.22), which can be derived in the same way. The transfer terms in the equation for $\chi_{\alpha'}^{\alpha}$ would, for the reaction $\alpha \to \beta$ that we are trying to calculate, be of second order in transfer reactions. Let us calculate to first order in transfer reactions. Therefore, we ignore couplings to other partitions, in which case $\chi_{\alpha'}^{\alpha}$ satisfies Eq. (17.9), that is,

$$(E_{\alpha'} - T_{\alpha'} - U_{\alpha'})\chi_{\alpha'}^{\alpha(+)} = \sum_{\alpha''} U_{\alpha'\alpha''}\chi_{\alpha''}^{\alpha(+)}. \tag{17.24}$$

As mentioned before, these equations, (17.23) and (17.24), which constitute the CCBA, are to be solved according to the physical boundary conditions (17.15) and (17.16).

The approximation to the problem given here is referred to as a Born approximation because the transfer process is treated in first order, as is evident in the coupled system (17.23), (17.24) or in (17.5). The inelastic transitions within partitions α and β are treated to all orders in the retained channels, which accounts for the designation "coupled channels," or, taken together, CCBA.

E. COUPLED REACTION CHANNELS (CRC)

It is usually true that transfer processes have smaller cross sections than, say, collective inelastic transitions, so, that in such cases the CCBA scheme makes sense. However, it is conceivable that certain special transfer reactions are large enough to require a higher-order treatment (Stamp, 1966; Rawitscher, 1967; Austern, 1969; Ascuitto *et al.*, 1971a). One such example is the case of a pairing vibrational 0^+ state in an even nucleus. Such states do not have a large inelastic cross section. On the contrary, it is hindered. In contrast, the (p, t) or (t, p) transfer reaction connects pairing vibrational states with a large cross section. In this case the relative importance of inelastic and transfer reactions is actually inverse to the importance given in the preceding discussion. The situation is depicted in Fig. 17.3, where two strong transfer reactions compete with an inelastic transition in a (p, p′) reaction.

Situations such as the preceding example can be handled by an obvious extension of the source-term equations, (17.23) and (17.24), by keeping the transfer source term in (17.24) (Stamp, 1966; Rawitscher, 1967; Austern, 1969; Ascuitto and Glendenning, 1970b; Ascuitto *et al.*, 1971a; Coker *et al.*, 1973). That equation, obtained from (17.19) in a way parallel to the derivation of (17.22), is

$$(E_{\alpha'} - T_{\alpha'} - U_{\alpha'})\chi_{\alpha'}^{\alpha(+)} = \sum_{\alpha''} U_{\alpha'\alpha''}\chi_{\alpha''}^{\alpha(+)} + (\Phi_{\alpha'}|V_\alpha - \hat{U}_\alpha|\Psi_\alpha^{(+)}). \quad (17.25)$$

This still contains the (unavailable) exact solution $\Psi_\alpha^{(+)}$. Because we especially want to represent those parts of the last term on the right-hand side that represent transitions from the partition β to α, we replace $\Psi_\alpha^{(+)}$ by

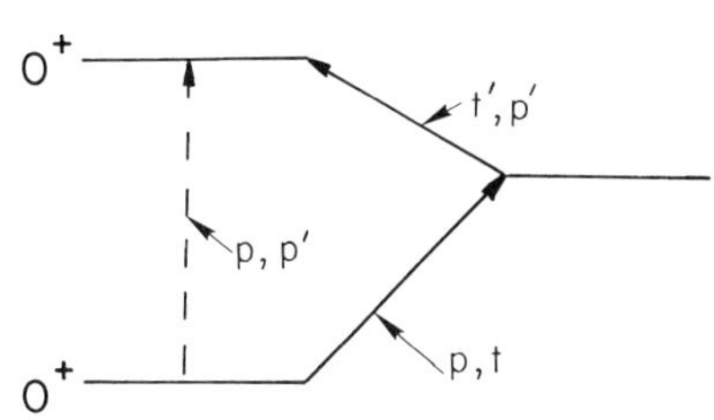

Fig. 17.3. The transfer reactions through an intermediate state in another nucleus can compete with weak inelastic transitions.

(17.20b). This gives us the solvable equations

$$(E_{\alpha'} - T_{\alpha'} - U_{\alpha'})\chi_{\alpha'}^{\alpha(+)} = \sum_{\alpha''} U_{\alpha'\alpha''}\chi_{\alpha''}^{\alpha(+)} + \sum_{\beta'} (\Phi_{\alpha'}|V_{\alpha} - \hat{U}_{\alpha}|\chi_{\beta'}^{\alpha(+)}\Phi_{\beta'}). \quad (17.26)$$

Equations (17.23) and (17.26) are the coupled reaction equations, being coupled to each other through their last term in each case. Whereas Eqs. (17.23) and (17.24) represent such transfer couplings as in Fig. 17.2 that go from $\alpha' \rightarrow \beta'$, as well as the inelastic transitions $\alpha' \rightarrow \alpha''$, $\beta' \rightarrow \beta''$, etc., Eqs. (17.23) and (17.26) include, in addition, the inverse transfer $\beta' \rightarrow \alpha'$. These latter equations treat all transitions, inelastic and transfer, to all orders among the retained channels. However, they are still approximate because, instead of using the complete (but unavailable) sets implied by (17.20), we recognize the practical limitation of solving only a finite number (indeed, few) of coupled equations representing the physically most interesting channels. The boundary conditions, again, are the physical ones (17.15) and (17.16).

Concerning the (CRC) equations, there is a great deal of discussion in the literature on what is referred to as the nonorthogonality problem. This problem is aggravated in much of the literature by making an expansion, not as in (17.20a) *or* (17.20b), but simultaneously represent Ψ as an expansion in several mass partitions

$$\Psi_{\alpha} = \sum_{\alpha'} \chi_{\alpha'}^{\alpha}\Phi_{\alpha'} + \sum_{\beta'} \chi_{\beta'}^{\alpha}\Phi_{\beta'} + \cdots. \quad (17.27)$$

In regions of configuration space where the nuclei are close to each other, the sets of function Φ_{α} and Φ_{β} are obviously not orthogonal; they are eigenfunctions of different Hamiltonians. Therefore, (17.27) is an expansion in an overcomplete set. As the discussion in this chapter shows, it is unnecessary to introduce (17.27) and all the resulting problems of nonorthogonality as long as only two partitions are thought to dominate the situation. However, if there are physical reasons to expect that a given partition, say, γ, is coupled to two (or more) other partitions, say α and β, then, in the equations for the γ channels,

$$(E_{\gamma} - T_{\gamma} - U_{\gamma})\chi_{\gamma}^{\alpha(+)} = \sum_{\gamma'} U_{\gamma\gamma'}\chi_{\gamma'}^{\alpha(+)} + (\Phi_{\gamma}|V_{\gamma} - \hat{U}_{\gamma}|\Psi_{\alpha}^{(+)}). \quad (17.28)$$

One would be tempted to insert an ansatz such as (17.27) into the source term. In this case, the need to address the overcompleteness of the expansion (17.27) becomes unavoidable. [Note that in our notation label α always refers to the definite partition a, A, and a sum over α' refers only to nuclear excitations in that partition. Some authors use a sum over a channel label such as α to sum over many partitions, so that at first sight the use of an overcomplete set (17.27) is not apparent.]

F. PARTIAL-WAVE EXPANSIONS AND DERIVATION OF THE CROSS SECTION

Because the CCBA or CRC equations involve the three-dimensional coordinates, the task of solving them is very much simplified by using partial-wave expansions of the wave functions. Then the power of the angular momentum algebra developed by Racah and Wigner can be employed to deal with the angular parts. The remaining computational problem concerns coupled equations in the radial coordinate only, and the boundary conditions (17.15, 17.16) are then easier to apply.

To this end we define, analogous to (8.32), channel functions

$$\phi_{\alpha I}^{M} = \Big[[Y_{\ell_\alpha}(\hat{\mathbf{r}}_\alpha)\Phi_{J_a}]_{j_\alpha}\Phi_{J_A}\Big]_I^M, \tag{17.29a}$$

where, in context, α stands for the set of quantum numbers

$$\alpha = \ell_\alpha J_a j_\alpha J_A \tag{17.29b}$$

and a similar such function $\phi_{\beta I}^{M}$. The two alternative ways of expanding $\Psi_\alpha^{(+)}$ in (17.20) are now replaced by eigenfunctions of total angular momentum and z projection, I and M, because these are conserved. Thus

$$\Psi_{\alpha I}^{(+)M} = \sum_{\alpha'} (1/r_\alpha) u_{\alpha'}^{\alpha I} \phi_{\alpha' I}^{M}, \tag{17.30a}$$

or

$$\Psi_{\alpha I}^{(+)M} = \sum_{\beta} (1/r_\beta) u_{\beta}^{\alpha I} \phi_{\beta I}^{M}. \tag{17.30b}$$

The superscript αI on the radial functions serves to remind us of the entrance channel, total angular momentum and parity π, mention of which is suppressed for brevity. In complete analogy with the derivation of (17.24), we obtain, using (17.30a) and the spherical coordinate representation of ∇^2 given in the Appendix,

$$(E_{\alpha'} - T_{\ell\alpha'} - U_{\alpha'\alpha'}^{I}) u_{\alpha'}^{\alpha I}(r_\alpha) = \sum_{\alpha''} U_{\alpha'\alpha''}^{I} u_{\alpha''}^{\alpha I}(r_\alpha), \tag{17.31}$$

where

$$T_{\ell\alpha} = \frac{\hbar^2}{2m_\alpha}\left(-\frac{d^2}{dr_\alpha^2} + \frac{\ell_\alpha(\ell_\alpha + 1)}{r_\alpha^2}\right). \tag{17.32}$$

Analogous to (8.47), we impose the boundary condition

$$u_{\alpha'}^{\alpha I}(r_\alpha) \to \delta_{\alpha\alpha'} F_{\ell_\alpha}(k_\alpha r_\alpha) + \left(\frac{v_\alpha}{v_{\alpha'}}\right)^{1/2} \frac{(S^I - 1)_{\alpha'\alpha}}{2i} O_{\ell_{\alpha'}}(k_{\alpha'} r_\alpha). \tag{17.33}$$

which can be recognized as the partial wave form of 17.15. Then a linear combination

$$\Psi_{\alpha}^{(+)} = \sum_{IM} \mathrm{A}_{\alpha I}^{M} \Psi_{\alpha I}^{(+)M}, \tag{17.34}$$

with

$$\mathrm{A}_{\alpha I}^{M_a + M_A} = (1/k_\alpha) C_{0\ M_a M_a}^{\ell_\alpha\ J_a\ j_a} C_{M_a M_A M_a + M_A}^{j_a\ J_A\ I} \sqrt{4\pi(2\ell_\alpha + 1)}\, i^{\ell_\alpha} e^{i\sigma_\alpha}, \tag{17.35}$$

yields a plane wave in the entrance channel α and outgoing waves in this and the open inelastic channels like (8.55), as is required by the situation that we wish to describe.

Now we derive the partial wave analog of (17.23). Substitute (17.30b) in the left side of (17.21) to obtain

$$(E - H_\beta - T_\beta - \hat{U}_\beta)\Psi = (1/r_\beta) \sum_{\beta'} (E_{\beta'} - T_{\ell_{\beta'}} - \hat{U}_\beta)\, \phi_{\beta' I}^{\ M} u_{\beta'}^{\alpha I}(r_\beta). \tag{17.36}$$

Multiply by $\phi_{\beta I}^{\ M}$ and integrate over the arguments of this function to obtain

$$(1/r_\beta)(E_\beta - T_{\ell\beta} - U_{\beta\beta}) u_\beta^{\alpha I}(r_\beta) - (1/r_\beta) \sum_{\beta'} U_{\beta\beta'}^{I} u_{\beta'}^{\alpha I}(r_\beta),$$

where

$$U_{\beta\beta'}^{I} = (\phi_{\beta I}^{\ M} | \hat{U}_\beta | \phi_{\beta' I}^{\ M}). \tag{17.37}$$

By the same steps, the right side of (7.21) becomes, upon using (17.30a),

$$\sum_{\alpha'} (\phi_{\beta I}^{\ M} | V_\beta - \hat{U}_\beta | (1/r_\alpha) u_{\alpha'}^{\alpha I} \phi_{\alpha' I}^{\ M}),$$

which remains a function of r_β, all other coordinates being integrated. Thus the coupled equations for given I are

$$\begin{aligned} &(E_\beta - T_{\ell\beta} - U_{\beta\beta}^{I}) u_\beta^{\alpha I}(r_\beta) \\ &\quad = \sum_{\beta'} U_{\beta\beta'}^{I} u_{\beta'}^{\alpha I}(r_\beta) + r_\beta \sum_{\alpha'} (\phi_{\beta I}^{\ M} | V_\beta - \hat{U}_\beta | \frac{1}{r_\alpha} u_{\alpha'}^{\alpha I} \phi_{\alpha' I}^{\ M}). \end{aligned} \tag{17.38}$$

Because there are only outgoing waves in the channel β for the reaction $\alpha \to \beta$, the boundary conditions on the solutions of this equation are[‡]

$$u_\beta^{\alpha I}(r_\beta) \to (1/2i)(v_\alpha/v_\beta)^{1/2} S_{\beta,\alpha}^{I} O_{\ell_\beta}(k_\beta r_\beta). \tag{17.39}$$

To find the scattering amplitude, use the asymptotic form (4.13) of the outgoing wave and examine

$$(1/r_\beta) u_\beta^{\alpha I}(r_\beta) \phi_{\beta I}^{\ M}.$$

[‡] The velocity is $v_\alpha = \hbar k_\alpha / m_\alpha$.

to find the amplitude of

$$\frac{1}{r_\beta}\left(\frac{v_\alpha}{v_\beta}\right)^{1/2} \Phi_{J_b}^{M_b}\Phi_{J_B}^{M_B} \exp\{i[k_\beta r_\beta - \eta_\beta \ln(2k_\beta r_\beta)]\}.$$

It is,

$$(1/2i)S_{\beta,\alpha}^{I} i^{-\ell_\beta} e^{i\sigma_\beta} \sum_{m_\beta \mu} C_{\mu\, M_B M}^{j_\beta\, J_B\, I} C_{m_\beta M_b \mu}^{\ell_\beta\, J_b\, j_\beta} Y_{\ell_\beta}^{m_\beta}(\hat{\mathbf{r}}_\beta).$$

This was computed using the function (17.30) as the entrance channel. As we saw previously, we actually must use the linear combination (17.34) to satisfy the initial condition of having a plane wave in the entrance channel plus outgoing waves. Thus the amplitude for the indicated transition is the same linear combination of these amplitudes, namely,

$$\begin{aligned} &F(J_a M_a J_A M_A \to J_b M_b J_B M_B; \hat{\mathbf{k}}_\beta) \\ &= \frac{1}{2ik_\alpha} \sum_{\ell_\alpha j_\alpha \ell_\beta j_\beta I M m_\beta} [4\pi(2\ell+1)]^{1/2} i^{\ell_\alpha - \ell_\beta} \exp[i(\sigma_\alpha + \sigma_\beta)] S_{\beta,\alpha}^{I} Y_{\ell_\beta}^{m_\beta}(\hat{\mathbf{k}}_\beta) \\ &\quad \times C_{0\, M_a M_a}^{\ell_\alpha J_a\, j_\alpha} C_{M_a M_A M}^{j_a\, J_A\, I} C_{m_\beta M_b m_\beta + M_b}^{\ell_\beta\, J_b\, j_\beta} C_{m_\beta + M_b M_B M}^{j_\beta\, J_B\, I}. \end{aligned} \tag{17.40}$$

Notice that $\hat{\mathbf{r}}_\beta \to \hat{\mathbf{k}}_\beta$ as $\mathbf{r}_\beta \to \infty$. This replacement was made in the spherical harmonic. The cross section, as in Chapter 8, is computed from

$$\frac{d\sigma}{d\Omega} = \frac{1}{(2J_a+1)(2J_A+1)} \sum_{M_a M_A M_b M_B} |F(\hat{\mathbf{k}}_\beta)|^2. \tag{17.41}$$

To recapitulate, the CCBA equations in partial wave expansion are (17.31) and (17.38) with boundary conditions (17.33) and (17.39). The amplitude for a transfer reaction is (17.40).

G. TWO-NUCLEON TRANSFER BETWEEN VIBRATIONAL NUCLEI

The inelastic effects on transfer reactions are second or higher order (i.e., one order in the transfer interaction and one or more in the inelastic interaction), whereas the direct single-step transfer is first order. Therefore, we anticipate that the inelastic effects will be most strongly seen in nuclei where there are enhanced inelastic transitions. The first theoretical indication that the effects might be not only large enough to observe but, in fact, that they might change cross sections by a factor of two or more, came in a study of vibrational nuclei, specifically the nickel isotopes (Glendenning, 1969b; Ascuitto and Glendenning, 1970a). The coupled-channel calculation of the inelastic scattering of nucleons was briefly discussed as an example in

Chapter 8. The development of the coupled equations for the two-nucleon transfer reaction, including the expressions for two-particle parentage factors for quasi-particle states, coherent super-positions of which can be used to represent vibrational states, is detailed in the literature (Ascuitto and

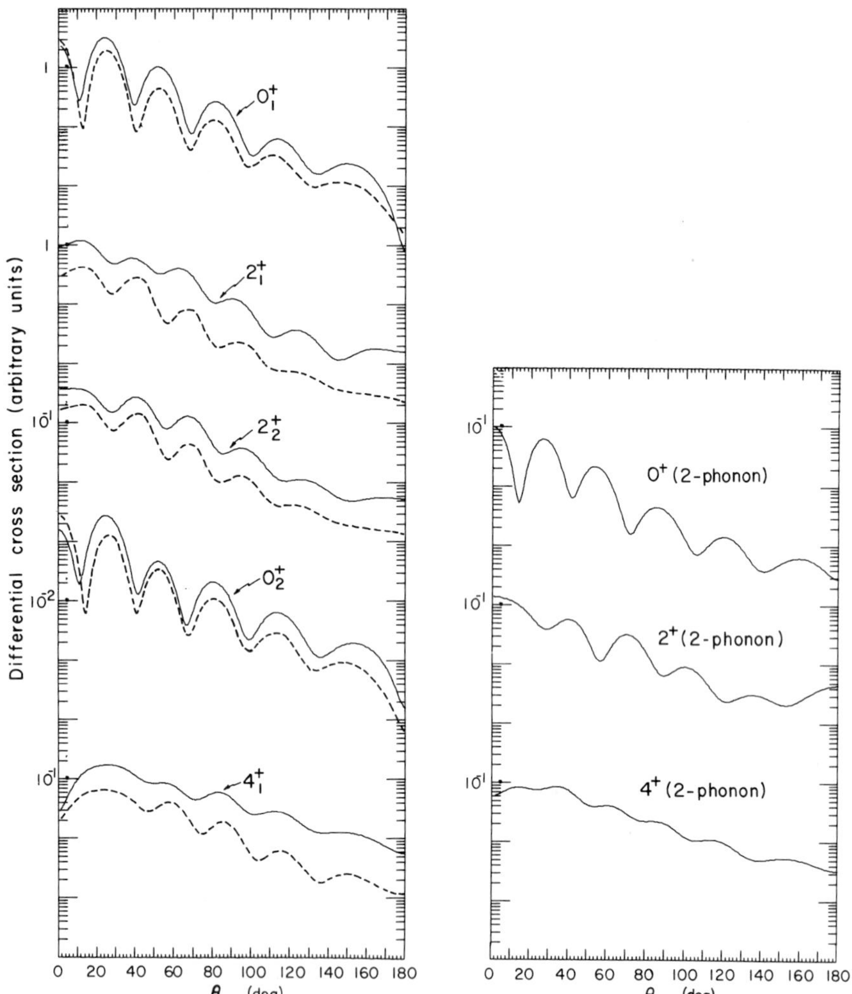

Fig. 17.4. Calculated cross sections of the $^{62}Ni(p, t)$ reaction at 30 MeV to ground and two-quasi-particle states are shown on the left, and "microscopic" two-phonon states on the right. Solid lines (———) include multiple-step processes. Dashed lines (- - -) show DWBA calculations, using optical-model parameters that reproduce the same triton and proton elastic cross sections as the coupled-channel calculation (From Ascuitto and Glendenning, 1970a.)

Glendenning, 1970). Typical results reported are illustrated in Fig. 17.4. Here we see changes in the cross section of the (p, t) reaction of up to a factor of five or more, caused by the indirect transitions.

H. TWO-NUCLEON TRANSFER BETWEEN ROTATIONAL NUCLEI

The first convincing evidence of strong inelastic effects resulted from an analysis of the (p, t) reaction data for deformed rare earth nuclei (Ascuitto *et al.*, 1971b, 1972). The deformations of such nuclei, including $L = 2, 4$, and 6 components in the shape

$$R = R_0\left(1 + \sum_{\lambda=0,2,4} \beta_\lambda Y_{\lambda 0}\right) \tag{17.42}$$

was discussed as an example of the coupling of channels in inelastic scattering in Chapter 13. As discussed, the transitions in deformed nuclei are so strong because the intrinsic structure within a band is unchanged and different states correspond merely to different rates of rotation of the entire

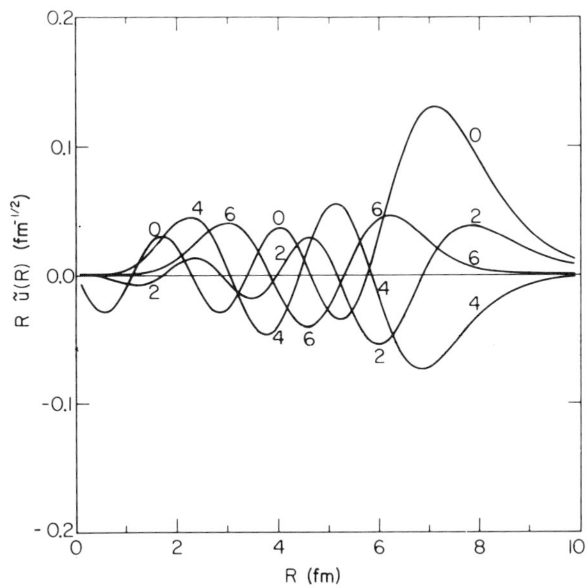

Fig. 17.5. The projected wave functions or form factor that show the strength (determined mainly by the surface region) of the various multipoles for transferring a pair of particles between ground-state members of the rotational nuclei ^{176}Yb and ^{174}Yb (From Ascuitto, *et al.*, 1972.)

nucleus. For example, the charge acted on by the electromagnetic field is Ze rather than e in the case of a single-particle state. Thus electromagnetic transitions are enhanced and, similarly, so are inelastic transitions. Therefore, strongly deformed nuclei are expected to exhibit the largest effects of inelastic transitions. The common intrinsic structure of the ground band was assumed in the calculations to be a BCS state (Ascuitto *et al.*, 1971b, 1972). The calculation of the form factors for the two-neutron transfer reaction is quite involved, depending as it does on the calculation of the quasi-particle states in a deformed potential and of the parentage amplitudes for the BCS states constructed for such quasi-particles. The form factors are shown in Fig. 17.5.

The calculated cross sections for the ^{176}Yb(p, t) reaction are shown in Fig. 17.6. There is a single normalization constant for all states (D_0 of Chapter

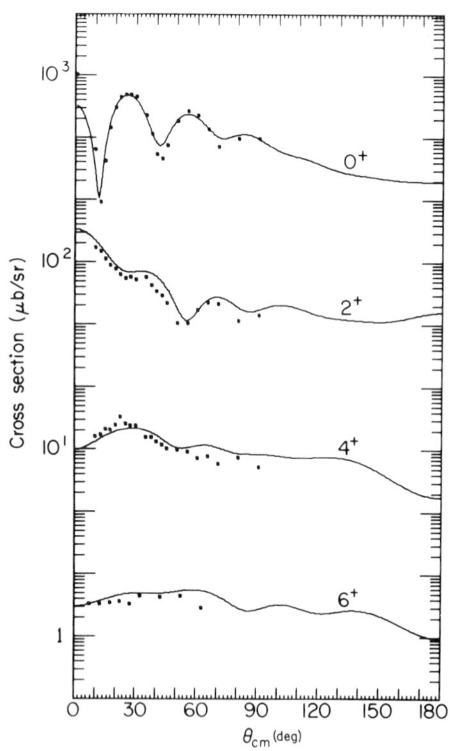

Fig. 17.6. Complete calculation for the ground-band members of ^{174}Yb produced in the (p, t) reaction at 19 MeV. Calculations include all inelastic and reaction transitions connecting all four states in both nuclei. The O^+ is normalized to the data and the same normalization was used for the others. [Data are from Oothoudt, *et al.* (1970), and the calculation from Ascuitto, *et al.* (1972)].

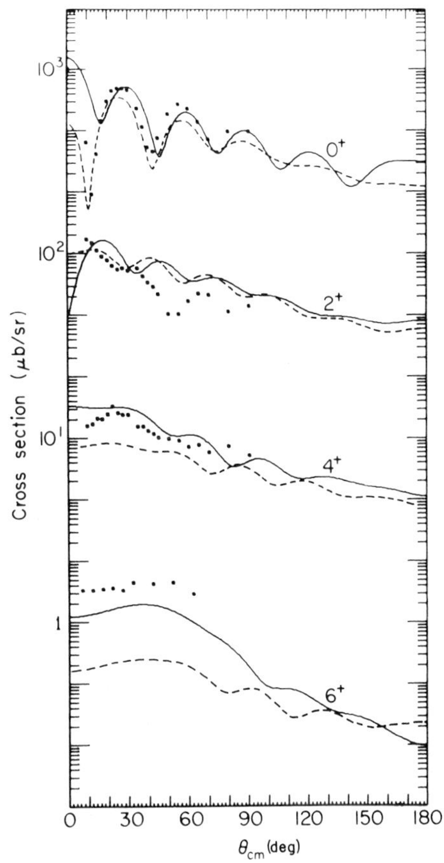

Fig. 17.7. Two DWBA calculations are shown for ^{176}Yb(p, t) reaction at 19 MeV. The solid curve (———) corresponds to proton and triton parameters *adjusted* to reproduce the elastic cross sections of the coupled-channel calculation. The dashed curve (– – –) uses the *same* parameters as for the coupled-channel calculation. The two sets of curves have the same normalization, which is reduced by a factor of three from the complete calculation shown in Fig. 17.6 [Data are from Oothoudt *et al.* (1970), and the calculations from Ascuitto *et al.* (1972)].

15). The nuclear shape was determined from α scattering, as discussed in Chapter 13, and the deformation parameters, β_2, β_4, so determined specify the inelastic form factors. The agreement of the relative cross sections with the data (Oothoudt *et al.*, 1970) is seen to be very good. This is all the more remarkable if comparison is made with the DWBA calculations of Fig. 17.7 which, of course, do not take into account the inelastic effects. An analysis of a particular cross section in terms of the individual cross sections arising

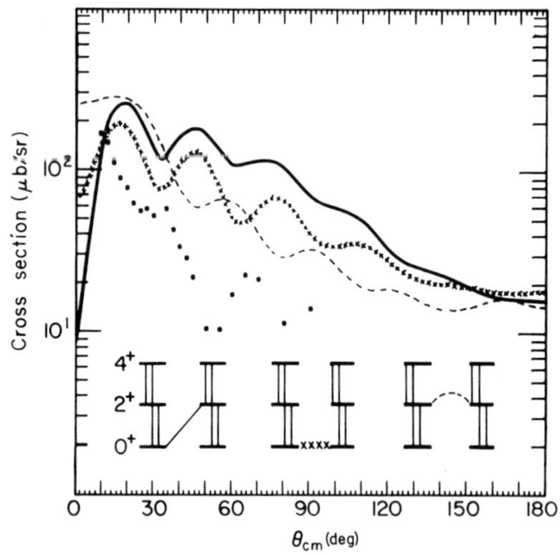

Fig. 17.8. Cross sections for the 2^+ state that correspond to the individual transfer processes shown for ^{176}Yb(p, t) reaction at 19 MeV. Note that the direct and indirect routes are comparable in magnitude [Data are from Oothoudt *et al.* (1970) and the calculations from Ascuitto *et al.* (1972)].

from specific transitions reveals the enormous sensitivity to an accurate description of all aspects of the reaction. This is illustrated in Fig. 17.8 for the 2^+ state. The two indirect transitions would each lead to cross sections that are comparable to the direct transition in strength. No single one of these three resembles the data. Rather, the agreement shown in Fig. 17.7 for this state results from the interference of these three amplitudes, each one different from the data but which, when interfering, bring about the agreement. This is very convincing evidence for the role played by inelastic transitions.

I. INDIRECT TRANSITIONS IN ANALOG REACTIONS

The excitation of analog states in charge-exchange reactions such as (p, n) is another situation in which indirect transitions play an important, or even dominant, role for vibrational or rotational nuclei (Madsen *et al.*, 1972). This is to be expected in view of the strength of the inelastic transitions. The 2^+ analog state shown in Fig. 17.9 can, of course, be excited directly from the ground state of the target nucleus. However, in comparison with the $0^+ \rightarrow 0^+$A and $2^+ \rightarrow 2^+$A transition, it is weak (Madsen *et al.*, 1972, 1976). The latter require no change of structure by virtue of the fact that the transition connects analog states. The inelastic transitions are also very strong in collec-

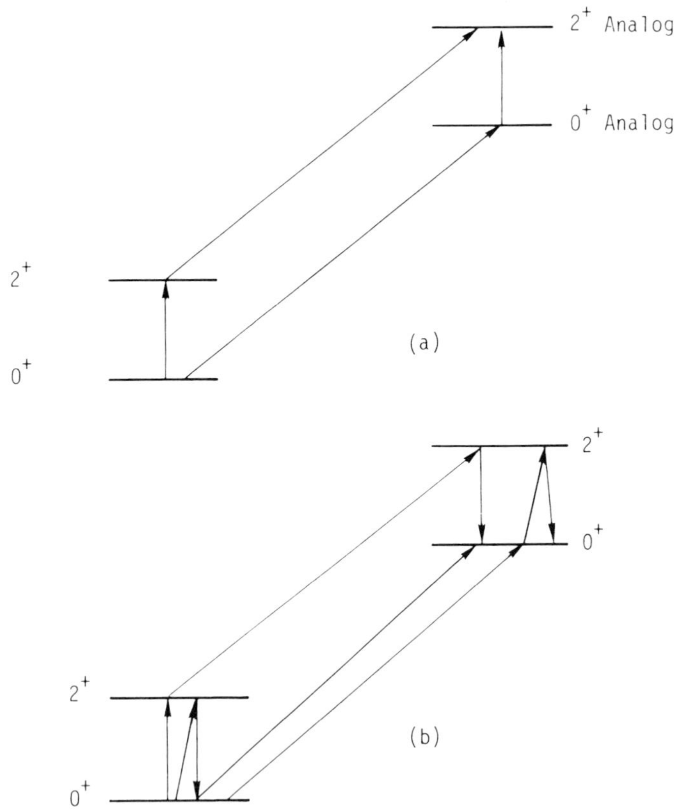

Fig. 17.9. Typical transitions to analog states showing direct and some indirect paths.

tive nuclei. For this reason the two-step transitions can be comparable in strength or stronger than the single-step direct transition.

Figure 17.10 shows a calculation (Wong *et al.*, 1979) of the cross section for analog states in ^{148}Eu. These levels lie in the range of 12- to 14-MeV excitation, but because they are analogs of the lowest states of the ^{148}Sm target, they are prominent features in the spectrum of the (p, n) reaction. Additional transitions to those shown in Fig. 17.9 were taken into account. For the 0^+A state, the cross section is very much reduced in comparison with the calculated cross section for the single direct transition. This indicates a destructive interference between the direct and the indirect transitions shown in Fig. 17.9. In contrast, the direct transition to the 2^+A is unimportant in comparison with the indirect transitions. Understanding these aspects of the reaction has been very important in understanding the systematics of analog transitions in charge-exchange reactions.

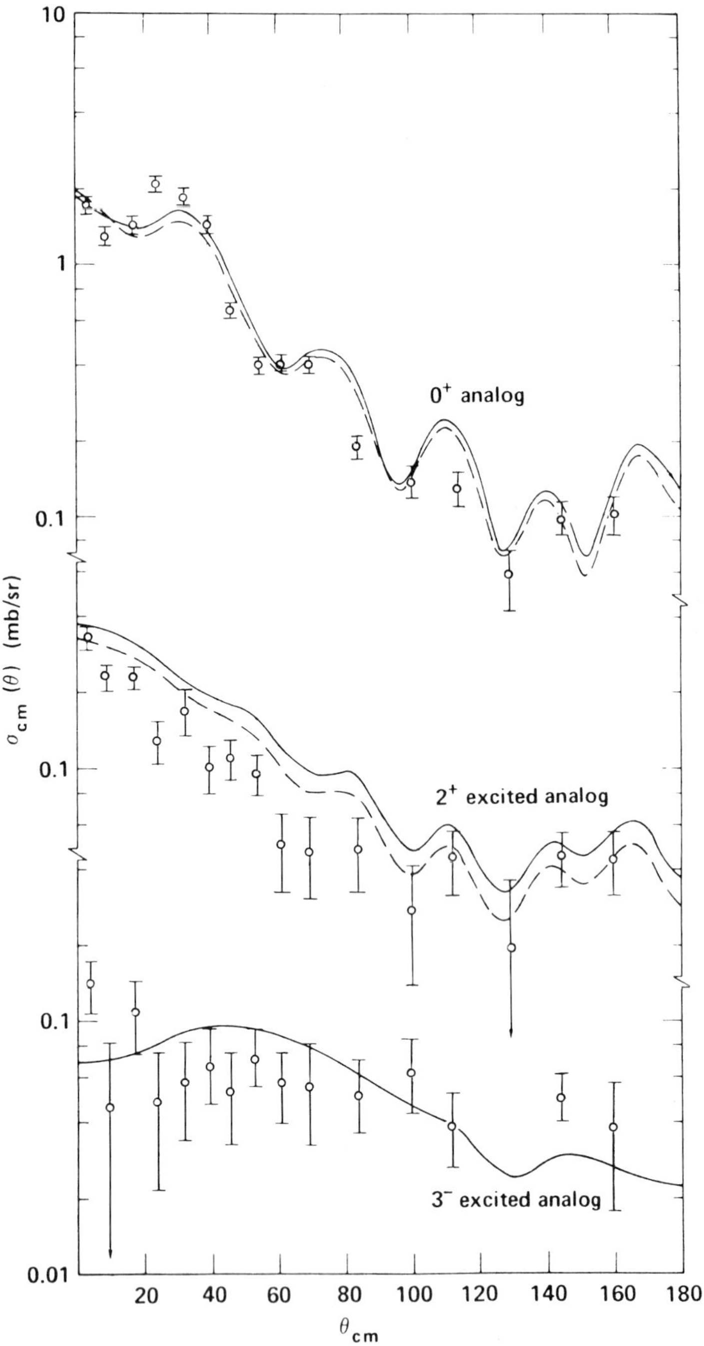

Fig. 17.10. Cross sections for analog transitions in which higher-order processes play a decisive role here for ^{148}Eu reaction at 26 MeV. In calculation (A) (———), the 0^+, 2^+ and 3^- states in the target, and their analogs in the final nucleus, were all coupled. In (B) (– – –), the 0^+, 2^+ and 2^+(2-photon) states in the target and their analogs were coupled (From Wong, *et al.*, 1979).

Chapter 18

Heavy-Ion Reactions

A. SPECIAL FEATURES

The development of the theory of direct reactions discussed in the preceding chapters applies also to the collisions between complex nuclei. These are sometimes referred to as heavy ions, although it is the nuclear interactions that are of interest not the ions, whose charges simply provide the means of acceleration.

Three special features can be noted immediately. The first is that the Coulomb force plays a major role. The second has to do with the approximate classical nature of nucleus–nucleus reactions which follows essentially from the fact that the angular momenta involved are large. This can be understood by noting that direct reactions between two complex nuclei surely result from grazing collisions in which there is very little overlap of matter from the two nuclei. Otherwise they would be involved in much more profound rearrangements of energy, angular momentum, and so on, than can be consistent with the notion of a direct reaction. The angular momentum of a grazing collision is

$$L = k_{\text{eff}} R,$$

$$R = r_0(A_1^{1/3} + A_2^{1/3}), \qquad r_0 \simeq 1.3 \quad \text{fm}, \tag{18.1}$$

where k_{eff} is the wave number corresponding to the energy of relative motion at the point of grazing. This is less than the collision energy because of the Coulomb repulsion. At the point of contact of the nuclear surfaces, the Coulomb energy is

$$B \simeq Z_1 Z_2 e^2/R, \tag{18.2}$$

so

$$k_{eff} = \sqrt{(2M/\hbar^2)(E - B)}, \tag{18.3}$$

where M is the reduced mass. Introducing the approximate values of the universal constants

$$\hbar c \simeq 197 \quad \text{MeV fm}, \qquad e^2 \simeq 1.44 \text{ MeV fm},$$

we find for a specific example of $^{18}O + {}^{120}Sn$ at $E_{lab} = 100$ MeV that $L \simeq 60$ units of $\hbar$. By comparison, a proton incident on Sn at 100 MeV carries approximately $L = 12$ for a grazing collision.

The approximate classical nature of heavy-ion reactions permits a more intuitive understanding of the reactions than is possible for light particles. This view has been elaborated in review articles elsewhere (Glendenning, 1974, 1975b) and we shall exploit only some of the notions.

The third special feature of heavy-ion reactions is a technical one. It has to do with the large size of the nuclei. This renders inappropriate the zero-range approximation that was used in some of the examples of the earlier chapters. The finite size must be taken into account. Related to this is the so-called recoil effect, which refers to how accurately the coordinate relationships are handled in the matrix elements involved in a reaction. This effect is sometimes of decisive importance in interpreting the data (Kovar *et al.*, 1972, 1973; Devries, 1973; Devries and Kubo, 1974), although many qualitative features of direct reactions between heavy ions can be understood without carrying the calculations to such accuracy (Ascuitto and Glendenning, 1973a, b; Glendenning and Wolschin, 1975, 1977). In Chapter 16 we discussed how to calculate the transition amplitude without neglecting the finite range and recoil effects.

B. THE POTENTIAL

The high angular momenta and the strong Coulomb force are peculiar aspects of direct reactions between heavy ions. The central potential acting between the centers of mass of the nuclei is usually assumed to consist of the nuclear optical potential and the Coulomb potential. It is useful to add to this the centrifugal potential in the radial equation coming from the

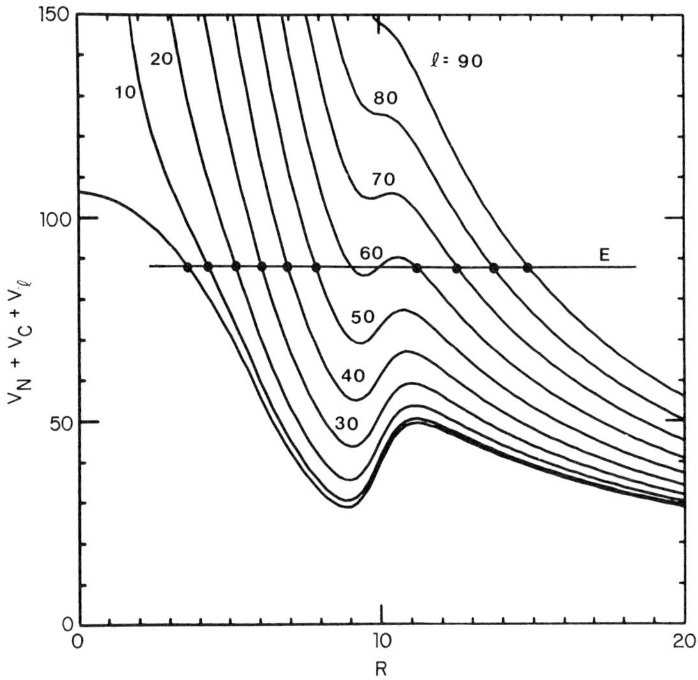

Fig. 18.1. For the system $^{18}O + {}^{120}Sn$ and the potential parameters (listed in Chapter 14), the sum of the nuclear, Coulomb, and centrifugal potentials are shown for the values of angular momentum indicated. The lowest curve, corresponding to $\ell = 0$, is the nuclear + Coulomb potential. The barrier is seen to be ~49 MeV. The horizontal line marks $E_{cm} = 87$ MeV, corresponding to a lab energy of 100 MeV. The turning points for various values of ℓ are marked by dots. Note the discontinuity occurring at the critical $\ell_C \simeq 57$ (Glendenning, 1975b).

kinetic energy operator (4.6). The real effective potential in the radial Schrödinger equation is therefore

$$V_{eff}(r) = \frac{V_0}{1 + \exp[(r - R)/a]} + \frac{Z_1 Z_2 e^2}{r} + \frac{\ell(\ell + 1)\hbar^2}{2Mr^2} \qquad (18.4)$$

for $r > R_c$. The Coulomb force for small radii is given by (4.35). This potential is plotted in Fig. 18.1 for the system $^{18}O + {}^{120}Sn$, which we shall use as a typical example at a typical laboratory energy of 100 MeV. The corresponding c.m. energy is 87 MeV and is marked on the graph. The classical turning points for the various angular momenta are marked (recall Chapter 4, Section L). The potential parameters for the nuclear part are

$$V_0 = -40 \text{ MeV}, \qquad r_0 = 1.31, \qquad a = 0.45, \qquad r_c = 1.2, \qquad (18.5)$$

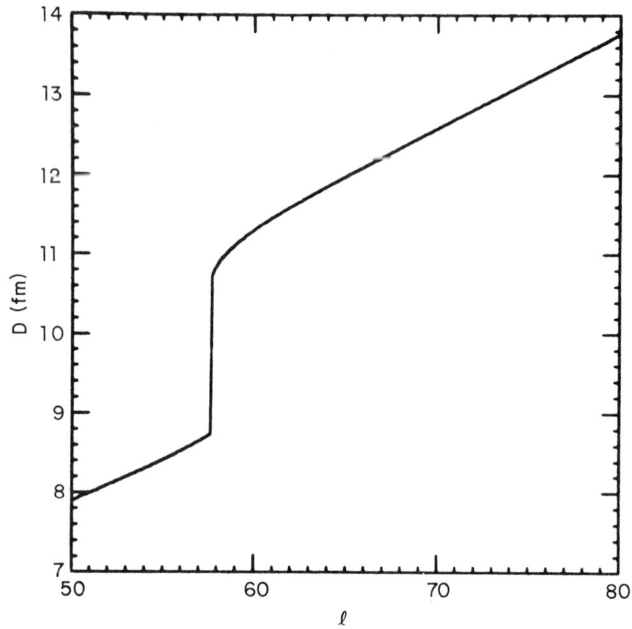

Fig. 18.2. The variation of the penetration depth or distance of closest approach D as a function of angular momentum for 100-MeV lab energy $^{18}O + ^{120}Sn$. The nuclear potential parameters are $V_0 = 40$ MeV, $r_0 = 1.31$, $a = 0.45$, and $r_c = 1.2$ (Glendenning, 1975b).

and the radius R is the sum of the nuclear radii

$$R = r_0(A_1^{1/3} + A_2^{1/3}). \tag{18.6}$$

Notice that the attractive nuclear potential is always weaker than the Coulomb potential. Therefore a classical particle in the field of such a potential would experience a marked deceleration as it approaches , until about 10 fm. The low ℓ-trajectory particles are then briefly accelerated.

The classical turning point or distance D of closest approach, as a function of ℓ, undergoes a discontinuity at that value of $\ell = \ell_c$ for which the line E in Fig. 18.1 is tangent to the potential (about 57 in the example; see also Fig. 18.2). This critical angular momentum is an important concept. Orbits with $\ell > \ell_c$ stay outside the nucleus; they barely touch. Those with $\ell < \ell_c$ plunge into the nucleus (Fig. 18.2). Several trajectories are shown in Fig. 18.3. Because of the large pathlength in the nuclear medium, those trajectories with $\ell < \ell_c$ are expected to be absorbed out of the direct reaction modes. The suddenness of the transition is reflected by the step-function-like be-

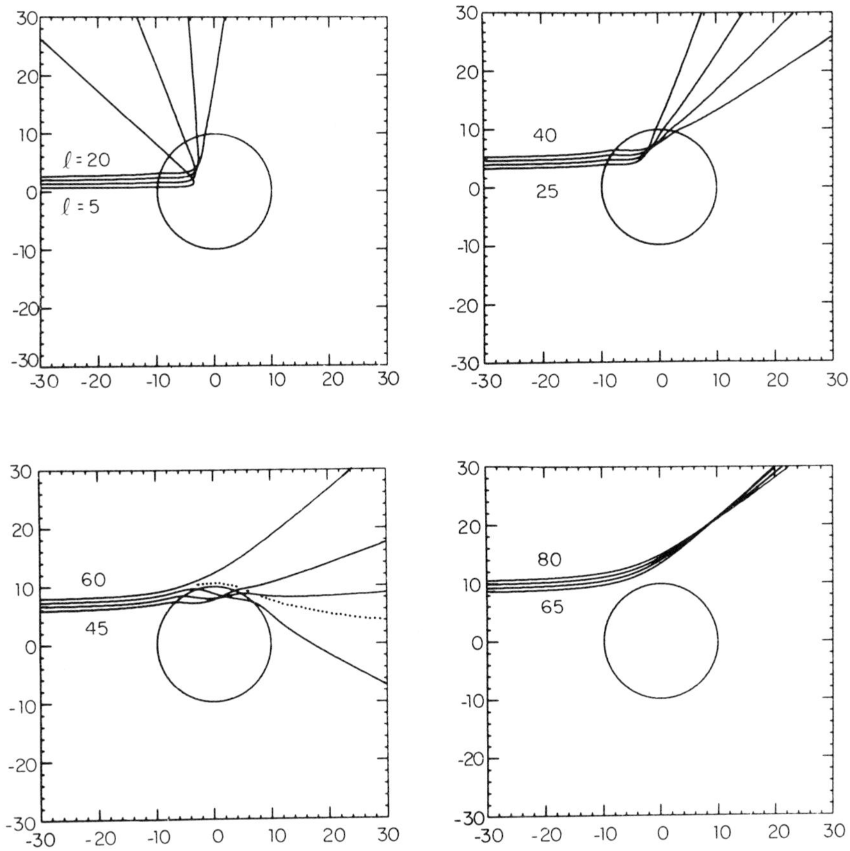

Fig. 18.3. Some trajectories in the potential (18.5) at intervals of $\ell = 5$ for 100-MeV ^{18}O scattered from ^{120}Sn. The circle marks the half-value of the nuclear potential. An additional skimming orbit is shown by dotted line (Glendenning, 1975b).

havior of the elastic S-matrix seen in Fig. 18.4. Note, however, that it is a smooth function of ℓ.

C. DEFLECTION FUNCTIONS, CLASSICAL CONDITIONS, PLUNGING ORBITS, GRAZING PEAK

We saw some typical trajectories in Fig. 18.3, and others are shown in Fig. 18.5. For some purposes it is useful to make a plot of their scattering angle versus the angular momentum or, equivalently, the impact parameter. A few points on such a plot, which is called a deflection function, could be

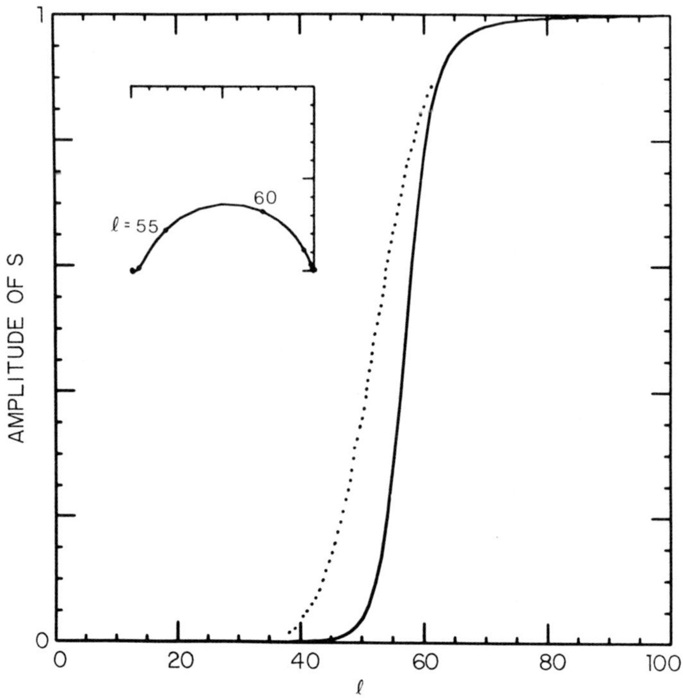

Fig. 18.4. The amplitude and phase (inset) of the S-matrix for elastic scattering of 100-MeV $^{18}O + {}^{120}Sn$ corresponding to the potential of (18.5) and $W = -15$ MeV. The effect of coupling to the 2^+ state in Sn is included. The half-value occurs at $\ell \simeq 57$, the critical angular momentum that is determined by the real potential. The dotted line corresponds to $V_0 = 0$ and shows, by comparison, how the real potential draws high angular momentum orbits into the absorbing region. The inset shows S plotted in the complex plane, its real part on the x axis. The points that represent it are joined by straight lines. The position of $\ell = 55$ and 60 are marked (Glendenning, 1975b).

read from the trajectories. We show such a deflection function in Fig. 18.6, which corresponds to 100-MeV laboratory-energy $^{18}O + {}^{120}Sn$ scattered from the potential shown at the bottom of Fig. 18.1. The location of the four orbits of Fig. 18.5 is marked. Three of them were especially chosen to show how different orbits in a potential such as that acting between heavy ions (Fig. 18.1) can scatter to the same angle (Ford and Wheeler, 1959). This introduces some very special features into the cross sections, as we shall discuss.

Classically, ℓ can have any value, integer and noninteger, but in quantum mechanics only integer values are allowed. We see that for oxygen the deflection function is sampled at very small intervals. This is in contrast to a typical proton reaction in which integer angular momenta are sampled very

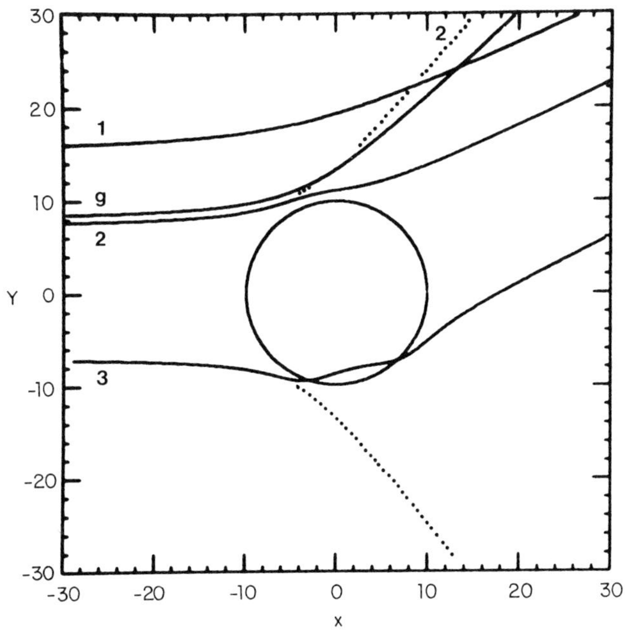

Fig. 18.5. Four classical orbits in the potential (18.5) are plotted, three of which, having impact parameters 1, 2, and 3 (3 < 2), scatter to the same angle θ. The orbit g is the grazing one. The circle marks the half-value of the Woods–Saxon nuclear potential. The orbits are for ^{18}O scattered by ^{120}Sn at 100-MeV lab energy. Dotted lines show pure Coulomb orbits. The scale is in fm. (Ascuitto and Glendenning, 1974).

sparsely. This is a necessary condition for making sense of a classical discussion of scattering. Only when the density of angular momentum states per unit impact parameter ($L/b = k$) is large does the classical discussion become useful. We use the word "useful" to avoid implying that a classical description is quantitatively valid. For deflection functions like those shown, classical mechanics is not quantitatively valid because large changes in θ occur between neighboring angular-momentum states in the vicinity of $\ell = \ell_c$. This means that a wave packet would be broken up by the action of the nuclear field near this angular momentum and scattered in all directions.

Thus classical, or semiclassical, methods cannot be expected to be more than qualitatively valid whenever the nuclear force is involved appreciably if it is strong enough to produce sharp changes in θ_ℓ. As we shall see in our later discussion, elastic, inelastic, one-particle transfer, and two-particle transfer are sensitive to increasingly smaller distances, that is, to the region where the nuclear force is strong. Semiclassical calculations, therefore,

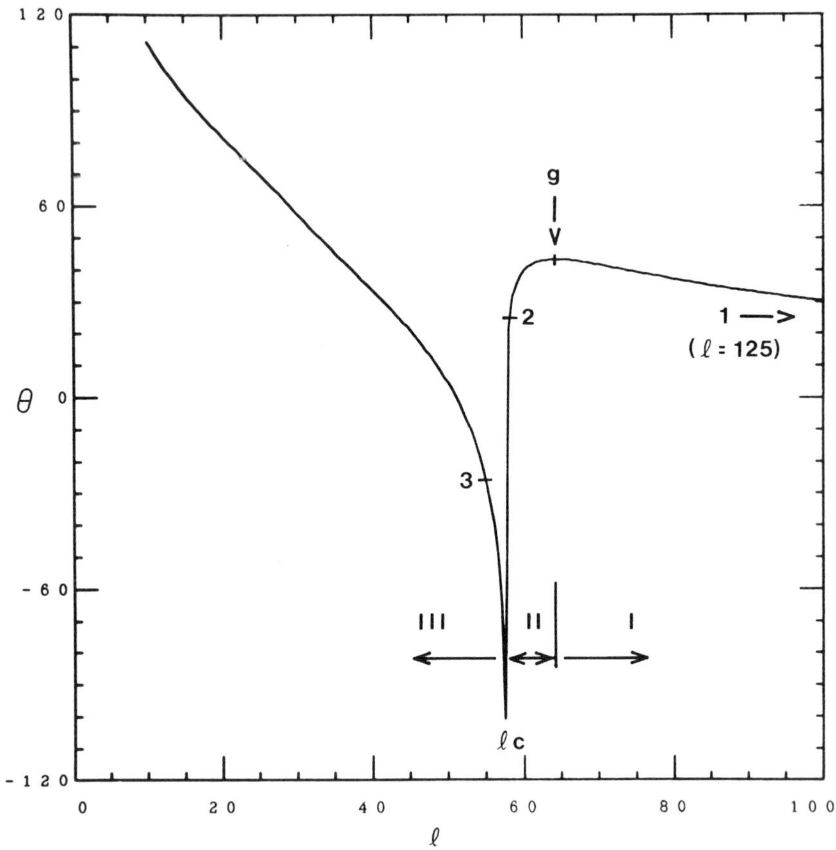

Fig. 18.6. Deflection function from which the classical scattering angle of a trajectory with angular momentum ℓ can be read for 100-MeV $^{18}O + {}^{120}Sn$. Points on the function corresponding to the four orbits of Fig. 18.5 are marked. The back scattering for small ℓ is caused by the repulsive Coulomb core (Fig. 18.1), which at this energy is not surmounted. The singularity occurs at the critical angular momentum ℓ_c. The three regions I, II, and III are discussed in the text. The grazing angle θ_g, which is the maximum to which non-penetrating orbits can scatter, is about 43° (Glendenning, 1975b).

become increasingly unreliable for the preceding sequence of interactions, except of course for energies below the Coulomb barrier. The exception is a drastic one because it excludes most of the interesting physics.

Earlier, we noted the *smooth* behavior of the S matrix as a function of ℓ when the de Broglie wavelength is short enough. The reason for this is clearly connected with the density of integer angular-momentum states on the deflection function.

We identified three classes of orbits and illustrated one from each class in Fig. 18.5, and we now identify the region of the deflection function (Fig. 18.6) to which they correspond. Region I corresponds to orbits that turn away from the nucleus because of the repulsive Coulomb field. However, the lower ℓ orbits of this region do come into the tail region of the nuclear field and are not deflected to as large an angle as they would have been in a pure Coulomb field. The opposite effects of the two fields produce the situation characterized by the grazing orbit ℓ_g, which scatters to the maximum angle possible for nonpenetrating orbits. This orbit marks the boundary with region II. The nuclear attraction causes the orbits of this region to *skim* the edge of the attractive region. The critical angular momentum ℓ_c, marks the point where the deflection function has a singularity. Because ℓ_c corresponds to the top of the barrier where the net radial force is zero, it corresponds to a capture (though unstable) orbit. Trajectories with ℓ very close to ℓ_c spiral around the nucleus. The skimming orbits of region II are in sharp contrast with the plunging orbits of region III below ℓ_c, such as 3 in Fig. 18.5. The plot of the penetration depth D in Fig. 18.2 clearly shows the discontinuity between these orbits and others.

The maximum in the deflection angle for nonpenetrating orbits, marked by g, is a prominent feature of the deflection function and provides an understanding of angular distributions. An early explanation for the grazing angular distribution was based on the notion of pure Coulomb orbits and is therefore deficient. According to that explanation, close orbits scatter to large angles but are absorbed, whereas distant orbits scatter forward but do not very successfully cause transitions because of the distance between the nuclei. Between these is an optimum orbit that defines the grazing peak. *Instead* we see that it is the attractive nuclear field, counterbalancing the repulsion of the Coulomb force, that produces a *maximum* scattering angle for nonpenetrating orbits. This we call the grazing angle. Some of the penetrating orbits can scatter *both* forward or back of this angle but suffer attenuation by absorption, whereas more distant orbits, which scatter only forward of the grazing angle contribute little to reactions because of the distance between the nuclei. The bunching of orbits is clearly visible in Fig. 18.3.

It might at first be imagined that the trajectories of region III do not contribute at all to heavy-ion reactions because of strong absorption. However, the absorption of such close orbits may be offset by other factors. Although the trajectories and deflection functions refer to classical *elastic* scattering, we consider them relevant to other quasi-elastic processes in which the exchange of energy, mass, charge, and angular momentum are all *relatively* small (e.g., $\Delta E/E \ll 1$). Such processes include inelastic scattering and one- or several-nucleon transfer reactions, all leading to low-lying states. For the purpose of this qualitative discussion, we suppose that the trajectory is undisturbed and that the cross section for any of the quasi-elastic processes can be estimated from three factors. One factor is the intrinsic elastic cross section, as estimated here from the classical deflection function. Flat regions of $\theta(\ell)$ indicate the scattering of many trajectories to nearly the same angle and give rise, therefore, to a large cross section. The other two factors are the absorption into other channels represented in the quantum calculations by the imaginary potential, and finally the form factor for the particular quasi-elastic process. We can gain an appreciation for the various regions of sensitivity of quasi-elastic processes from their form factors in r space. Thus for inelastic, one- and two-nucleon transfer, these form factors behave, at large distance, as

$$F_{\text{inel}}(r) = -\beta R \frac{\partial V}{\partial r} + \beta_c \frac{3Z_1Z_2e^2R_c^L}{(2L+1)r^{L+1}}, \qquad r > R_c, \tag{18.7}$$

$$F_1(r) \to \exp[-(2mB_1/\hbar^2)^{1/2}r], \tag{18.8}$$

$$F_2(r) \to \exp[-(4mB_2/\hbar^2)^{1/2}r]. \tag{18.9}$$

In these expressions, β and β_c are the nuclear and charge deformations, respectively, R the radius of the vibrating nucleus, and B_1 and B_2 the binding energies of one and two nucleons. The last two expressions approximate the behavior in the tail region. Actual computed form factors for two-nucleon transfer are shown in Fig. 18.7 (Ascuitto and Glendenning, 1973a.) In the order listed, the preceding processes are increasingly concentrated in the nuclear region and thus will exhibit an increasing sensitivity to the interaction at small distance. Although the penetrating orbits of region III are subject to strong loss to other more complicated channels (for which we sometimes use the word "absorption," not meaning to imply, however, compound nucleus formation), nonetheless, a process such as two-nucleon transfer to a coherent state *strongly* prefers the deep penetration. Whether or not this region manifests itself in a particular case depends on how the three factors balance out.

We now make a very crude estimate of the relative contributions of orbits 2 and 3 of Fig. 18.5 to two-nucleon transfer. To estimate the absorption we note that the decay length in a semiinfinite medium is given by the reciprocal of the imaginary part of the wave number. In this application, we compute a typical *local* value for the orbit in question and use a rough estimate of the path length in the absorptive region, read from Fig. 18.5. The wave number is

$$K \simeq \left[\frac{2M}{\hbar^2}(E-V)\right]^{1/2}\left(1 - \frac{i}{2}\frac{W}{E-V}\right), \tag{18.10}$$

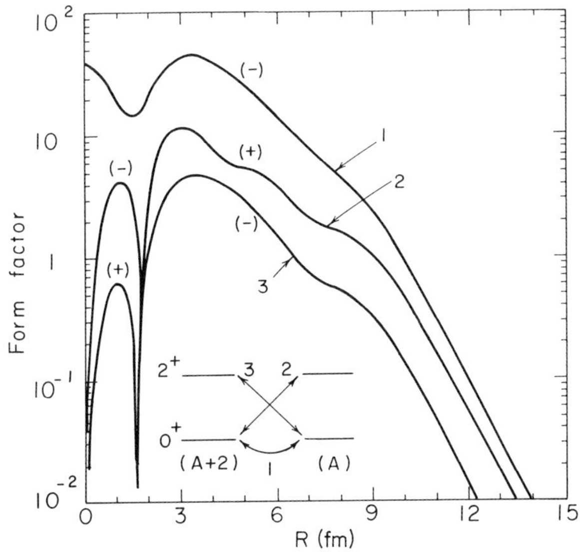

Fig. 18.7. The $\ell = 0$ and $\ell = 2$ form factors for two-nucleon transfer in the reaction $^{120}Sn(^{18}O, ^{16}O)^{122}Sn$ or its inverse. Absolute value is plotted, and signs are indicated (Glendenning, 1975b).

which is roughly $6(1 + i/8)\ \text{fm}^{-1}$. Thus the amplitude decreases as $\exp(-3r/4)$. The pathlength of orbit 3 is almost 8 fm. Orbit 1 is outside in the tail region of $W(r)$; we give its amplitude to be attenuated by about 1 fm pathlength or e^{-1}. From Fig. 18.7, the form factor is approximately e^{-r} in the region in question, which appears squared in the cross section. Finally, the elastic cross section for angular momentum ℓ is given classically by

$$\left(\frac{d\sigma}{d\Omega}\right)_\ell = \lambda\!\!\!^{-2} \frac{(\ell + \frac{1}{2})}{\sin\theta} \left|\frac{d\theta}{d\ell}\right|^{-1}. \tag{18.11}$$

The three factors for each orbit are collected in Table I. We see that for two-nucleon transfer the two close orbits are within a factor e in importance,

TABLE I

Estimate of the Relative Contribution of Orbits 2 and 3 of Fig. 18.5 to Two-Particle Transfer

Orbit	Probability of survival	Transfer probability	Elastic cross section	Product
2	e^{-2}	e^{-8}	1	e^{-10}
3	e^{-12}	1	e	e^{-11}

revealing that the contribution of penetrating orbits is by no means negligible. Of course, we must rely in the final analysis on a quantum calculation. The quantity that will illustrate our point is the absolute value of the S matrix plotted as a function of angular momentum. The S matrix gives the amplitude in the outgoing channel to which it corresponds, given that the flux in the incident channel is unity. Thus, for example, the partial cross section for given ℓ is

$$\sigma_\ell = (\pi/k^2)(2\ell + 1)|S_\ell|^2, \qquad \text{nonelastic.} \tag{18.12}$$

Using the potential of Fig. 18.1 (with $W = -15$ MeV) and the form factor of Fig. 18.7, we compute the S matrix for two-nucleon transfer to the ground state for the reaction ^{120}Sn(^{18}O, ^{16}O) ^{122}Sn, $E_{\text{lab}} = 100$ MeV. The result, shown in Fig. 18.8, shows a very smooth narrow function of ℓ centered at

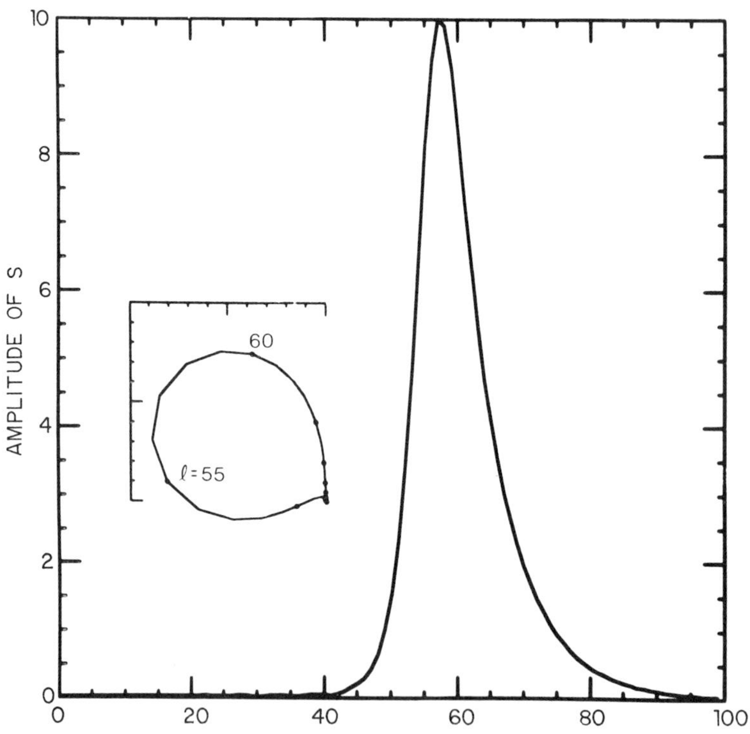

Fig. 18.8. The amplitude of the S-matrix for the ground-state transition in the reaction ^{120}Sn(^{18}O, ^{16}O)^{122}Sn at E = 100 MeV. The inset shows S plotted in the complex plane, its real part on the x axis. Points representing S are joined by straight lines, and its position at $\ell = 55$ and 60 are marked. It begins and ends at the origin (Amplitude is multiplied by 509.34) (Glendenning, 1975b).

$\ell = 57$, which is the critical angular momentum for this pontential at this energy, as was seen in Figs. 18.1, 18.2, and 18.6. The S Matrix at the values of $\ell = 55$ and 58, which correspond approximately to orbits 2 and 3 (the skimming and plunging orbits of Fig. 18.5), are seen to be comparable, confirming the expectation formed by the preceding crude estimate.

This case, in which the reaction region is confined to a *very narrow* band of orbits, is in marked contrast to the inelastic scattering to the collective 2^+ state in the entrance channel of the preceding reaction. Its S matrix, shown in Fig. 18.9, reveals the very large contribution of the high-ℓ region, corresponding to the very slow fall off of the electric quadrupole part of the field. It will be noted that the S matrices for the two cases are cut off at the same

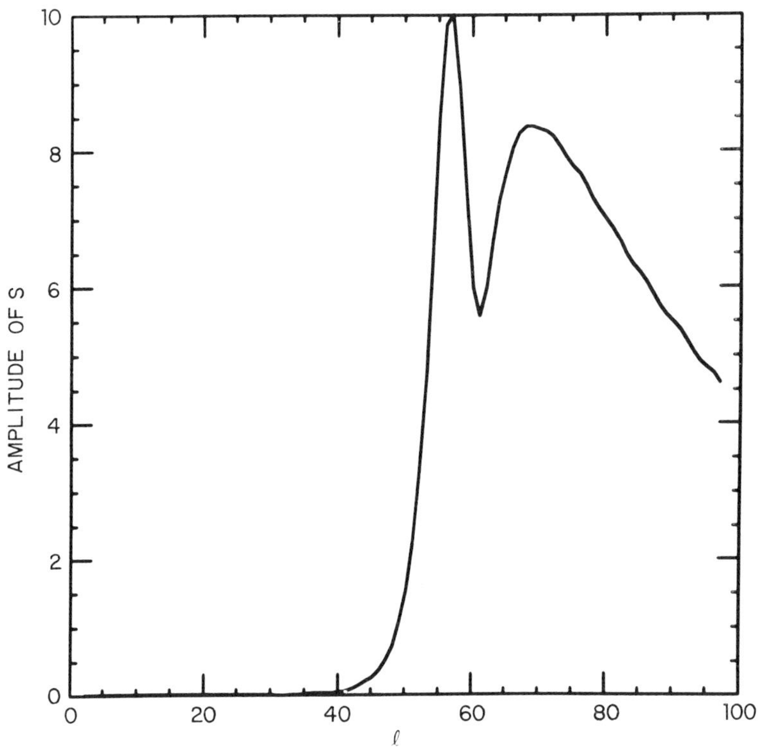

Fig. 18.9. The amplitude of the S-matrix corresponding to the inelastic excitation of the 2^+ vibrational state of Sn in the reaction $^{18}O + {}^{120}Sn$ at 100 MeV. Actually, three S's occur for each incident ℓ, corresponding to the three possible values of the final angular momentum ℓ, $\ell \pm 2$. The one for $\ell' = \ell$ is shown, being typical. The lower peak is dominated by the nuclear field, whereas the upper one, which extends to very high ℓ, is dominated by the Coulomb field. The minimum arises from a destructive interference between the two for ℓ in the vicinity of 62 (Amplitude is multiplied by 102.53) (Glendenning, 1975b).

$\ell_{1/2} = 54$, at least for the absorption strength used here, which is consistent but not very well determined by elastic scattering. Nevertheless, the two-nucleon transfer is more sensitive to close distances, perhaps up to the penetration depth (Fig. 18.2), for $\ell_{1/2}$. This is so because the *relative* contribution of this orbit is bigger than in the case of inelastic scattering.

The preceding discussion reveals that, depending on the particular reaction, we can select the sensitivity to various orbits and hence to reaction regions. In addition, further fine tuning for a given reaction can be achieved by selecting the angle region.

D. ELASTIC AND INELASTIC SCATTERING AND THE OPTICAL POTENTIAL

It is often said that elastic scattering is sensitive only to the tail region of the potential. Here we attempt to make this statement more precise. To do so we are motivated by the importance of the grazing angle, as revealed in the deflection function. Classically, this is the maximum angle to which non-penetrating orbits can scatter. Presumably, then, this is the invariant that is possessed by all potentials that can be found to yield the same quantum elastic scattering. Therefore, we focus on the penetration depth $D(\ell_g)$ of the grazing orbit ℓ_g. For the potential, listed earlier (Chapter 14), we read from Fig. 18.6 that $\ell_g \simeq 64$, and from Fig. 18.2 that $D(\ell_g) \simeq 12$ fm.

This is the distance at which we choose the real part of all our potentials to have the same value, which is only 0.36 MeV. The radius and diffuseness in each case are chosen so that in a 1 fm neighborhood of this point the potentials do not differ radically, as illustrated in Fig. 18.10. The deflection

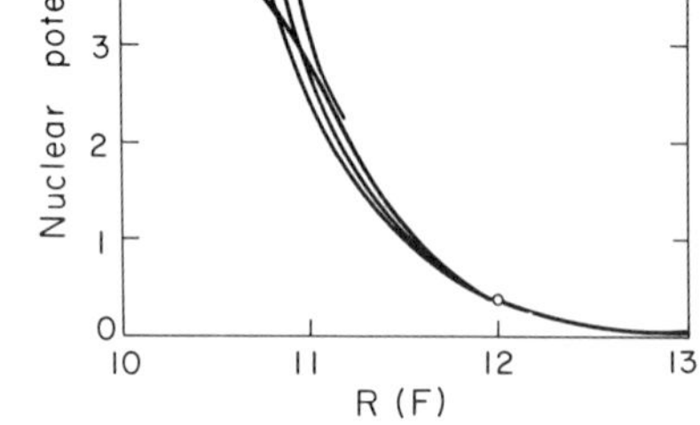

Fig. 18.10. The tail region of the real part of potentials that yield elastic cross sections that are indistinguishable. The central value of the depth is in each case indicated. These are members of a continuum of potentials. Here V can have any value lying between these curves, and possibly beyond (Glendenning, 1974, 1975b).

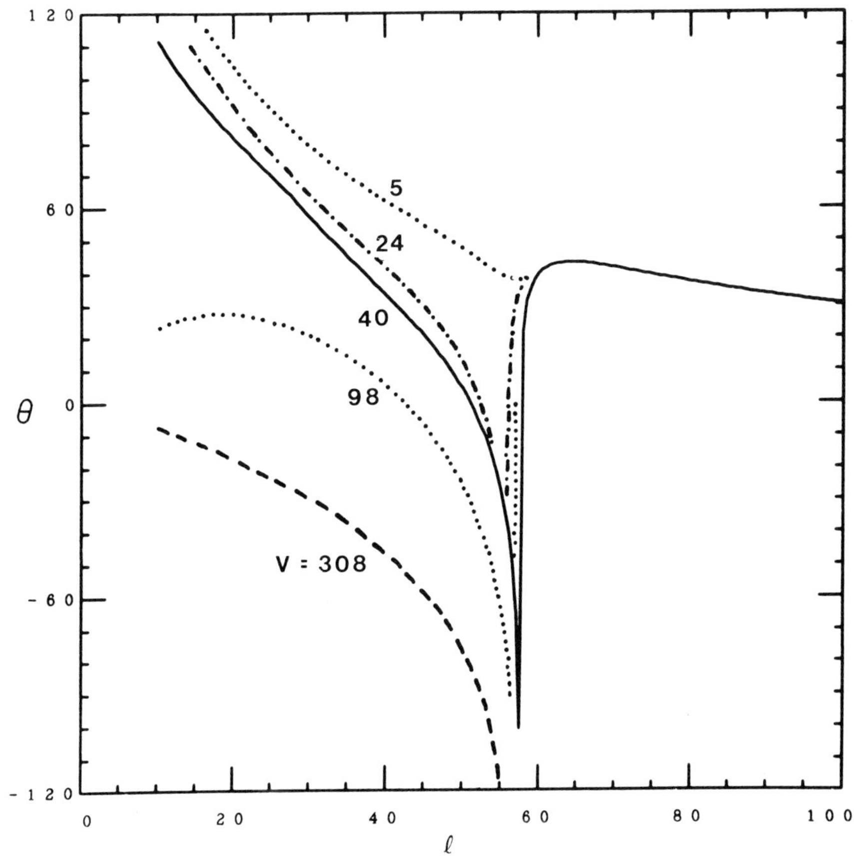

Fig. 18.11. Deflection function for 100-MeV $^{18}O + ^{120}Sn$ corresponding to the potentials whose tails are plotted in Fig. 18.10. They are identical beyond $\ell \simeq 60$. (Glendenning, 1975b).

functions for these potentials are shown in Fig. 18.11 and are all seen to have the same form in the neighborhood of ℓ_g and beyond. They do differ radically for smaller ℓ. Nevertheless, in *each* case an imaginary potential can be found that yields quantum elastic cross sections, which are indistinguishable by typical experiments (Fig. 18.12). We emphasize that these potentials are only a selection from a continuum (see Fig. 18.11). From this fact we conclude that elastic scattering at best can be used to determine the potential only roughly in the vicinity of the penetration depth of the grazing orbit. How rough the determination is can be surmised from Fig. 18.10. Were it not for the fact that we need to know the interaction to calculate more complicated processes, we would dismiss the elastic optical potential with the comment that it is a very

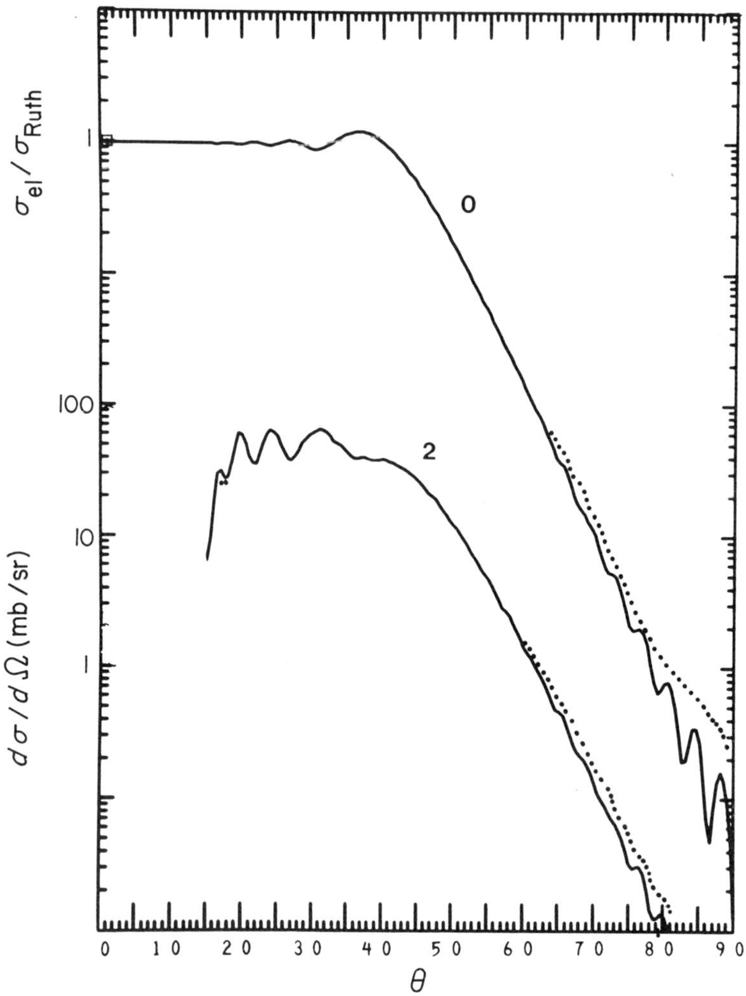

Fig. 18.12. Elastic and inelastic 2^+ cross sections for 100-MeV $^{18}O + {}^{120}Sn$. The two scales refer to the two curves. Solid curves is for V = 308 and dotted for V = 5 (Glendenning, 1975b). For the inelastic cross section, our calculation for $\theta \leq 15°$ is inaccurate because only the contributions for $\ell \leq 215$ and $R \leq 30$ fm were computed at this energy. Cross sections throughout are in mb/s.

complicated and nonunique way of generating a very simple S matrix (Fig. 18.4).

At this point we note, in case the hope is entertained that the cross section determined at several energies could be used to distinguish among the potentials, that $D(\ell_g)$ is almost constant, ranging from 11.6 to 11.8 fm for E between

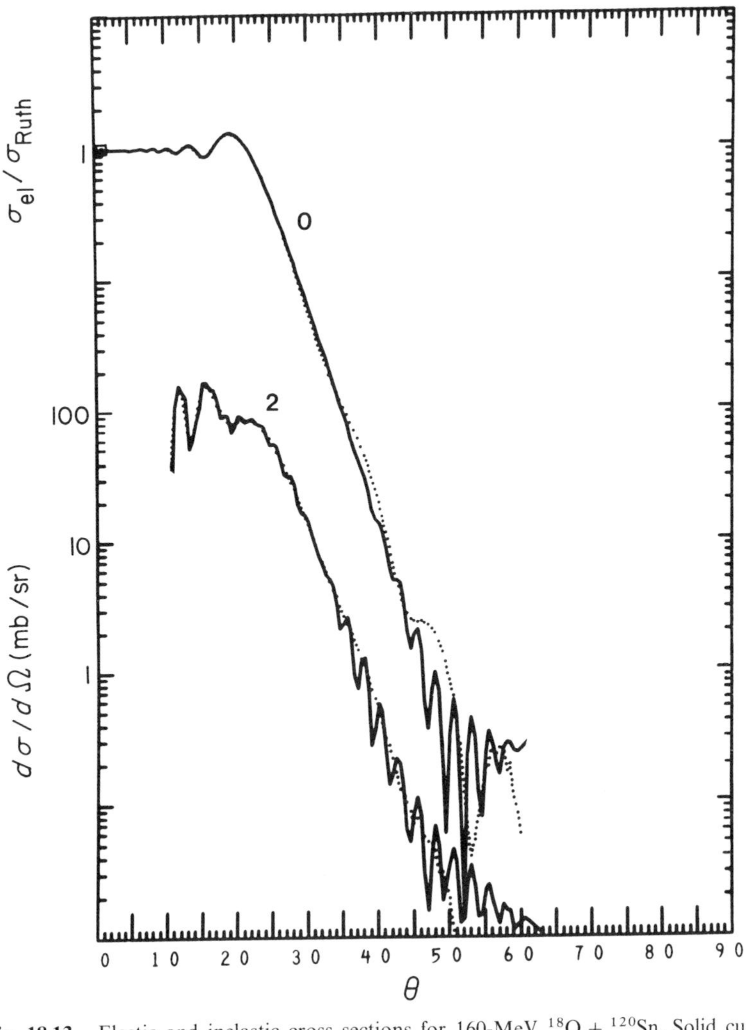

Fig. 18.13. Elastic and inelastic cross sections for 160-MeV $^{18}O + {}^{120}Sn$. Solid curve is for V = 308 and dotted curve is for V = 5 (Glendenning, 1975b).

60 and 160 MeV. Therefore, over the range of energies from the barrier up to several times its value, the same region of configuration space dominates the elastic cross section. This is confirmed by the calculations shown in Fig. 18.12 and Fig. 18.13 for $E = 100$ and 160 MeV, where elastic and inelastic scattering are compared for the two extreme potentials of Fig. 18.10.

The inelastic cross sections to the collective 2^+ state are also shown in these figures. It is disappointing to note that they are hardly more sensitive than elastic scattering. Moreover, the differences, small as they are, are probably well within the uncertainty of the model for the vibrational state.

E. INDIRECT TRANSITIONS IN TWO-NUCLEON TRANSFER, OPPOSITE BEHAVIOR FOR STRIPPING AND PICKUP

Nucleon transfer reactions were originally conceived as involving a direct transfer from the ground state of the target to the final state of the residual nucleus. More recently, we have learned that indirect transitions involving an inelastic transition as an intermediate step, as illustrated in Fig. 18.14, sometime compete in importance with the direct transition and, in certain cases, can strongly modify the shape of the angular distribution. Such processes were predicted to be strong in calculations of the (p, t) pickup reaction on vibrational nuclei, but their first convincing appearance in experiment was observed in deformed nuclei (see Chapter 17).

Interestingly, the signature of indirect reactions connecting *vibrational nuclei* is weak for two-nucleon pickup but strong for stripping. The reason for this is that the parentage amplitude (15.49) connecting the ground to a vibrational state has opposite sign for the reverse reaction (transitions 2 and 3 in Fig. 18.14) (Ascuitto and Glendenning, 1973b). This implies that the direct transition (2 or 3) will interfere in the opposite sense with the second-order indirect transitions for pickup as compared with stripping. It turns out to be constructive for pickup, such as the frequently studied (p, t) reaction, and the signature is therefore weak compared with the less commonly studied stripping reactions where the destructive interference alters the angular distribution shape. This probably accounts for the very late discovery of these processes in light-nuclide transfer reactions.

In heavy-ion reactions the effect of indirect transitions on two-nucleon stripping connecting vibrational nuclei is illustrated in Fig. 18.15 (Ascuitto and Glendenning, 1973a). There, the cross sections computed separately for the direct and the two second-order indirect transitions to the 2^+ state are

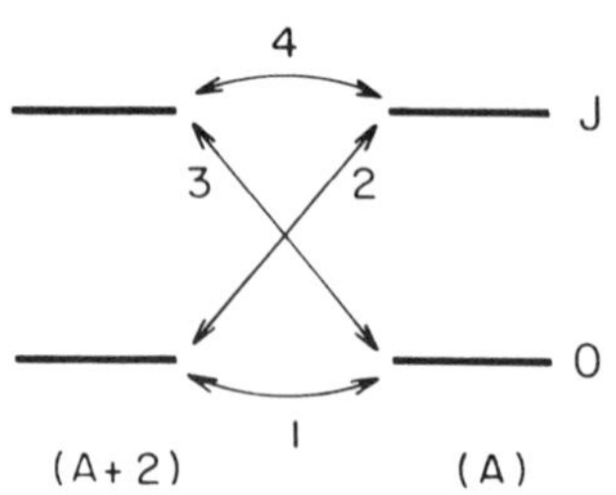

Fig. 18.14. The various direct and indirect amplitudes connecting a pair of states in the initial and final nucleus. (Glendenning, 1975b).

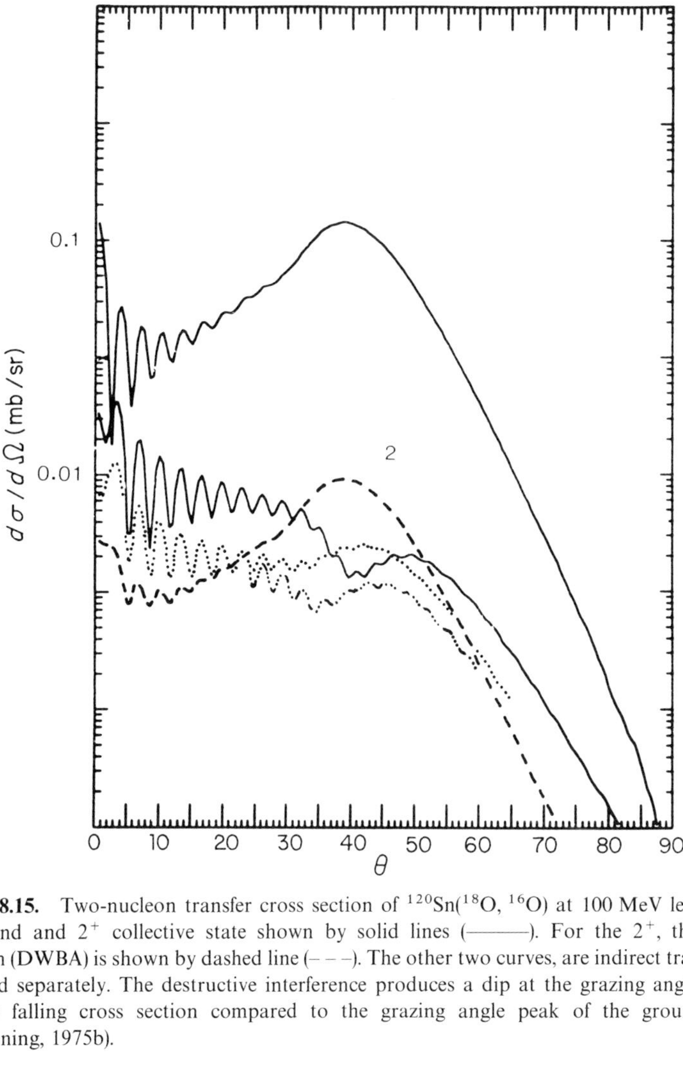

Fig. 18.15. Two-nucleon transfer cross section of $^{120}Sn(^{18}O, ^{16}O)$ at 100 MeV leading to the ground and 2^+ collective state shown by solid lines (———). For the 2^+, the direct transition (DWBA) is shown by dashed line (– – –). The other two curves, are indirect transitions calculated separately. The destructive interference produces a dip at the grazing angle and a generally falling cross section compared to the grazing angle peak of the ground state (Glendenning, 1975b).

shown. When all amplitudes are taken together, the cross section is very strongly altered from the grazing peaked angular distribution characteristic of the direct transition. In particular, for the strength of the inelastic transitions used here, the interference produces a dip at the grazing angle.

The ground state is affected very little by the indirect transitions because branches 2 and 3 of Fig. 18.14 are parts of the second-order indirect transitions leading to the ground state and, as mentioned earlier, have opposite

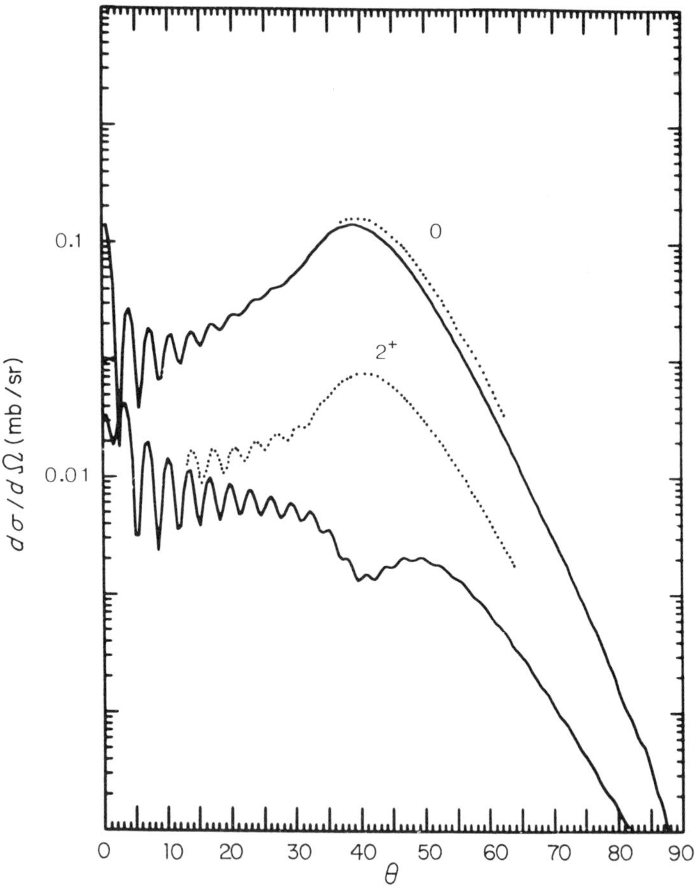

Fig. 18.16. The two-nucleon stripping (———) and pickup (· · ·) reactions are compared for $^{120}\text{Sn} + {}^{18}\text{O} \leftrightarrows {}^{122}\text{Sn} + {}^{16}\text{O}$ at 100 MeV. Note the marked difference of the cross sections to the two 2^+ states arising from the opposite sense in which the direct transition interferes with the indirect in stripping compared to pickup. (Ground-state transitions would be time reversed if calculated at the time-reversed energies) (Ascuitto and Glendenning, 1973b).

signs. As a result, they tend to cancel each other so that the ground-state transition in a good vibrational nucleus will be almost purely direct.

The inverse pickup reaction leading to a vibrational state contrasts (Ascuitto and Glendenning, 1973b) sharply with the stripping reaction, as seen in Fig. 18.16. (The ground-state transitions would be time-reversed and identical if carried out at the reverse energies. Here they were done at the same bombarding energy.) The collective state was, for these illustrative calculations, taken to be a superposition of two quasi-particle states (Arvieu and

Veneroni, 1960a, b) represented by eigenstates of a Woods–Saxon potential. The dependence on the nuclear wave functions is amplified by the fact that the several transfer amplitudes are comparable in magnitude and are coherent, which provides unusual sensitivity. The predicted difference in pickup and stripping reactions was subsequently confirmed by experiment (Scott *et al.*, 1975, Bond *et al.*, 1977).

F. FORWARD-ANGLE RIPPLES, UNCERTAINTY PRINCIPLE

A prominent feature of all the two-nucleon cross sections appearing here are the forward-angle ripples, or high-frequency oscillations. This phenomenon is likely to be very general for heavy-ion reactions, for it depends on a sufficiently high localization in ℓ. Such a localization occurs in heavy-ion transfer reactions because the reaction in r space is a ring, bounded on the inner side by absorption and on the outer side by the decay of the bound-state wave functions. The frequency of the ripples is determined by the angular momentum ℓ_0 corresponding to the peak in S. The angle between ripples is then $\delta\theta \simeq \pi/\ell_0$. We have noticed that ℓ_0 generally corresponds to the critical angular momentum ℓ_c. If the S matrix were a delta function $\delta(\ell - \ell_0)$, then the angular distribution would be given by $|Y_{\ell_0}(\theta)|^2$. The ripples will be damped as the scattering angle increases from zero according to the width $\Delta\ell$ of S.

One can support this picture by computing the angle $\delta\theta$ corresponding to an extra path length of one wavelength between two orbits scattered to the same angle from opposite sides of the nucleus (such as orbits 2 and 3 of Fig. 18.5):

$$\delta\theta = \sin^{-1}(\lambda/2R) \simeq \lambda/2R = \pi/KR = \pi/\ell_0. \tag{18.13}$$

Of course, one must use the *local* wavelength for λ because this refers to an orbit scattered through the nuclear edge (e.g., orbit 3 of Fig. 18.5). Doing so, one computes $\Delta\theta \simeq 3°$ for the $E = 100$ MeV case, in agreement with the ripples.

Thus the frequency of the ripples, if it corresponds to ℓ_c, provides no new information, the critical angular momentum being the one parameter that elastic scattering unambiguously provides. The existence of the ripples depends on the localization in ℓ space.

It is worth noting that in the neighborhood of $\theta = 0$ the first one or two oscillations show a dependence on the angular-momentum transfer (in our case $L = 2$). It would be hard to detect the difference in cases where the frequency of the ripples is so small, but at lower energy or for lighter targets it is feasible.

It is interesting to relate the width of the grazing angular distribution (Fig. 18.15) to the width of the corresponding S matrix (Fig. 18.8). Reading these values at one-half the maximum value, we have $\Delta\theta \simeq 18^\circ = 18\pi/180$ radians and $\Delta\ell \simeq 10$. This gives

$$\Delta\theta\,\Delta\ell \simeq \pi. \tag{18.14}$$

Thus we see that the grazing peak could not be much narrower and still be consistent with the uncertainty principle.

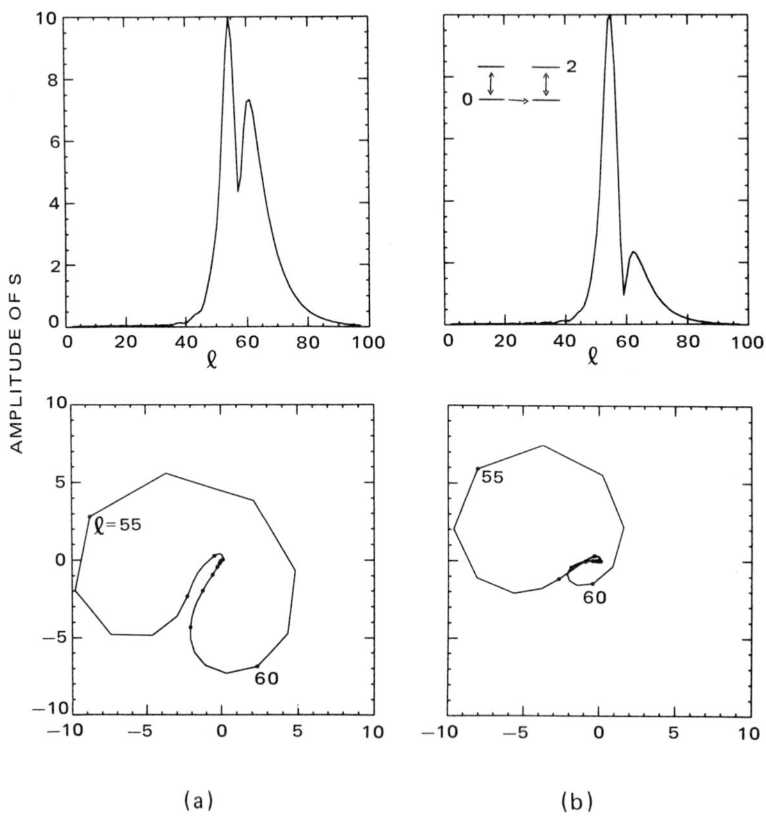

Fig. 18.17. For two-nucleon transfer to the 2^+ state for $^{120}Sn(^{18}O, ^{16}O)$ at 100 MeV the amplitude of the S-matrix is shown together with the contributions from the direct and two second-order indirect routes (Fig. 18.14). The minimum near $\ell = 57$ is produced by the destructive interference between the two indirect terms and the direct. This is evident from the plots

G. STRUCTURE OF THE *S* MATRIX, SPECULATIONS ON FUTURE DEVELOPMENTS

In this section we shift emphasis from the cross sections and their dependence on the physical parameters to the S matrices underlying the cross sections. There are several reasons why the S matrix may prove a useful entity to study. Rather trivially, it portrays the important regions of ℓ space and therefore tells us the interaction region. But beyond this, we are motivated by two other possibilities. One concerns the direct parametrization of S, especially its high-ℓ domain. The other concerns the development of semiquantal

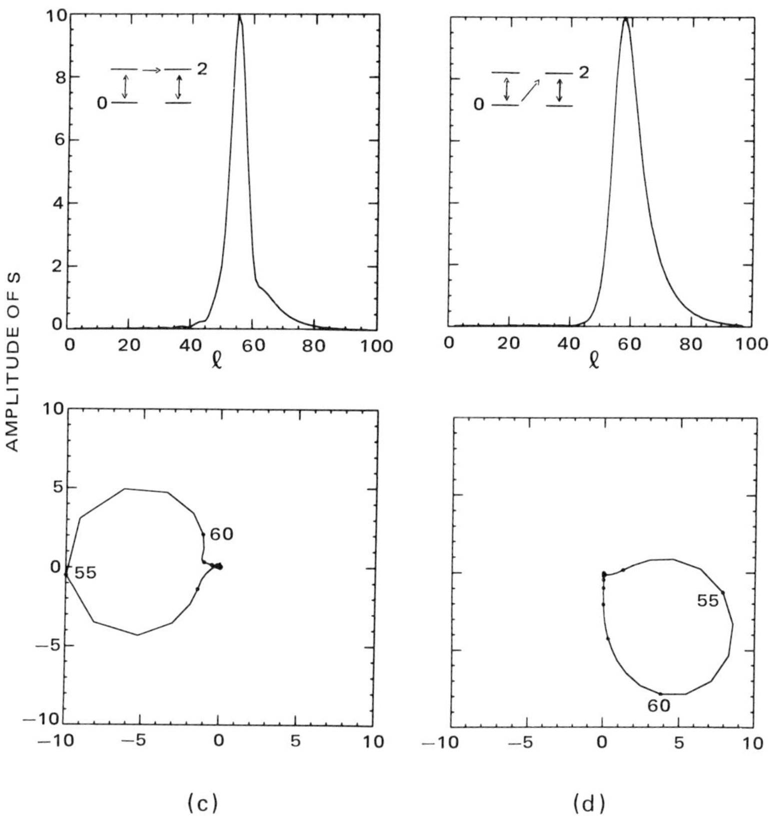

showing S in the complex plane. The complete S is (to high accuracy) the vector sum of the last three. Points corresponding to $\ell = 55$ and 60 are indicated. (Amplitude magnification: (a) $\times 4054$, (b) $\times 5986$, (c) $\times 3735$, (d) $\times 3957$) (Glendenning, 1975b) methods of computing S (Miller, 1970, 1974; Broglia and Winther, 1972; Harvey *et al.*, 1974; Fuller, 1975; Koeling and Malfliet, 1975; Frahn and Gross, 1976; Rowley and Marty, 1976; Brink and Takigawa, 1977; Lee *et al* 1978). In heavy-ion reactions we will sometimes be faced with the need to carry large numbers (perhaps thousands) of partial waves. This is especially so for inelastic scattering because of the slow fall-off of the electric field. It is also true for transfer reactions when both target and projectile are truly heavy and the energy exceeds the Coulomb barrier (hundreds of MeV). One can then imagine that it may be useful to combine a quantum calculation of S up to the region of the critical angular momentum, where the nuclear field causes rapid changes in the deflection function (Fig. 18.6), with one of the other methods for the high-ℓ region, where the conditions for classical scattering are more closely fulfilled. In the high-ℓ region, the phase and the amplitude of S are very smooth and slowly changing functions of ℓ, so that the possibility of interpolating S between calculated values at widely spaced intervals also exists. Although these particular calculations were no problem, for truly heavy projectiles the density of ℓ points on these figures is enormously magnified, and a fully quantum calculation of each contributing ℓ would be very costly.

Nucleon transfer reactions, where multiple-step processes are important, present a special problem because of interference. The amplitude of S for the 2^+ state produced in the two-nucleon transfer reaction is shown in the first part of Fig. 18.17. The minimum, occurring near ℓ_c, arises from the interference of one direct and two second-order amplitudes, also shown. This interference produces the unusual two-lobed figure when S is plotted in the complex plane. (Third- and higher-order contributions are included in the complete calculation but are much smaller than second and first order.) The indirect amplitudes, which peak at a slightly lower ℓ than the direct ones, plunge precipitously above the peak, but then exhibit a shoe. This mirrors the Coulomb excitation amplitude (Fig. 18.9) reduced in strength and cutoff much more rapidly by the particle-transfer process, which cannot occur with appreciable probability when the distance separating the nuclei is too large. The rapid convergence of the particle-transfer process, even when receiving contributions from Coulomb excitation, was correctly predicted prior to the calculations (Ascuitto and Glendenning, 1972) and was decisive in encouraging us to carry out fully quantal calculations. To a high degree of accuracy, the complete S matrix is the sum of the individual amplitudes. (The inaccuracy enters at the third order of the inelastic transitions.) Using whatever methods seem appropriate for calculating the separate components of S and their low- and high-ℓ regions, the unusual two-lobed S matrix characterizing the complete transition can be obtained.

H. SUMMARY

We have reviewed the aspects of classical scattering that are useful in understanding the dependence of heavy-ion reactions on the physical quantities governing the interaction. We emphasized how classical scattering can be used to suggest how to choose experiments that may be decisive in answering specific questions. When classical scattering is used in conjunction with quantum calculations, we are in a position to understand reactions in detail not previously attained in nuclear physics.

The structure of the S matrix, both as a function of ℓ and decomposed into separate components corresponding to specific processes, provides valuable insights. For reactions in which many hundreds of angular momenta are involved, a two-part approach to the calculation of the S matrix was suggested, employing quantum mechanics for the region where the nuclear force is dominant and semiclassical or semiquantal methods for the high-ℓ region, where the smooth behavior of S even permits an interpolation between widely spaced values of ℓ.

One chapter can hardly do justice to heavy-ion reactions. The reader is referred to recent conference proceedings as a guide to the literature.

Notes to Chapter 18

There are a number of interesting articles and reviews other than those referred to in this chapter as well as conference proceedings, such as: Frahn and Venter (1963); Bromley (1969); Frahn (1972); Norenberg and Weidenmuller (1976); Kahana and Baltz (1977); "Proceedings of the International Conference on Reactions between Complex Nuclei," Robinson *et al.* (1974); "Proceedings of the Symposium on Classical and Quantum-Mechanical Aspects of Heavy-Ion Collisions," Harvey *et al.* (1974); "Proceedings of the Macroscopic Features of Heavy-Ion Collisions," Argonne, (1976); and "Heavy Ion Collisions," Vol. 1 and 2, Boch, (1980).

Chapter 19

Polarizability of Nuclear Wave Functions in Heavy-Ion Reactions

A. INTRODUCTION

In this chapter we present some evidence for the polarizability of the wave function of one nucleus by the field of another during their passage in a collision (Delic *et al.*, 1977, 1978; Pruess, 1977, 1979). It is plausible that such a distortion or polarization can occur. However, its effects have never been observed during all the years that nuclear reactions initiated by light particles have been studied. Such reactions have been the primary tool of the nuclear spectroscopist, perhaps because of the very fact that they can be interpreted in terms of the wave functions of the isolated participants.

Many heavy-ion reactions have been interpreted successfully in terms of the wave functions describing states of the isolated nuclei and the distorted-wave Born approximation, which yields the transition amplitude between the assumed states. Such reactions, which proceed in a single step from initial to final state, have a very characteristic angular distribution at energies moderately above the Coulomb barrier. They have a broad peak centered at an angle referred to as the grazing angle. This angle can be inferred from an analysis of the elastic scattering. Although the elastic cross sections do not carry much detailed information about the nucleus–nucleus interaction, they

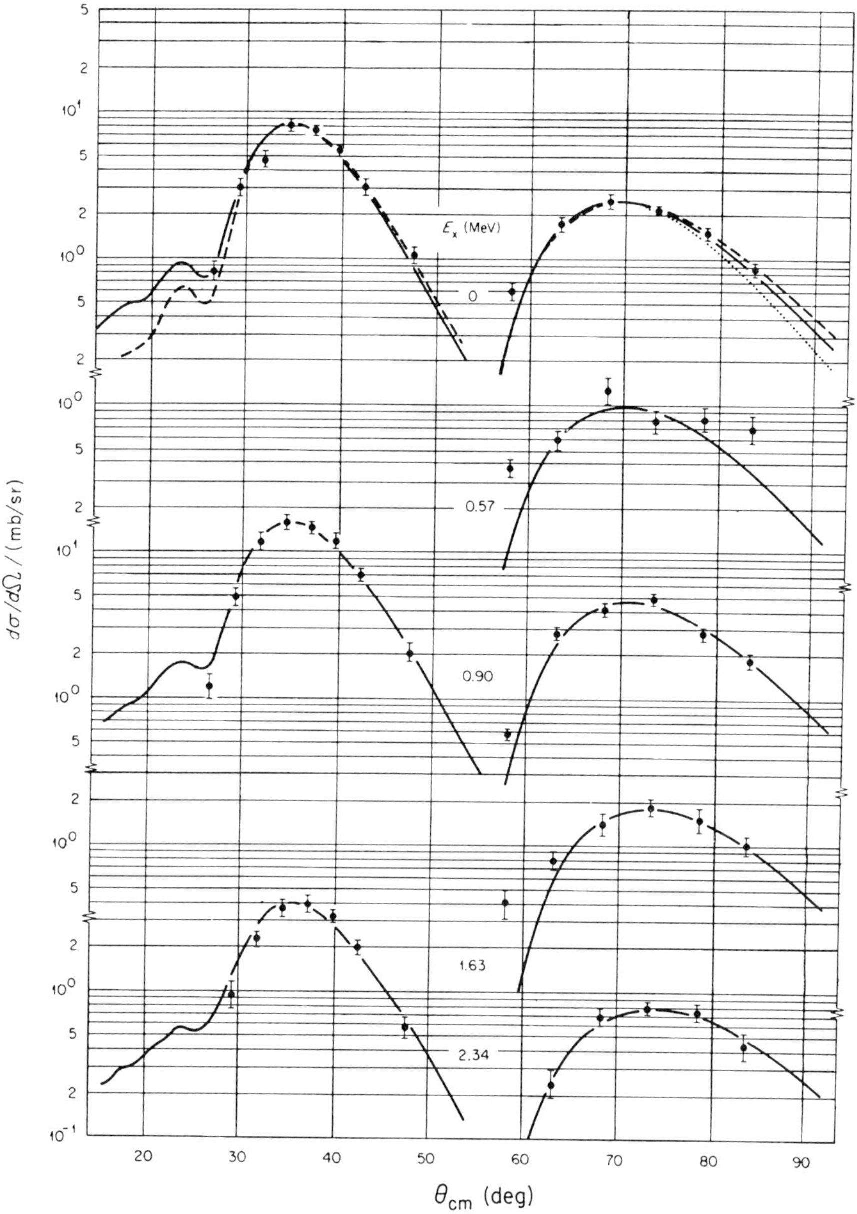

Fig. 19.1. Typical agreement of the DWBA with the data for uncomplicated single-step transitions involving strongly bound states. Here the reaction is $^{208}Pb(^{12}C, ^{13}C)^{207}Pb$ at 116.4 MeV on the left, and 77.4 MeV on the right. [From Toth *et al.* (1976).]

unambiguously determine the grazing maximum in the corresponding deflection function. For truly quasi-elastic processes, those for which all changes ($\Delta E/E$, $\Delta Z/Z$, etc.) between initial and final channels are relatively small, the grazing angle in both elastic channels are equal, and the reaction has its maximum at this angle. When they are unequal, owing to some inelasticity, the reaction peaks at an angle between the two grazing angles of the initial and the final channel. The distorted-wave Born approximation is presumed to locate correctly this intermediate angle for the reaction, if the conditions for the validity of the DWBA are satisfied. We show in Fig. 19.1 examples in which the description of the reaction in terms of a direct single-step transition between shell-model states of the isolated nuclei agrees very well with the experimental data (Toth *et al.*, 1976). This is the quality of agreement that is to be expected if the preceding description of the reaction is accurate. Significant deviations are a signal that the nature of the reaction is different than assumed.

By now we are familiar with some reactions that do have substantially different angular distributions than those having the characteristic grazing maximum. These are reactions in which transitions through intermediate states compete with or dominate the direct transition spoken of previously (Glendenning and Wolschin, 1977 and references there in). In cases in which the final state is reached by several different transitions, the phase of their interference and their relative strengths are determined by the wave functions of the various states involved. Because the two-step transitions have a broad angular distribution compared with the one-step transitions, the distribution of reactions having both distributions will vary widely, depending on the phase and relative strengths of the contributing transitions (Fig. 19.2).

There is a third body of data that appears to have the normal peaked angular distribution, but for which shifts of the peaks from the positions expected from DWBA calculations occur *systematically* for a number of levels in the same nucleus (Kovar *et al.*, 1972). It is the systematic evolution of the discrepancy in several levels of the same nucleus that suggests that some explanation must be sought other than the interference between direct and indirect transitions spoken of previously.

B. THE EXPERIMENTAL FACTS

At comparable energies of the reactions exhibited in Fig. 19.1 and for the same target and projectile, Fig. 19.3 shows reactions leading to six states in the final nucleus. A systematic discrepancy evolves from the ground to the highest-lying level shown when one compares the experiment with the usual DWBA. The data peak at a more forward angle than expected. We note that

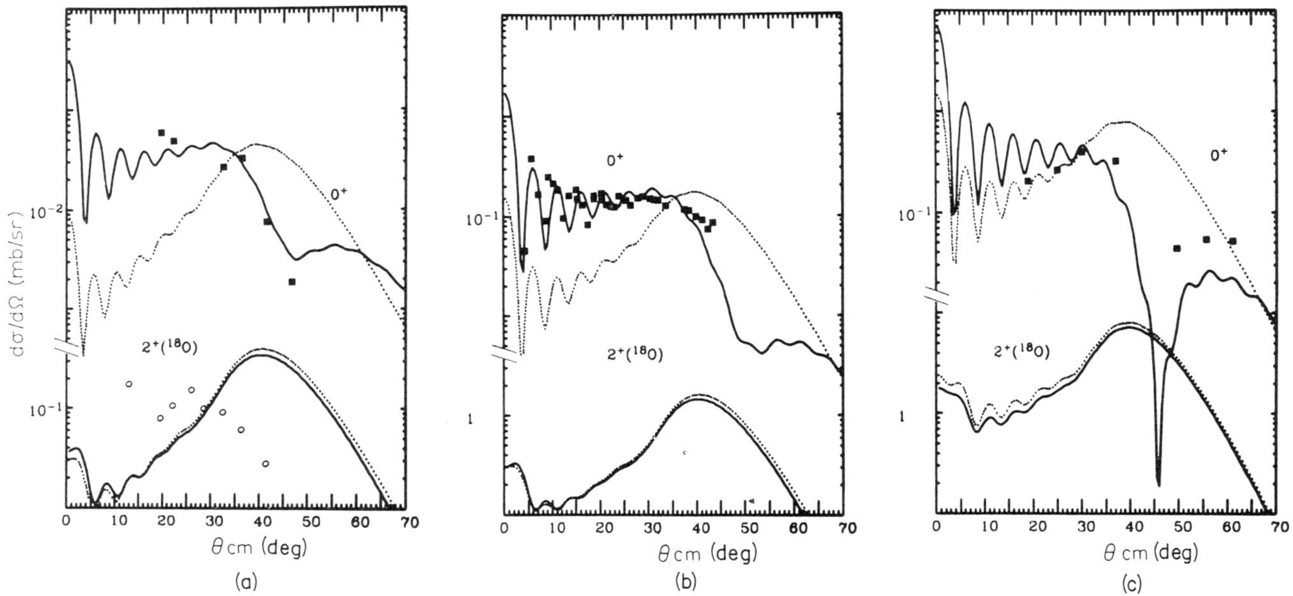

Fig. 19.2. We illustrate how, for the ground state, the interference between direct and indirect transitions alters the angular distribution, and how the ultimate result depends on the nuclear wave functions, as shown for the three different cases. In the graphs solid curve is for direct plus indirect transitions and dotted curve for direct transitions. Graph (a) for $^{60}Ni(^{16}O, ^{18}O)$ at 73.2 MeV (From Berkeley), (b) For $^{62}Ni(^{16}O, ^{18}O)$ at 70.7 MeV (From Brookhaven), and (c) For $^{64}Ni(^{16}O. ^{18}O)$ at 68 MeV (From Brookhaven) [Glendenning and Wolschin, (1977)].

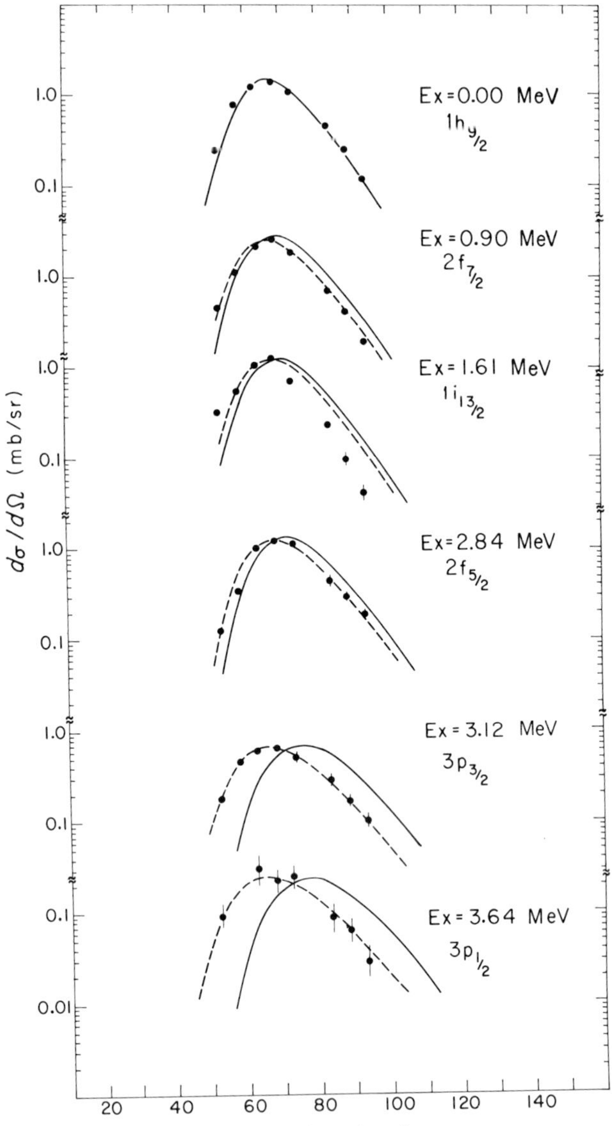

Fig. 19.3. DWBA calculations (——) become systematically out of alignment with data (– – –) as binding energy decreases from 3.8 MeV at the ground state for low bombarding energy ($E = 78$ MeV) for $^{208}Pb(^{12}C, ^{11}B)^{209}Bi$ [Kovar *et al.* (1972)].

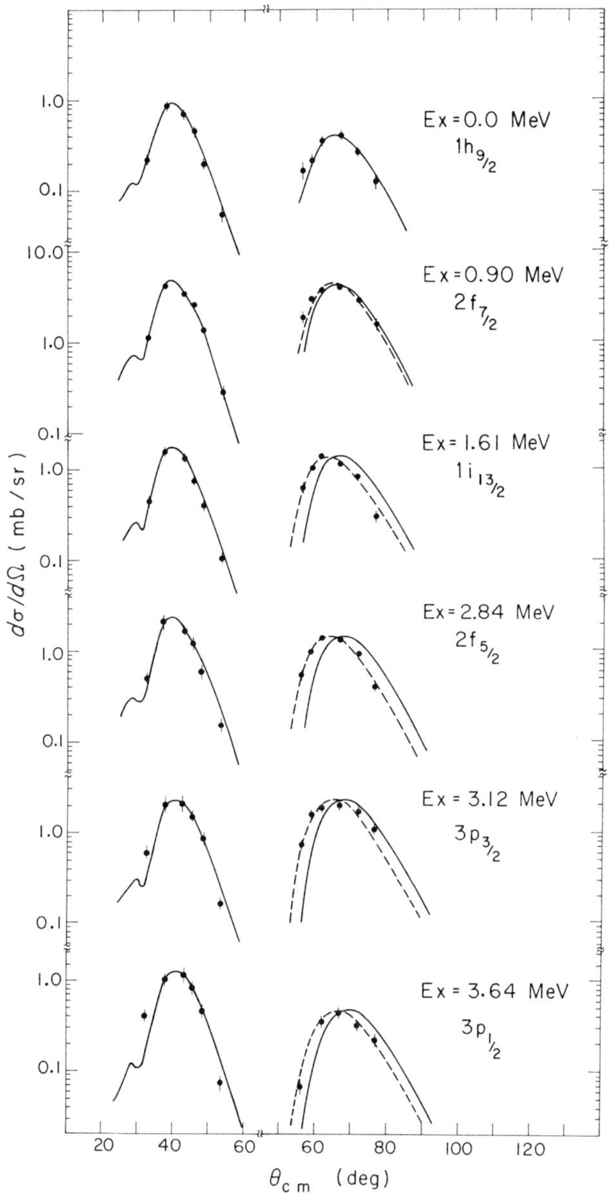

Fig. 19.4. Similar reaction leading to same states as Fig. 19.2 at two higher energies at left 140 MeV and at right 104 MeV for $^{208}Pb(^{16}O, {}^{15}N)^{209}Bi$. The discrepancy between DWBA (———) and data (– – –) vanishes at highest energy [Kovar *et al.* (1972)].

the discrepancy appears to be correlated with how weakly the transferred particle is bound. The states of ^{209}Bi are presumed to be the shell-model states of a single proton lying outside the magic ^{208}Pb core. The ground state is bound by only 3.8 MeV, and the binding decreases for the more highly excited states. This is in contrast with the binding of the levels produced in the reaction illustrated in Fig. 19.1. The ground state is bound by 7.4 MeV, and the higher-lying levels are even more strongly bound because they correspond to removing deeper-lying particles from the ^{208}Pb core.

Figure 19.4 shows similar results for a related reaction leading to the same ^{209}Bi states at two higher energies. We note a second correlation: the discrepancy appears to depend on the bombarding energy.

The experimental facts can be summarized as follows:

(1) For the weakly bound states, the grazing peak in the angular distribution is shifted forward in angle from its expected location.

(2) The shift grows systematically with decreasing binding.

(3) The shifts decrease with increasing bombarding energy.

C. A POSSIBLE EXPLANATION

As suggested in the introduction, we believe that the preceding observations may indicate that the wave function of a weakly bound particle in one nucleus may be polarized by the field of the other nucleus that is participating in the reaction and that the transition occurs to or from the polarized state rather than from the state of the isolated nucleus, as has been assumed.

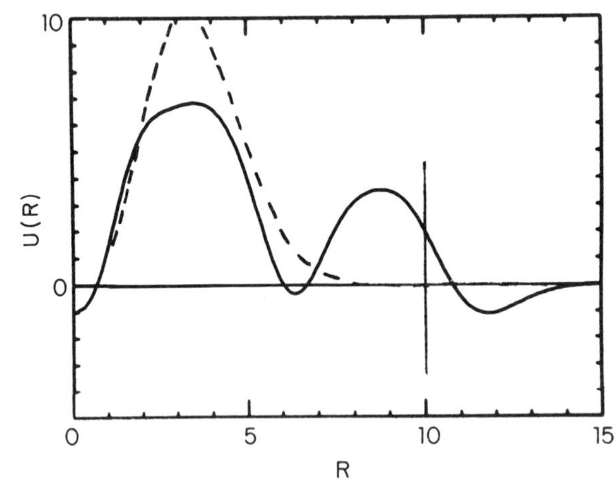

Fig. 19.5. Wave function along axis joining ^{40}Ca to ^{16}O for a pure $1F_{7/2}$ state (- - -) and polarized state (———). The center of ^{40}Ca is at the origin and of ^{16}O at 10 fm. (From Delic *et al.*, 1978)].

In Fig. 19.5 we show a two-center shell-model wave function corresponding to a fixed distance between the nuclei. It shows how the wave function can be polarized from its original form and drawn toward the other nucleus.

Whether such a polarization can occur during a reaction time depends on the two characteristic times, the collision time, and the single-particle relaxation time,

$$\tau(\text{coll.}) \simeq 2R/V = 2R/\sqrt{2E/m\text{A}},$$

$$\tau(\text{s.p. relaxation}) \gtrsim 2R/V_{\text{F}} = 2R/\sqrt{2\mathscr{E}_{\text{F}}/m},$$

where A is the projectile mass number, E the bombarding energy, V_{F} and $\mathscr{E}_{\text{F}}$, the Fermi velocity and energy in the nucleus (target or projectile) that is polarized (i.e., that has a weakly bound state). This ratio is

$$\frac{\tau(\text{coll.})}{\tau(\text{s.p.})} \simeq \sqrt{[\mathscr{E}_{\text{F}}/(E/\text{A})]} \simeq \sqrt{(30/5)} = \sqrt{6}.$$

Thus in typical collisions the collision time is longer than the single-particle relaxation time, but not much longer. Therefore, polarization may have time to develop, although the characteristic times are not decisive.

D. ADIABATIC ESTIMATE

Having satisfied ourselves that polarization may occur, we now make what we expect will be an overestimate of the effect. If the overestimate corresponds to too small an angular shift in comparison with experiment, then polarizability is clearly not the explanation of the anomalies described earlier. If it does produce shifts comparable or larger than observed, then we will feel encouraged to pursue a dynamical description of the reaction (outlined later).

The scheme we use for our estimate can be called adiabatic, because it assumes infinite time for the development of the polarization at each separation distance. In the usual DWBA amplitude, we replace the wave function of the weakly bound state $\phi_{\text{B}}(r)$ by a two-center shell-model state $\phi_{\text{B}}(r, R)$, which depends on the separation distance R of the two nuclei. Schematically, we can write the amplitude in the adiabatic approximation as,

$$T_{\text{adi}} \propto \langle \chi_\beta^{(-)}(\mathbf{R})\phi_{\text{B}}(\mathbf{r}, R) | \hat{V}(r) | \phi_\alpha(\mathbf{r}')\chi_\alpha^{(+)}(\mathbf{R})\rangle, \tag{19.1}$$

$$\begin{aligned} \phi_{\text{B}}(\mathbf{r}, R) &\equiv \phi_{(n\ell j)\Omega}(\mathbf{r}, R) \\ &= \sum_{n\ell j} A_{n\ell j}^{n\ell j\Omega}(R)\phi_{n\ell j\Omega}(\mathbf{r}) \\ &\xrightarrow[R\to\infty]{} \phi_{n\ell j\Omega}(\mathbf{r}). \end{aligned} \tag{19.2}$$

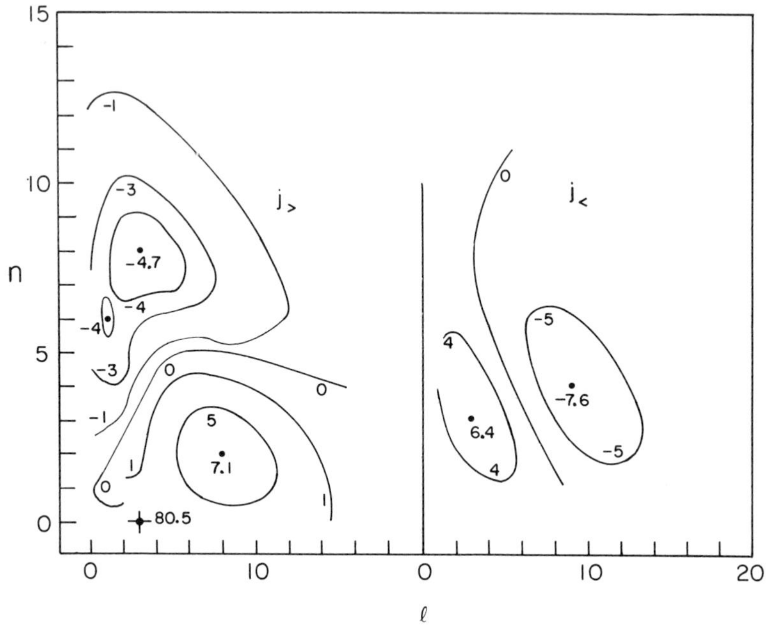

Fig. 19.6. Polarized wave function represented by contours of the amplitudes $A(R)$ in the n, ℓ plane for R = 10 fm), Eq. 19.2. Conditions are the same as in Fig. 19.5. The original shell-model state ($n = 1$, $\ell = 3$) has only an 80% amplitude in the polarized state. (Asymptotic quantum numbers are $n = 0$, $\ell = 3$, $j = \frac{7}{2}$, $\Omega = \frac{1}{2}$.) (From Delic *et al.*, 1978.)

The projection Ω of j on the axis joining the centers of nuclei is conserved. Otherwise, high components of angular momentum $\bar{\ell}$ and radial nodes $\bar{n}$ are introduced by the localization in space of the polarizing source. These are illustrated in Fig. 19.6 by contours in the $n\ell$ plane. This figure is a complete description of the polarized wave function for a state in scandium polarized by an oxygen nucleus at a distance of 10 fm. The same wave function viewed along the axis joining their centers is a more familiar visualization and was shown in Fig. 19.5.

The preceding amplitude has been specified more precisely elsewhere (Delic *et al.*, 1977, 1978; Pruess, 1977, 1979). We calculate it with a finite-range interaction without neglecting recoil. Two examples are shown in Figs. 19.7 and 19.8. The magnitudes of the forward shifts are approximately reproduced by our adiabatic estimate. Also, the cross section in the case of the Ca target is overestimated, which, as noted, might be expected from an adiabatic calculation if indeed the polarizability is the explanation of the anomalies. Thus we conclude that polarization of weakly bound states probably does occur in heavy-ion collisions.

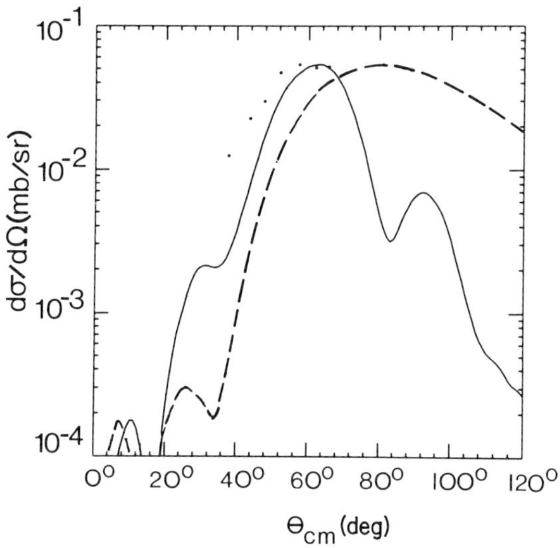

Fig. 19.7. Dashed curve (— — — —) is the usual DWBA multiplied by 5.6, and the solid curve (———) is the result of our adiabatic polarizability calculation divided by 43. They are compared with data (Korner *et al.* (1973)) for the reaction ^{40}Ca (^{16}O, ^{15}N)^{41}Sc at $E = 48$ MeV. (From Delic *et al.*, 1978.)

E. DYNAMICAL DESCRIPTION OF POLARIZABILITY

We now outline a dynamical description of the polarization of nuclear wave functions during a heavy-ion reaction and show how the adiabatic model described previously can be theoretically justified. Suppose the state β produced in a transfer reaction is weakly bound. This state, because of its weak binding, can be easily excited to both the bound and continuum states of the nucleus. The many second-order transitions leading from an initial state α to β through these excited states contributes second-order terms to the amplitude:

$$T_{\alpha\to\beta} = \langle\chi_\beta^{(-)}|(\phi_\beta|\hat{V}_\alpha|\phi_\alpha) + \sum_{\beta'}(\phi_\beta|\hat{V}_\beta|\phi_{\beta'})\frac{1}{E_{\beta'}^{(+)} - H_\beta}(\phi_{\beta'}|\hat{V}_\alpha|\phi_\alpha)|\chi_\alpha^{(+)}\rangle, \quad (19.3)$$

where the first term is the usual DWBA. We neglect rescattering among the intermediate states β' by replacing the interaction in H_β by an optical potential, so that $H_\beta \approx T_\beta + U_\beta$ is the Hamiltonian and $E_{\beta'} = E - \mathscr{E}_{\beta'}$ the energy of relative motion, with $\mathscr{E}_{\beta'}$ the excitation energy of the nucleus. The long-range and smooth behavior of the nucleus–nucleus interaction $\hat{V}$ and the narrow Q window, especially at low energy, ensure that intermediate scattering states of H_β with energy far from E_β can be neglected. Defining a

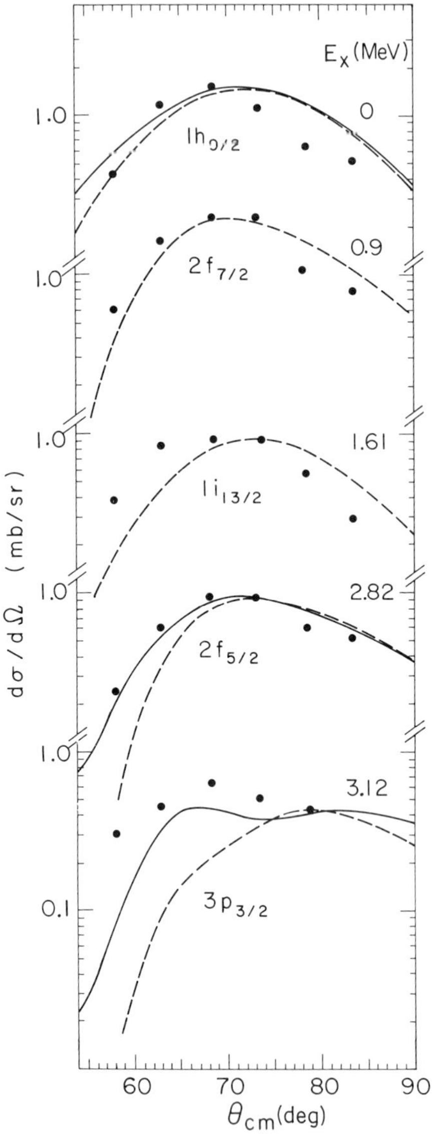

Fig. 19.8. At $E = 78$ MeV, for the reaction populating various states in ^{209}Bi, the usual DWBA (— —) and our adiabatic polarizability (———) cross sections are compared with the data [Kovar *et al.* (1972)].

projection operation P, which projects on eigenstates of H_β with energy E_i such that $|E_i - E_\beta| < |\mathscr{E}_\beta - \mathscr{E}_{\beta'}|$, the Green's function can be expanded as

$$P \frac{1}{E_{\beta'}^{(+)} - H_\beta} = P \frac{1}{\mathscr{E}_\beta - \mathscr{E}_{\beta'}} \sum_{n=0}^{\infty} \left(\frac{H_\beta - E_\beta}{\mathscr{E}_\beta - \mathscr{E}_{\beta'}} \right)^n. \tag{19.4}$$

If we were to keep only the $n = 0$ term, we would obtain

$$T_{\alpha\to\beta} = \left\langle \chi_\beta^{(-)} \left(\phi_\beta + \sum_{\beta'} \frac{(\phi_\beta | \hat{V}_\beta | \phi_{\beta'})}{\mathscr{E}_\beta - \mathscr{E}_{\beta'}} \phi_{\beta'} \right) \middle| \hat{V}_\alpha \middle| \phi_\alpha \chi_\alpha^{(+)} \right\rangle. \tag{19.5}$$

This shows that the transition is from the state α to a polarized state β, with the polarization amplitude corresponding to first-order perturbation theory. This is the theoretical justification for our schematic calculation. However, the bombarding energy dependence is contained in the $n > 0$ terms. Therefore, it is only when the full amplitude (19.3) is calculated that the dynamics of the reaction will be registered in the results.

Appendix

Some Useful Reminders

A. LAPLACE OPERATOR

The operator ∇^2 can be written in spherical coordinates as

$$\nabla^2 = \frac{1}{r^2}\frac{\partial}{\partial r}\left(r^2\frac{\partial}{\partial r}\right) - \frac{1}{r^2}\frac{1}{\hbar^2}L^2(\theta, \phi), \tag{A1}$$

$$L^2(\theta, \phi) = -\hbar^2\left[\frac{1}{\sin\theta}\frac{\partial}{\partial\theta}\left(\sin\theta\frac{\partial}{\partial\theta}\right) + \frac{1}{\sin^2\theta}\frac{\partial^2}{\partial\phi^2}\right]. \tag{A2}$$

B. SPHERICAL HARMONICS

The spherical harmonics $Y_\ell^m(\theta, \phi)$ are eigenfunctions of the preceding operator L^2 and of L_Z,

$$L_Z = (\mathbf{r} \times \mathbf{p})_Z = -i\hbar\,\partial/\partial\phi, \tag{A3}$$

such that

$$L^2 Y_\ell^m(\theta, \phi) = \ell(\ell + 1)\hbar^2 Y_\ell^m(\theta, \phi), \tag{A4}$$

$$L_Z Y_\ell^m = m\hbar Y_\ell^m. \tag{A5}$$

They are related to the Legendre functions, for example,

$$Y_\ell^0(\theta, \phi) = [(2\ell + 1)/4\pi]^{1/2} P_\ell(\cos\theta), \tag{A6}$$

where $P_0(x) = 1$, $P_1(x) = x$, $P_2(x) = \frac{1}{2}(3x^2 - 1)$, etc. They are orthonormal,

$$\int Y_\ell^{m*}(\theta, \phi) Y_{\ell'}^{m'}(\theta, \phi) \sin\theta \, d\theta \, d\phi = \delta_{\ell\ell'}\delta_{mm'}. \tag{A7}$$

Phases used here are the so-called Condon and Shortley phases,

$$\begin{aligned} Y_\ell^{-m}(\theta, \phi) &= (-)^m Y_\ell^{m*}(\theta, \phi) \\ &= (-)^\ell Y_\ell(\pi - \theta, \phi + \pi). \end{aligned} \tag{A8}$$

The spherical harmonic addition theorem states that

$$P_\ell(\cos\omega) = 4\pi/(2\ell + 1) \sum_m Y_\ell^{m*}(\theta, \phi) Y_\ell^m(\theta', \phi'), \tag{A9}$$

where ω is the angle between vectors pointing, respectively, in the directions (θ, ϕ) and (θ', ϕ'). We often denote the angle θ, ϕ by $\hat{\mathbf{r}}$, it being understood that θ, ϕ are polar coordinates of the vector $\mathbf{r}$. Consequently,

$$\cos\omega = \hat{\mathbf{r}} \cdot \hat{\mathbf{r}}' = \mathbf{r} \cdot \mathbf{r}'/rr'. \tag{A10}$$

There is a very useful theorem for the product of two spherical harmonics of the *same* angles:

$$Y_\ell^m(\theta, \phi) Y_{\ell'}^{m'}(\theta, \phi) = \sum_{LM} \left(\frac{\hat{\ell}\hat{\ell}'\hat{L}}{4\pi}\right)^{1/2} \begin{pmatrix} \ell & \ell' & L \\ m & m' & M \end{pmatrix} \begin{pmatrix} \ell & \ell' & L \\ 0 & 0 & 0 \end{pmatrix} Y_L^{M*}(\theta, \phi), \tag{A11a}$$

or

$$[Y_\ell(\theta, \phi) Y_{\ell'}(\theta, \phi)]_L^M = (-)^L \left(\frac{\hat{\ell}\hat{\ell}'}{4\pi}\right)^{1/2} \begin{pmatrix} \ell & \ell' & L \\ 0 & 0 & 0 \end{pmatrix} Y_L^M(\theta, \phi), \tag{A11b}$$

where the large parentheses ( ) are called 3-j symbols (described later) and $\hat{\ell} = 2\ell + 1$. Note that the 3-j symbol with the zeros in the lower position requires that

$$\ell + \ell' + L = \text{even}. \tag{A12}$$

C. ANGULAR-MOMENTUM COUPLING; CLEBSCH–GORDAN COEFFICIENTS

Let $\phi_{\ell_1}^{m_1}$ and $\psi_{\ell_2}^{m_2}$ be eigenfunctions of angular momentum of, say, two quantum systems with the eigenvalues indicated. Sometimes they are denoted by

$$|\ell_1 m_1\rangle \qquad \text{and} \qquad |\ell_2 m_2\rangle.$$

Denote by **L**, the vector sum

$$\mathbf{L} = \ell_1 + \ell_2. \tag{A13}$$

Eigenfunctions of L^2 and L_Z with eigenvalues $L(L+1)$ and M are

$$|\ell_1\ell_2 LM\rangle = \sum_{m_1 m_2} \langle \ell_1 m_1 \ell_2 m_2 | \ell_1 \ell_2 LM\rangle |\ell_1 m_1\rangle |\ell_2 m_2\rangle. \tag{A14a}$$

This is referred to as the vector coupling of $|\ell_1 m_1\rangle$ and $|\ell_2 m_2\rangle$. The coefficients are the very important Clebsch–Gordan coefficients, often denoted variously by

$$\begin{aligned}\langle \ell_1 m_1 \ell_2 m_2 | \ell_1 \ell_2 LM\rangle &= C^{\ell_1 \ell_2 L}_{m_1 m_2 M} = \langle \ell_1 m_1 \ell_2 m_2 | LM\rangle \\ &= (-)^{\ell_1 - \ell_2 + M}\sqrt{2L+1}\begin{pmatrix} \ell_1 & \ell_2 & L \\ m_1 & m_2 & -M \end{pmatrix},\end{aligned} \tag{A15}$$

Sometimes commas or parenthesis are used to indicate associations such as $|(\ell_1\ell_2)LM\rangle$ or $|\ell_1\ell_2, LM\rangle$. In the last line we indicate the relation to the 3-j symbols. They are real and are nonvanishing only if

$$\begin{aligned} m_1 + m_2 &= M, \\ |\ell_1 - \ell_2| \le L &\le \ell_1 + \ell_2. \end{aligned} \tag{A16}$$

Symmetries are easiest to state in terms of $\left(\begin{smallmatrix} \ell_1 & \ell_2 & L \\ m_1 & m_2 & M \end{smallmatrix}\right)$, which is

(1) invariant to cyclic permutation of columns,
(2) multiplied by $(-)^{\ell_1+\ell_2+L}$ if two columns are interchanged,
(3) multiplied by the preceding phase if the signs of m_1, m_2, and M are simultaneously changed.

They obey orthogonality relations which follow from the completeness relation $\sum_B \langle A|B\rangle\langle B|A'\rangle = \delta_{AA'}$,

$$\sum_{m_1=-\ell_1}^{\ell_1} \sum_{m_2=-\ell_2}^{\ell_2} \langle \ell_1 m_1 \ell_2 m_2 | \ell_1 \ell_2 LM\rangle \langle \ell_1 m_1 \ell_2 m_2 | \ell_1 \ell_2 L'M'\rangle = \delta_{LL'}\delta_{MM'}, \tag{A17a}$$

$$\sum_L \sum_M \langle \ell_1 m_1 \ell_2 m_2 | \ell_1 \ell_2 LM\rangle \langle \ell_1 m'_1 \ell_2 m'_2 | \ell_1 \ell_2 LM\rangle = \delta_{m_1 m'_1}\delta_{m_2 m'_2}. \tag{A18a}$$

Written in terms of the 3-j, these appear as

$$\sum_{m_1 m_2} \begin{pmatrix} \ell_1 & \ell_2 & \ell_3 \\ m_1 & m_2 & m_3 \end{pmatrix}\begin{pmatrix} \ell_1 & \ell_2 & \ell'_3 \\ m_1 & m_2 & m'_3 \end{pmatrix} = \frac{1}{2\ell_3 + 1}\delta_{\ell_3\ell'_3}\delta_{m_3 m'_3}, \tag{A17b}$$

$$\sum_{\ell_3 m_3} (2\ell_3 + 1)\begin{pmatrix} \ell_1 & \ell_2 & \ell_3 \\ m_1 & m_2 & m_3 \end{pmatrix}\begin{pmatrix} \ell_1 & \ell_2 & \ell_3 \\ m'_1 & m'_2 & m_3 \end{pmatrix} = \delta_{m_1 m'_1}\delta_{m_2 m'_2}. \tag{A18b}$$

It will be noted in the text that we use both the Clebsch–Gordan and the 3-j symbols rather than choosing always to use one or the other. The reason is that vector coupling relations such as (A14a) are more concise and easier to remember in the Clebsch notation. However, the simplicity of the 3-j with regard to symmetries (which of course can be translated to the Clebsch notation) makes the 3-j simpler to manipulate under some circumstances.

Using (A18a), the inverse of (A14a) is

$$|\ell_1 m_1\rangle|\ell_2 m_2\rangle = \sum_{LM} \langle \ell_1\ell_2 LM|\ell_1 m_1 \ell_2 m_2\rangle|\ell_1\ell_2 LM\rangle. \qquad \text{(A14b)}$$

Frequently, a square bracket notation is used for (A14a), especially when one wants to write a more specific notation for the particular wave functions involved; thus

$$[\phi_{\ell_1}\phi_{\ell_2}]_L^M = \sum_{m_1 m_2} C_{m_1 m_2 M}^{\ell_1 \ell_2 L}\phi_{\ell_1}^{m_1}\phi_{\ell_2}^{m_2}. \qquad \text{(A14c)}$$

Special values are

$$C_{m\,-m\,0}^{\ell\;\ell\;0} = (-1)^{\ell-m}(2\ell+1)^{-1/2}, \qquad \text{(A19)}$$

and

$$C_{0\,0\,0}^{\ell_1\ell_2\ell_3} = 0 \qquad \text{if} \quad \ell_1+\ell_2+\ell_3 = \text{odd}. \qquad \text{(A20)}$$

D. ANGULAR-MOMENTUM RECOUPLING

The recoupling of angular momentum is an exceedingly important tool for the evaluation of matrix elements. The notation $|(j_1 j_2)j_{12}, j_3; J\rangle$ implies that eigenfunctions of $j_1 j_2 j_3$ are coupled in the order

$$\mathbf{j}_1 + \mathbf{j}_2 = \mathbf{j}_{12}, \qquad \mathbf{j}_{12} + \mathbf{j}_3 = \mathbf{J}. \qquad \text{(A21)}$$

This is related to a different ordering through the transformation

$$|(j_1 j_2)j_{12}, j_3; JM\rangle = \sum_{j_{23}} \langle j_1, (j_2 j_3)j_{23}; J|(j_1 j_2)j_{12}, j_3; J\rangle|j_1, (j_2 j_3)j_{23}; JM\rangle. \qquad \text{(A22)}$$

The recoupling coefficient is related to the 6-j *symbol*,

$$\begin{Bmatrix} j_1 & j_2 & j_{12} \\ j_3 & J & j_{23} \end{Bmatrix}(\hat{j}_{12}\hat{j}_{23})^{1/2}(-)^{j_1+j_2+j_3+J} = \langle j_1, (j_2 j_3)j_{23}; J|(j_1 j_2)j_{12}, j_3; J\rangle. \qquad \text{(A23)}$$

The arguments of the 6-j symbol can assume only values as are allowed by the vector coupling relations implied by the right-hand side. Otherwise they vanish. The labels connected by such vector relationships can be can be

indicated by

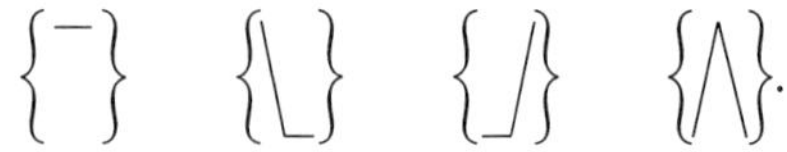

There exist tables of the 3-j and 6-j symbols although in applications they are usually computed with a subroutine in a computer. The 6-j symbol is also related by a factor to the Racah coefficient. The properties of these important quantities can be found in several texts (Edmonds, 1957; Brink and Satchler, 1968). In particular, as transformation coefficients between complete sets of states, many relations between them, such as orthogonality and completeness, can be written immediately. For example,

$$\sum_{j_{12}} \langle j_1, (j_2 j_3) j_{23}; J | (j_1 j_2) j_{12}, j_3; J \rangle \langle (j_1 j_2) j_{12}, j_3; J | j_1, (j_2 j_3) j'_{23}; J \rangle = \delta_{j_{23} j'_{23}} \tag{A24}$$

$$\sum_{j_{12}} \langle j_2, (j_1 j_3) j_{13}; J | (j_1 j_2) j_{12}, j_3; J \rangle \langle (j_1 j_2) j_{12}, j_3; J | j_1, (j_2 j_3) j_{23}; J \rangle = \langle j_2, (j_1 j_3) j_{13}; J | j_1 (j_2 j_3) j_{23}; J \rangle \tag{A25}$$

are ways of writing $\sum_B \langle A|B\rangle\langle B|A'\rangle = \langle A|A'\rangle$. Substitution of (A23) yields relations among the 6-j symbols. An important special value is

$$\begin{Bmatrix} j_1 & j_2 & j_3 \\ j_2 & j_1 & 0 \end{Bmatrix} = (-)^{j_1+j_2+j_3} (\hat{j}_1 \hat{j}_2)^{-1/2}. \tag{A26}$$

The symmetries are as follows. The 6-j symbol is invariant to *any* permutation of columns, for example,

$$\begin{Bmatrix} a & b & c \\ d & e & f \end{Bmatrix} = \begin{Bmatrix} b & a & c \\ e & d & f \end{Bmatrix}, \tag{A27}$$

and interchange of the upper and lower arguments in each of *any two* columns, for example,

$$\begin{Bmatrix} a & b & c \\ d & e & f \end{Bmatrix} = \begin{Bmatrix} d & e & c \\ a & b & f \end{Bmatrix}. \tag{A28}$$

There are also relations connecting 3-j and 6-j symbols. One of them can be derived, as an example, from

$$\begin{aligned} |(j_1 j_2) j_{12}, j_3; JM\rangle &\equiv \sum_{m_1 m_2} \langle j_1 m_1 j_2 m_2 | j_1 j_2 J M\rangle \\ &\quad \times \sum_{m_{12} m_3} \langle j_{12} m_{12} j_3 m_3 | j_{12} j_3 J M\rangle |j_1 m_1\rangle |j_2 m_2\rangle |j_3 m_3\rangle \\ &= \sum_{j_{23}} \langle j_1, (j_2 j_3) j_{23}; J | (j_1 j_2) j_{12}, j_3; J\rangle | j_1, (j_2 j_3) j_{23}; JM\rangle, \end{aligned} \tag{A29}$$

which simply expresses the meaning of the symbols. Now, however, note the trivial relation, obtained by writing the implied couplings,

$$\langle j_1 m_1 j_2 m_2 j_3 m_3 | j_1, (j_2 j_3) j_{23}; JM \rangle = \sum_{m_{23}} \langle j_1 m_1 j_{23} m_{23} | j_1 j_{23}, JM \rangle \langle j_2 m_2 j_3 m_3 | j_2 j_3, j_{23} m_{23} \rangle. \quad \text{(A30)}$$

(The m_{23} sum is a dummy sum because the only nonvanishing value of the last bracket, from (A16), is the one for which $m_{23} = m_2 + m_3$.) Using this in (A29), it follows at once that

$$\sum_{m_{12}} \langle j_1 m_1 j_2 m_2 | j_1 j_2, JM \rangle \langle j_{12} m_{12} j_3 m_3 | j_{12} j_3, JM \rangle = \sum_{j_{23}, m_{23}} \langle j_1 (j_2 j_3) j_{23}; J | (j_1 j_2) j_{12}, j_3; J \rangle \times \langle j_1 m_1 j_{23} m_{23} | j_1 j_{23}, JM \rangle \langle j_2 m_2 j_3 m_3 | j_2 j_3, j_{23} m_{23} \rangle. \quad \text{(A31)}$$

This yields one of the relations between 3-j and 6-j symbols,

There are also special notations for the coupling for four angular momenta. Given $j_1 j_2 j_3 j_4$, there are several ways that they can be coupled to a total angular momentum J, for example,

$$j_1 + j_2 = j_{12}, \qquad j_3 + j_4 = j_{34}, \qquad j_{12} + j_{34} = J,$$

or

$$j_1 + j_3 = j_{13}, \qquad j_2 + j_4 = j_{24}, \qquad j_{13} + j_{24} = J. \quad \text{(A32)}$$

The transformation connecting the two coupling schemes is denoted by

$$\langle (j_1 j_2) j_{12}, (j_3 j_4) j_{34}; J | (j_1 j_3) j_{13}, (j_2 j_4) j_{24}; J \rangle \equiv \begin{Bmatrix} j_1 & j_2 & j_{12} \\ j_3 & j_4 & j_{34} \\ j_{13} & j_{24} & J \end{Bmatrix} (\hat{j}_{12} \hat{j}_{34} \hat{j}_{13} \hat{j}_{24})^{1/2} \equiv \begin{bmatrix} j_1 & j_2 & j_{12} \\ j_3 & j_4 & j_{34} \\ j_{13} & j_{24} & J \end{bmatrix}. \quad \text{(A33)}$$

The curly brace is known as a 9-j symbol. Orthogonality and completeness relations can be derived in an analogous fashion to the 6-j, as well as relations connecting the 9-j and 6-j symbols. For finite values of these 9-j symbols, the six vector relations (A32) must be obeyed.

Two important special values of the 9-j symbols are

$$\begin{Bmatrix} a & b & e \\ c & d & e \\ f & f & 0 \end{Bmatrix} = \frac{(-)^{b+c+e+f}}{(\hat{e}\hat{f})^{1/2}} \begin{Bmatrix} a & b & e \\ d & c & f \end{Bmatrix},$$

$$\begin{Bmatrix} a & a & 0 \\ b & b & 0 \\ c & c & 0 \end{Bmatrix} = \frac{1}{(\hat{a}\hat{b}\hat{c})^{1/2}},$$

where $\hat{a} = 2a + 1$. The symmetry is such that an even permutation of rows or columns leaves the value of the symbol unchanged, but an odd permutation of rows or columns produces a change of sign

$$\begin{Bmatrix} a & b & c \\ d & e & f \\ g & h & i \end{Bmatrix} = (-)^{\sigma} \begin{Bmatrix} b & a & c \\ e & d & f \\ h & g & i \end{Bmatrix},$$

where σ is the sum of all nine arguments.

The orthogonality relation and other sum rules can be inferred from (A33). For example, suppressing $j_1 j_2 j_3$, etc. in (A33), we can write

$$\begin{aligned} \delta_{j_{13}j'_{13}}\delta_{j_{24}j'_{24}} &= \langle j'_{12}j'_{24}; J | j_{13}j_{24}; J\rangle \\ &= \sum_{j_{12}j_{34}} \langle j'_{13}j'_{24}; J | j_{12}j_{34}; J\rangle \langle j_{12}j_{34}; J | j_{13}j_{24}; J\rangle, \end{aligned}$$

which can now be rewritten in terms of the 9-j.

There are other important symmetry relations for 3-j, 6-j, and 9-j symbols, both among themselves and connecting each other, that the reader should know exist (see Edmonds (1957)).

E. SPHERICAL TENSORS

An operator in quantum mechanics can be characterized by how much angular momentum it carries when applied to a state of the system. A spherical tensor $T_{\lambda\mu}$ is a set of $2\lambda + 1$ operators $\mu = \lambda, \lambda - 1, \ldots, -\lambda$, which, if it acts on a state of zero angular momentum, produces one with angular momentum λ and projection μ:

$$T_{\lambda\mu}|\alpha, J = 0\rangle = |\beta, J = \lambda, M = \mu\rangle. \tag{A34}$$

Examples are the spinor $\boldsymbol{\sigma}$ ($\lambda = 1$) or the quadrupole operator ($\lambda = 2$). The projection μ is onto the z axis of a specified coordinate system. The relationship to a rotated coordinate system is frequently of interest (e.g., laboratory frame and intrinsic frame of a deformed nucleus). Denote by $T'_{\lambda\mu}$ the tensor operators obtained from $T_{\lambda\mu}$ by transforming the dynamical variables such as position and spin from the unprimed coordinate system to the primed (rotated) one. The functions that express the transformation are the rotation functions $D^{\lambda}_{\mu\mu'}$,

$$T'_{\lambda\mu} = \sum T_{\lambda q} D^{\lambda}_{q\mu}(\omega), \tag{A35}$$

where ω denotes the rotation (Euler angles). Unfortunately, there is very little consistency in the definitions of the rotations and the D functions, and it is important in carrying out a calculation with them to use consistently the properties as derived by a single convention (e.g., Edmonds). An irreducible tensor operator can be defined by the commutation relations

$$\begin{aligned} [J_{\pm}, T_{\lambda\mu}] &= [(\lambda \mp \mu)(\lambda \pm \mu + 1)]^{1/2} T_{\lambda\mu\pm 1}, \\ [J_Z, T_{\lambda\mu}] &= \mu T_{\lambda\mu}, \end{aligned} \tag{A36}$$

where J is the total angular momentum of the system and $J_{\pm} = J_x \pm iJ_y$.

Two tensor operators $U_{\lambda_1\mu_1}$ and $V_{\lambda_2\mu_2}$ can be multiplied according to the vector coupling rule

$$T_{\lambda\mu} = \sum_{\mu_1\mu_2} \langle \lambda_1\mu_1\lambda_2\mu_2 | \lambda_1\lambda_2, \lambda\mu \rangle U_{\lambda_1\mu_1} V_{\lambda_2\mu_2} \equiv [U_{\lambda_1} V_{\lambda_2}]^{\mu}_{\lambda} \tag{A37}$$

(*tensor product*), and the result $T_{\lambda\mu}$ is still a tensor operator. The *scalar product* of two tensors of the *same* rank is defined as

$$\begin{aligned} U_{\lambda} \cdot V_{\lambda} &= \sum_{\mu} (-)^{\mu} U_{\lambda\mu} V_{\lambda-\mu} \\ &= (-)^{\lambda}(2\lambda + 1)^{1/2}[U_{\lambda} V_{\lambda}]^{0}_{0}, \end{aligned} \tag{A38}$$

where the second line is obtained through use of (A19).

F. REDUCED MATRIX ELEMENTS— WIGNER–ECKART THEOREM

The matrix element of a tensor operator $T_{\lambda\mu}$ has the very special property

$$\langle j'm' | T_{\lambda\mu} | jm \rangle = (-)^{j'-m'} \begin{pmatrix} j' & \lambda & j \\ -m' & \mu & m \end{pmatrix} (j' \| T_{\lambda} \| j), \tag{A39}$$

where the object $(j'\|T_{\lambda}\|j)$ is called the reduced matrix element of $T_{\lambda\mu}$. It is independent of the value of the z projections. The dependence on these projections of the left side is expressed completely by the 3-j symbol on the right. This is an expression of the Wigner–Eckart theorem.

G. SCALAR PRODUCT OF TWO COMMUTING TENSORS

Let the tensor operators U_k and V_k act on different parts, 1 and 2, of a physical system (e.g., U on target, V on projectile in a collision). If the angular momenta of the two parts of the system are j_1 and j_2 and the total angular momentum is J, then it can be proven (Edmonds) that the matrix element of the scalar product is

$$\langle j'_1 j'_2, J'M' | U_k \cdot V_k | j_1 j_2 JM \rangle$$
$$= (-)^{j_1 + j'_2 + J} \delta_{JJ'} \delta_{MM'} \begin{Bmatrix} J & j'_2 & j'_1 \\ k & j_1 & j_2 \end{Bmatrix} (j'_1 \| U_k \| j_1)(j'_2 \| V_k \| j_2). \quad \text{(A40)}$$

As an example, consider a two-body potential $V(\mathbf{r}_1 - \mathbf{r}_2)$. It can be expanded in multipoles because of the completeness of the Legendre functions and the dependence on $|\mathbf{r}_1 - \mathbf{r}_2|^2 = r_1^2 + r_2^2 - 2r_1 r_2 \cos\omega$,

$$V(\mathbf{r}_1 - \mathbf{r}_2) = \sum_\lambda g_\lambda(r_1, r_2) P_\lambda(\cos\omega)$$
$$= \sum_{\lambda\mu} \frac{4\pi}{2\lambda + 1} g_\lambda(r_1, r_2)(-)^\mu Y_\lambda^{-\mu}(\hat{\mathbf{r}}_1) Y_\lambda^\mu(\hat{\mathbf{r}}_2)$$
$$= \sum_\lambda \frac{4\pi}{2\lambda + 1} g_\lambda(r_1, r_2) Y_\lambda(\hat{\mathbf{r}}_1) \cdot Y_\lambda(\hat{\mathbf{r}}_2), \quad \text{(A41)}$$

where we used (A8), (A9), and (A38). In a shell-model problem, one may want to compute the matrix elements between two particle states,

$$|\ell_1 \ell_2, LM\rangle \equiv u_{n_1\ell_1}(r_1) u_{n_2\ell_2}(r_2) [Y_{\ell_1}(\hat{\mathbf{r}}_1) Y_{\ell_2}(\hat{\mathbf{r}}_2)]_L^M, \quad \text{(A42)}$$

where we use the square bracket to denote vector coupling when it is desired to indicate more detail than is convenient in the angular bracket (A14a). From (A40),

$$\langle \ell'_1 \ell'_2, L'M' | V(\mathbf{r}_1 - \mathbf{r}_2) | \ell_1 \ell_2, LM \rangle = \sum_\lambda \frac{4\pi}{2\lambda + 1} R_\lambda(n'_1 \ell'_1 n'_2 \ell'_2 | n_1 \ell_1 n_2 \ell_2)$$
$$\times (-)^{\ell_1 + \ell'_2 + L} \delta_{LL'} \delta_{MM'} \begin{Bmatrix} L & \ell'_2 & \ell'_1 \\ \lambda & \ell_1 & \ell_2 \end{Bmatrix}$$
$$\times (\ell'_1 \| Y_\lambda \| \ell_1)(\ell'_2 \| Y_\lambda \| \ell_2), \quad \text{(A43)}$$

where

$$R_\lambda \equiv \int u_{n'_1\ell'_1}(r_1) u_{n'_2\ell'_2}(r_2) g_\lambda(r_1, r_2) u_{n_1\ell_1}(r_1) u_{n_2\ell_2}(r_2) r_1^2\, dr_1\, r_2^2\, dr_2 \quad \text{(A44)}$$

denotes a radial integral. The reduced matrix elements of the spherical harmonics can be obtained from (A11) and (A39):

$$(\ell'\|Y_\lambda\|\ell) = (-)^{\ell'}\left(\frac{\hat{\ell}'\hat{\lambda}\hat{\ell}}{4\pi}\right)^{1/2}\begin{pmatrix}\ell' & \lambda & \ell\\ 0 & 0 & 0\end{pmatrix}. \tag{A45}$$

H. VECTOR PRODUCT OF TWO COMMUTING TENSORS

The reduced matrix element of the tensor product (A37) of two commuting tensors U_k and V_k acting on different parts, 1 and 2, of a system is given by

$$(j_1'j_2'J'\|[U_{k_1}(1)V_{k_2}(2)]_K\|j_1j_2J)$$
$$= (\hat{J}'\hat{J}\hat{K})^{1/2}(j_1'\|U_{k_1}\|j_1)(j_2'\|U_{k_2}\|j_2)\begin{Bmatrix}j_1' & j_1 & k_1\\ j_2' & j_2 & k_2\\ J' & J & K\end{Bmatrix}. \tag{A46}$$

Both this important result and (A40) can be derived by application of the same method used in the derivation of earlier results in the Appendix, namely, the straightforward application of the completeness property

$$\langle A|B\rangle = \sum_x \langle A|x\rangle\langle x|B\rangle. \tag{A47}$$

I. VECTOR AND VECTOR PRODUCT

The spherical components of a vector **a** are

$$a_{\pm 1} = \mp(a_x \pm ia_y)/\sqrt{2}, \qquad a_0 = a_3. \tag{A48}$$

We may also write the spherical components as

$$a_\mu = aY_1^\mu(\hat{\mathbf{a}}), \qquad a = |\mathbf{a}|, \tag{A49}$$

where, as an argument of Y, $\hat{\mathbf{a}}$ denotes the direction of **a**. The dot product (A.38) for a vector is

$$\mathbf{a}\cdot\mathbf{b} = \sum_\mu (-)^\mu a_\mu b_{-\mu} = -\sqrt{3}[\mathbf{ab}]_0^0. \tag{A50}$$

The cross product is

$$(\mathbf{a}\times\mathbf{b})_\mu = \frac{i}{\sqrt{2}}[\mathbf{ab}]_1^\mu \tag{A51}$$

J. REDUCED MATRIX ELEMENT OF SPIN OPERATOR

In connection with multipole expansion of the two-nucleon potential, it was convenient to couple the spin operator σ to the spherical harmonics, or more generally to define

$$\Sigma_S^\mu = \begin{cases} \sigma^\mu, & S = 1 \\ 1, & S = 0 \end{cases} \tag{A52}$$

where the spherical components are given by (A48). Construct the vector product

$$T_{LSJ}^{\ M} = [Y_L \Sigma_S]_J^M. \tag{A53}$$

The reduced matrix element of this

$$((\ell'\tfrac{1}{2})j' \| T_{LSJ} \| (\ell\tfrac{1}{2})j),$$

is given by Eq. (A46) in terms of the reduced matrix elements of Σ_S and Y_L, the latter being given by Eq. (A45). To find the reduced matrix element of Σ_S, write the Wigner-Eckart theorem, Eq. (A39), for the operator

$$\langle \tfrac{1}{2}m' | \Sigma_S^\mu | \tfrac{1}{2}m \rangle = (-)^{1/2-m'} \begin{pmatrix} \frac{1}{2} & S & \frac{1}{2} \\ -m' & \mu & m \end{pmatrix} (\tfrac{1}{2} \| \Sigma_S \| \tfrac{1}{2}), \tag{A54}$$

and choose, for simplicity of calculation, $\mu = 0$. Note, in addition, the special case of the 3-j symbol:

$$\begin{pmatrix} j & k & j \\ -j & 0 & j \end{pmatrix} = (2j + k + 1)^{1/2} \frac{(2j)!}{[(2j + k)!(2j - k)!]^{1/2}} \tag{A55}$$

whence

$$\begin{pmatrix} \frac{1}{2} & 0 & \frac{1}{2} \\ -\frac{1}{2} & 0 & \frac{1}{2} \end{pmatrix} = 1/\sqrt{2}, \qquad \begin{pmatrix} \frac{1}{2} & 1 & \frac{1}{2} \\ -\frac{1}{2} & 0 & \frac{1}{2} \end{pmatrix} = 1/\sqrt{6}.$$

Note also that

$$\langle \tfrac{1}{2}m' | \Sigma_1^0 | \tfrac{1}{2}m \rangle = \langle \tfrac{1}{2}m' | \sigma_z | \tfrac{1}{2}m \rangle = 2m\delta_{m'm}$$

$$\langle \tfrac{1}{2}m' | 1 | \tfrac{1}{2}m \rangle = \delta_{m'm}.$$

Hence, from the Wigner–Eckart theorem just given,

$$(\tfrac{1}{2} \| \boldsymbol{\sigma} \| \tfrac{1}{2}) = \sqrt{6}, \qquad (\tfrac{1}{2} \| 1 \| \tfrac{1}{2}) = \sqrt{2}, \tag{A56}$$

or, in terms of Σ_S,

$$(\tfrac{1}{2} \| \Sigma_S \| \tfrac{1}{2}) = \sqrt{2(2S + 1)} \tag{A57}$$

More general than the result for the spin $\frac{1}{2}$ case is

$$(j\|\mathbf{J}\|j') = \sqrt{j(j+1)(2j+1)}\,\delta_{jj'}, \tag{A58}$$

which can be derived in the same way.

K. ROTATION FUNCTIONS

The reader needs to be alert to the fact that there are many different conventions used for D functions and Euler angles. We use Edmonds's convention. Several important properties are

$$\frac{1}{8\pi^2}\int_0^{2\pi} d\alpha \int_0^{\pi} \sin\beta\, d\beta \int_0^{2\pi} d\gamma\, \mathscr{D}^{j*}_{m'm}(\alpha\beta\gamma)\mathscr{D}^{J}_{M'M}(\alpha\beta\gamma)$$

$$= \frac{1}{\hat{j}}\,\delta_{jJ}\delta_{m'M'}\delta_{mM}, \tag{A59}$$

$$\frac{1}{8\pi^2}\int_0^{2\pi} d\alpha \int_0^{\pi} \sin\beta\, d\beta \int_0^{2\pi} d\gamma\, \mathscr{D}^{j_1}_{m'_1m_1}\mathscr{D}^{j_2}_{m'_2m_2}\mathscr{D}^{j_3}_{m'_3m_3}$$

$$= \begin{pmatrix} j_1 & j_2 & j_3 \\ m'_1 & m'_2 & m'_3 \end{pmatrix}\begin{pmatrix} j_1 & j_2 & j_3 \\ m_1 & m_2 & m_3 \end{pmatrix} \tag{A60}$$

L. OSCILLATOR FUNCTIONS

The normalized harmonic oscillator function in three dimensions is

$$\phi_{n\ell m}(\sqrt{\nu}\,r) = u_{n\ell}(\sqrt{\nu}\,r)\, Y^m_\ell(\hat{\mathbf{r}}), \tag{A61}$$

where the radial function is orthonormal and real,

$$\int_0^\infty u_{n\ell}u_{m\ell}r^2\, dr = \delta_{nm}, \tag{A62}$$

and is

$$u_{n\ell}(\sqrt{\nu}r) = N_{n,\ell}(\sqrt{\nu}r)^\ell e^{-1/2\nu r^2} L_n^{\ell+1/2}(\nu r^2), \qquad n \geq 0, \tag{A63}$$

where the normalization constant can be obtained by recurring

$$N_{n,\ell} = \left(\frac{n}{n+\ell+\frac{1}{2}}\right)^{1/2} N_{n-1,\ell} = \left(\frac{1}{n+\ell+\frac{1}{2}}\right)^{1/2} N_{n,\ell-1},$$

$$N_{0,0} = 2\left(\frac{\nu^3}{\pi}\right)^{1/4}, \qquad N_{0,\ell} = \left(\frac{2}{2\ell+1}\right)^{1/2} N_{0,\ell-1}, \tag{A64}$$

and $L_n^{\ell+1/2}$ is a Laguerre polynomial

$$L_n^{\ell+1/2}(x) = \sum_{k=0}^{n} \binom{n+\ell+\frac{1}{2}}{n-k} \frac{(-x)^k}{k!} \tag{A65}$$

which can also be expressed as

$$\begin{aligned} L_n^{\ell+1/2}(x) &= \sum_{p=0}^{n} T_p, \\ T_p &= -\frac{n+\ell+\frac{3}{2}-p}{p}\,\frac{n+1-p}{x}\,T_{p-1}, \\ T_0 &= (-x)^n/n! \end{aligned} \tag{A66}$$

Note that there are two conventions used in the literature. In one $n \geq 0$ and in the other $n \geq 1$. The preceeding formulas are for the former convention but of course can be converted to the latter by the appropriate replacement of n with $n-1$. There are also two conventions as to phase.

For nuclei, a rough value of v is given by $v = A^{-1/3}\ \text{fm}^{-2}$. There is an important transformation for the vector-coupled product of two single-particle wave functions of the shell-model type (A61), due to Talmi (1952) and Moshinsky (1959). It relates the particle to the relative and c.m. coordinates:

$$\begin{aligned} [\phi_{n_1\ell_1}(\sqrt{v}\mathbf{r}_1)\phi_{n_2\ell_2}(\sqrt{v}\mathbf{r}_2)]_L^M &= \sum_{n\lambda N\Lambda} \langle n\lambda N\Lambda; L|n_1\ell_1 n_2\ell_2; L\rangle \\ &\quad \times [\phi_{n\lambda}(\sqrt{v/2}\mathbf{r})\phi_{N\Lambda}(\sqrt{2v}\mathbf{R})]_L^M. \end{aligned} \tag{A67}$$

with $\mathbf{r} = \mathbf{r}_1 - \mathbf{r}_2$, $2\mathbf{R} = \mathbf{r}_1 + \mathbf{r}_2$.

The functions (A61) are eigenfunctions of angular momentum, z projection and parity (A4, A5, A8) and of the oscillator Hamiltonian,

$$\begin{aligned} (-(\hbar^2/2m)\nabla^2 + \tfrac{1}{2}m\omega^2 r^2 - E_{n\ell})\phi_{n\ell m}(\sqrt{v}r) &= 0 \\ E_{n\ell} = (2n+\ell+\tfrac{3}{2})\hbar\omega, \qquad \omega &= \hbar v/m \end{aligned} \tag{A68}$$

M. SPHERICAL BESSEL FUNCTIONS

The Schrödinger equation for a free particle of energy E is

$$[(\hbar^2/2m)\nabla^2 + E]\phi(\mathbf{r}) = 0 \tag{A69}$$

or

$$(\nabla^2 + k^2)\phi(\mathbf{r}) = 0, \qquad k^2 = 2mE/\hbar^2, \tag{A70}$$

where k is a real for positive energy (scattering states).

A solution is

$$\phi = e^{i\mathbf{k}\cdot\mathbf{r}} = e^{ikr\cos\theta}. \tag{A71}$$

Because of the completeness of the Legendre functions P_ℓ or spherical harmonics Y_ℓ, this can be expanded as

$$\begin{aligned}\phi(\mathbf{r}) = e^{i\mathbf{k}\cdot\mathbf{r}} &= \frac{1}{kr}\sum_{\ell=0}^{\infty}(2\ell+1)i^\ell F_\ell(kr)P_\ell(\cos\theta)\\ &= \frac{1}{kr}\sum_\ell \sqrt{4\pi(2\ell+1)}\,i^\ell F_\ell(kr)Y_\ell^0(\theta),\end{aligned} \tag{A72}$$

where we used the relations between P and Y given by (A6), and θ in the angle between r and k.

Now substitute the spherical coordinate representations (A1) of ∇^2 and the properties of the spherical harmonics (A4) and (A7), to find

$$\left(-\frac{d^2}{dr^2} + \frac{\ell(\ell+1)}{r^2} - k^2\right)F_\ell(kr) = 0. \tag{A73}$$

Defining another function j_ℓ by

$$F_\ell(kr) = krj_\ell(kr), \tag{A74}$$

substitution into (A73) gives the equation satisfied by j_ℓ. It is the equation for the spherical Bessel function. There is, of course, another solution, which is irregular at the origin. Call it

$$G_\ell(kr) = -krn_\ell(kr). \tag{A75}$$

The minus sign is merely a convention, and n_ℓ is the spherical Neumann function.

In the case $\ell = 0$, the solution are obvious:

$$F_0(kr) = \sin kr, \qquad G_0(kr) = \cos kr. \tag{A76}$$

Also, for very large r it is evident that for any ℓ the solution is oscillatory, with wavelength

$$\lambda = 2\pi/k. \tag{A77}$$

The precise asymptotic properties are given in the text (4.12).

For small values of r, k can be ignored in the differential equation by comparison with $\ell(\ell+1)/r^2$ (except for $\ell = 0$). Then we see that

$$F_\ell(kr) \to C(kr)^{\ell+1}, \qquad G_\ell(kr) \to C'(kr)^{-\ell}, \qquad r \to 0. \tag{A78}$$

The plane-wave expansion (A72) is often written in terms of the directions of k and r instead of the angle between them, as in (A72). Use (A9) for this purpose to obtain

$$e^{i\mathbf{k}\cdot\mathbf{r}} = 4\pi \sum_{\ell=0}^{\infty} i^{\ell} j_{\ell}(kr) Y_{\ell}(\hat{\mathbf{k}}) \cdot Y_{\ell}(\hat{\mathbf{r}}). \tag{A79}$$

Equation (A73) is a special case of

$$\left(-\frac{d^2}{dr^2} + \frac{2k\eta}{r} + \frac{\ell(\ell+1)}{r^2} - k^2\right) F_{\ell}(\eta, kr) = 0, \tag{A80}$$

which is the radial Schrödinger equation in the case that a Coulomb potential with η defined by (4.10) is present. The solutions in this case are the Coulomb functions, $F_{\ell}(\eta, kr)$ and $G_{\ell}(\eta, kr)$, which are respectively regular and irregular at the origin, with asymptotic behavior for large r defined by (4.12).

Notes to Appendix

For a more complete review of the theory of angular momentum algebra see Edmonds (1957) and Brink and Satchler (1968), as well as those references cited in the Notes to Chapter 3.

References

Abul-Magd A. Y., and El-Nadi, M. (1966). *Prog. Theoret. Phys.* (*Japan*) **35**, 798.

Adams, G. S. *et al.*, 1977. *Phys. Rev. Lett.* **38**, 1387.

Adams, G. S., Bacher, A. D., Emery, G. T., Jones, W. P., Muller, D. W., Love, W. G., and Petrovich, F. (1980). *Phys. Lett.* **91B**, 23.

Alder, K., and Winther, A. (1966). "Coulomb Excitation." Academic Press, New York.

Alder, K., Bohr, A., Huus, T., Mottelson, B., and Winther, A. (1956). *Rev. Mod. Phys.* **28**, 432.

Amos, K. A., Madsen, V. A., and McCarthy, I. E. (1967). *Nucl. Phys.* **A94**, 103.

Aponick, A. A., Chesterfield, C. M., Bromley, D. A., and Glendenning, N. K. (1970). *Nucl. Phys.* **A159**, 367.

Arvieu, R., and Veneroni, M. (1960a). *C. R. Acad. Sci.* (*Paris*) **240**, 992.

Arvieu, R., and Veneroni, M. (1960b). *C. R. Acad. Sci.* (*Paris*) **252**, 670.

Arvieu, R., and Veneroni, M. (1963). *Phys. Lett.* **8**, 407.

Arvieu, R., Baranger, E., Veneroni, M., Baranger, M., and Gillet, V. (1963). *Phys. Lett.* **4**, 119.

Arvieu, R., Salusti, E., and Veneroni, M. (1964). *Phys. Lett.* **8**, 334.

Ascuitto, R. J., and Glendenning, N. K. (1969). *Phys. Rev.* **181**, 1396.

Ascuitto, R. J., and Glendenning, N. K. (1970a). *Phys. Rev.* **C2**, 1260.

Ascuitto, R. J., and Glendenning, N. K. (1970b). *Phys. Rev.* **C2**, 415.

Ascuitto, R. J., and Glendenning, N. K. (1972). *Nucl. Phys.* **A188**, 185.

Ascuitto, R. J., and Glendenning, N. K. (1973a). *Phys. Lett.* **45B**, 85.

Ascuitto, R. J., and Glendenning, N. K. (1973b). *Phys. Lett.* **47B**, 332.

Ascuitto, R. J., and Glendenning, N. K. (1974). *Phys. Lett.* **48B**, 6.

Ascuitto, R. J., Glendenning, N. K. and Sorensen, B. (1971a). *Nucl. Phys.* **A170**, 65.

Ascuitto, R. J., Glendenning, N. K. and Sorensen, B. (1971b). *Phys. Lett.* **34B**, 17.

Ascuitto, R. J., Glendenning, N. K. and Sorensen, B. (1972). *Nucl. Phys.* **A183**, 60.
Austern, N. (1969). *Phys. Rev.* **188**, 1595.
Austern, N. (1970). "Direct Nuclear Reaction Theories." Wiley, New York.
Austern, N., Butler, S. T., and McManus, H. (1953). *Phys. Rev.* **92**, 350.
Austern, N., Drisco, R. M., Halbert, E. C., and Satchler, G. R. (1964). *Phys. Rev.* **133B**, 3.
Ball, G. C., and Cerny, J. (1969). *Phys. Rev.* **177**, 1466.
Baltz, A. J., and Kahana, S. (1972). *Phys. Rev. Lett.* **29**, 1267.
Baltz, A. J., and Kahana, S. (1974). *Phys. Rev.* **C9**, 2243.
Baltz, A. J., Kauffmann, S. K., Glendenning, N. K., and Pruess, K. (1978). *Phys. Rev. Lett.* **40**, 20.
Baltz, A. J., Kauffmann, S. K., Glendenning, N. K., and Pruess, K. (1979). *Nucl. Phys.* **A327**, 221.
Baranger, M. (1960). *Phys. Rev.* **120**, 957.
Baranger, M. (1963). *In* "Cargese Lectures on Theoretical Physics" (M. Levy, ed.). Benjamin, New York.
Baranger, M. (1969). *In* "Proceedings of the International School of Physics, 'Enrico Fermi'" (M. Jean, ed.), Course XL, 1967, p. 511. Academic Press, New York.
Bardeen, J., Cooper, L. N., and Schrieffer, R. (1957). *Phys. Rev.* **108**, 1175.
Bayman, B. F. (1974). *Phys. Rev. Lett.* **32**, 71.
Belyaev, S. T. (1959). *K. Danske Vidensk. Selsk. Mat-fys. Medd.* **31** (11) 1.
Bernard V. and van Giai, N. (1979). *Nucl. Phys.* **A327**, 397.
Bertsch, G., Borysowicz, J., McManus, H., and Love, W. G. (1977). *Nucl. Phys.* **A284**, 399.
Bethe, H. A. (1956). *Phys. Rev.* **103**, 1353.
Bethe, H. A., and Butler, S. T. (1952). *Phys. Rev.* **85**, 1045.
Blair, J. S. (1959). *Phys. Rev.* **115**, 928.
Bloom, S. D., Glendenning, N. K., and Mozkowski, S. A. (1959). *Phys. Rev. Lett.* **3**, 542.
Bock, R., and Yoshida, H. (1972). *Nucl. Phys.* **A189,** 177.
Bogolyubov, N. N. (1958). *Sov. Phys. JETP* **7**, 41.
Bohr, N. (1936). *Nature* **137**, 344.
Bohr, A., and Mottelson, B. R. (1969 and 1975). "Nuclear Structure," Vols. I and II. Benjamin, New York.
Bohr, A., Mottelson, B. R., and Pines, D. (1958). *Phys. Rev.* **110**, 936.
Bond, P. D., Barret, J. D., Hansen, O., Kahana, S., Levine, M. J., and Schwartzchild, A. Z. (1973). *Phys Lett.* **47B**, 231.
Bond, P. D., Korner, H. J., Lemaire, M.-C., Pisano, D. J., and Thorn, C. E. (1977). *Phys. Rev.* **C16**, 177.
Bowen, P. H., Scanlon, J. P., Stafford, G. H., Thresher, J. J., and Hodgson, P. E. (1961). *Nucl. Phys.* **22**, 640.
Breit, G., and Wigner, E. P. (1936). *Phys. Rev.* **49**, 519, 642.
Brieva, F. A., and Rook, J. R. (1977). *Nucl. Phys.* **A291**, 317.
Brieva, F. A., and Rook, J. R. (1978). *Nucl. Phys.* **A297**, 206.
Brink, D. M., and Satchler, G. R. (1968). "Angular Momentum," 2nd ed. Oxford University Press, Oxford.
Brink, D. M., and Takigawa, N. (1977). *Nucl. Phys.* **A279**, 159.
Brody, T. A., and Moshinsky, M. (1960). "Tables of Transformation Brackets." Monografias del Instituto de Fisica, Mexico.
Bromley, D. A. (1969). in "International School of Physics, Enrico Fermi," Course XL. (Jean, M., ed.) Academic Press, New York.
Broek, H. W., Yntema, J. L., Buck, B., and Satchler, G. R. (1965). *Nucl. Phys.* **64**, 259.
Broglia, R. A., and Winther, A. (1972). *Phys. Rep.* **4C**, 155.
Brown, G. (1971). "Unified Theory of Nuclear Models and Forces" North-Holland, Amsterdam.
Brown, V. R., and Madsen, V. A. (1975). *Phys. Rev.* **C11**, 1298.
Brueckner, K. A., Levinson, C. A., and Mahmond, H. M. (1954). *Phys. Rev.* **95**, 217.

Brueckner, K. A., Eden, R. J., and Francis, N. C. (1955). *Phys. Rev.* **98**, 1445.
Buck, B. (1962). *Phys. Rev.* **127**, 940.
Burrows, H. B., Gibson, W. H., and Rotblat, J. (1950). *Phys. Rev.* **80**, 1095.
Buttle, P. J. A., and Goldfarb, L. J. B. (1966). *Nucl. Phys.* **78**, 409.
Buttle, P. J. A., and Goldfarb, L. J. B. (1968). *Nucl. Phys.* **A115**, 461.
Butler, S. T. (1950). *Phys. Rev.* **80**, 1095.
Butler, S. T. (1951). *Proc R. Soc. (London)* **A208**, 36.
Butler, S. T. (1952). *Phys Rev.* **88**, 685
Butler, S. T., and Austern, N. (1954). *Phys. Rev.* **93**, 355.
Butler, S. T., and Hitmain, O. H., Nuclear Stripping Reactions, (Horwitz Publications, Sydney, Australia 1957).
Cerny, J., Harvey, B. G., and Pehl, R. H. (1962). *Nucl. Phys.* **20**, 136.
Charlton, L. A. (1973). *Phys. Rev.* **C8**, 146.
Clegg, A. B., and Satchler, G. R. (1961). *Nucl. Phys.* **27**, 431.
Cohen, B. L. (1957). *Phys. Rev.* **105**, 1549.
Cohen, B. L. and Price, R. E. (1961). *Phys. Rev.* **121**, 1441.
Cohen, B. L., and Rubin, A. G. (1958). *Phys. Rev.* **111**, 1568.
Cohen, B. L., Fulmer, R. H., and McCarthy, A. L. (1962). *Phys. Rev.* **126**, 698.
Coker, W. R., Udagawa, T., and Wolter, H. H. (1973). *Phys. Rev.* **C7**, 1154.
Cooper, L. N., Mills, R. L., and Sessler, A. M. (1959). *Phys. Rev.* **114**, 1377.
Coulter, P. W., and Satchler, G. R. (1977). *Nucl. Phys.* **A293**, 209.
Dalitz, R. (1952). *Proc. Phys. Soc. (London)* **A65**, 175.
Day, B. D. (1967). *Rev. Mod. Phys.* **39**, 719.
Delic, G., and Robson, B. A. (1970). *Nucl. Phys.* **A155**, 97.
Delic, G., Pruess, K., Charlton, L. A., and Glendenning, N. K. (1977). *Phys. Lett.* **69B**, 20.
Delic, G., Pruess, K., Charlton, L. A., and Glendenning, N. K. (1978). *J. Phys. Soc. (Japan)* **44**, 272.
de-Shalit, A., and Feshbach, H. (1974). "Theoretical Nuclear Physics." Wiley, New York.
de-Shalit, A., and Talmi, I. (1963). "Nuclear Shell Theory." Academic Press, New York.
Devries, R. M. (1973). *Phys. Rev.* **C8**, 951.
Devries, R. M., and Kubo, K. I. (1973). *Phys. Rev. Lett.* **30**, 325.
Devries, R. M., and Kubo, K. I. (1974). *Phys. Lett.* **48B**, 285.
Dickens, J. K., Perey, F. G., Silva, R. J., and Tamura, T. (1963). *Phys. Lett.* **6**, 53.
Dirac, P. A. M. (1947). "The Principles of Quantum Mechanics," 3rd ed. Clarendon Press, Oxford.
Dodd, L. R., and Greider, K. R. (1969). *Phys. Rev.* **180**, 1187.
Doering, R. R., Galonsky, A., Patterson, D. M., and Bertsch, G. (1975). *Phys. Lett.* **35**, 1891.
Drisco, R. M., and Satchler, G. R. (1964). *Phys. Lett.* **9**, 342.
Edmonds, A. R. (1957). "Angular Momentum in Quantum Mechanics." Princeton University Press, Princeton, New Jersey.
Eisenberg, J. M. and Greiner, W. (1976). "Nuclear Theory" Vol. 1–3, North-Holland Publ., Amsterdam.
Elliott, J. P., Jackson, A. D., and Mavromates, H. A. (1968). *Nucl. Phys.* **A121**, 241.
El-Nadi, M. (1957). *Proc. Phys. Soc. (London)* **70**, 62.
Faessler, A., Glendenning, N. K., and Plastino, A. (1967). *Phys. Rev.* **159**, 846.
Feng, Da Hsuan, Nagarajan, M. A., Strayer, M. R., Vallieres, M., and Pinkston, W. T. (1981). *Phys. Rev. Lett.* **44**, 1037.
Feshbach, H. (1962). *Ann. Phys. (New York)* **19**, 287.
Feshbach, H. (1958). *Annu. Rev. Nucl. Sci. (Palo Alto)* **8**, 49.
Feshbach, H., and Weisskopf, V. F. (1949). *Phys. Rev.* **76**, 1550.
Feshbach, H., Peaslee, D. C., and Weisskopf, V. F. (1947). *Phys. Rev.* **71**, 145.
Feshbach, H., Porter, C. E. and Weisskopf, V. F. (1954). *Phys. Rev.* **96**, 448.
Flynn, E. R., Bary, J. G., and Blair, A. G. (1974). *Nucl. Phys.* **A218**, 285.

Ford, K. W., and Wheeler, J. A. (1959). *Ann. Phys. (New York)* **7**, 259
Frahn, W. E. (1972). *Ann. Phys. (New York)* **72**, 524.
Frahn, W. E., and Gross, D. H. E. (1976). *Ann. Phys. (New York)* **101**, 520.
Frahn, W. E., and Venter, R. H. (1963). *Ann. Phys. (New York)* **24**, 243
Frances, N. C., and Watson, K. M. (1953). *Phys. Rev.* **92**, 291.
Fuller, R. C. (1975). *Phys. Rev.* **C12**, 1561.
Fulmer, R. H., McCarthy, A. L., Cohen, B. L., and Middleton, R. (1964). *Phys. Rev.* **133B**, 955.
Gell-Mann, M., and Goldberger, M. L. (1953). *Phys. Rev.* **91**, 398.
Gerjuoy, E. (1958). *Phys. Rev.* **109**, 1806.
Gibbs, W. R., and Tobocman, W. (1961). *Phys. Rev.* **124**, 1496.
Gibson, E. F., Ridley, B. W., Kranshvar, J. J., Rickey, M. E., and Bassel, R. H. (1967). *Phys. Rev.* **155**, 1194.
Glendenning, N. K. (1959). *Phys. Rev.* **114**, 1297.
Glendenning, N. K. (1962). *Nucl. Phys.* **29**, 109.
Glendenning, N. K. (1963). *Annu. Rev. Nucl. Sci. (Palo Alto)* **13**, 191.
Glendenning, N. K. (1965). *Phys. Rev.* **137**, B102.
Glendenning, N. K. (1967). *Phys. Rev.* **156**, 1344.
Glendenning, N. K. (1968a). *Nucl. Phys.* **A117**, 49.
Glendenning, N. K. (1968b). University of California Radiation Laboratory Report UCRL-18225 [text of invited Washington APS paper].
Glendenning, N. K. (1969a). *In* "Proceedings of the International School of Physics, 'Enrico Fermi'" (M. Jean, ed.), Course XL, 1967, p. 332. Academic Press, New York.
Glendenning, N. K. (1969b). "Proceedings of the International Conference on the Properties of Nuclear States" (M. Harvey, ed.) p. 332. University of Montreal Press, Montreal.
Glendenning, N. K. (1974). *In* "Reactions between Complex Nuclei," (R. L. Robinson, F. K. McGowan, J. B. Ball, and J. H. Hamilton, eds.). North-Holland, Amsterdam.
Glendenning, N. K. (1975a) "Atomic Data and Nuclear Data Tables," Vol. 16, No. 1. Academic Press, New York.
Glendenning, N. K. (1975b). *Rev. Mod. Phys.* **47**, 659.
Glendenning, N. K. (1975c). *In* "Nuclear Spectroscopy and Reactions," Part D, p. 319, (J. Cerny, ed.) Academic Press, New York.
Glendenning, N. K., and Nagarajan, M. A. (1974). *Nucl. Phys.* **A236**, 13.
Glendenning, N. K., and Veneroni, M. (1965). *Phys. Lett.* **14**, 228.
Glendenning, N. K., and Veneroni, M. (1966). *Phys. Rev.* **144**, 839.
Glendenning, N. K., and Wolschin, G. (1975). *Phys. Rev. Lett.* **34**, 1642.
Glendenning, N. K., and Wolschin, G. (1977). *Nucl. Phys.* **A281**, 486.
Glendenning, N. K., Hendrie, D. L., and Jarvis, O. N. (1968). *Phys. Lett.* **26B**, 131.
Goldberger, M. L., and Watson, K. M. (1964). "Collision Theory." Wiley, New York.
Goldstone, J. (1957). *Proc. R. Soc. (London)* **A239**, 267.
Gomes, L. C., Walecka, J. D., and Weisskopf, V. F. (1958). *Ann. Phys. (New York)* **3**, 241.
Goodman, C. D. (1980). *In* "The (*p,n*) Reaction and the Nucleon–Nucleon Force" (C. D. Goodman, S. M. Austin, S. D. Bloom, J. Rapaport, and G. R. Satchler, eds.), p. 1490. Plenum, New York.
Goodman, C. D., Austin, S. M., Bloom, S. D., Rapaport, J., and Satchler, G. R. eds. (1980). "The (*p,n*) Reaction and the Nucleon–Nucleon Force." Plenum, New York.
Hamada, T., and Johnson, I. D. (1962). *Nucl. Phys.* **34**, 382.
Harvey, B. G., and Cerny, J. (1960). *Phys. Rev.* **120**, 2162.
Harvey, B. G., Cerny, J., Pehl, R. H., and Rivet, R. (1962). *Nucl. Phys.* **39**, 160.
Harvey, B. G., Hendrie, D. L., Jarvis, O. N., Mahoney, J., and Valentin, J. (1967). *Phys. Lett.* **24B**, 43.
Harney, H. L., Braun-Munzinger, P., and Gelbke, C. K. (1974). *Z. Phys.* **269**, 339.

Haybron, R. M., and McManus, H. (1964). *Phys. Rev.* **136B**, 1730.
Haybron, R. M., and McManus, H. (1965). *Phys. Rev.* **140B**, 638.
Haxel, O., Jensen, J. H. D., and Suess, H. E. (1949). *Phys. Rev.* **75**, 1766.
"Heavy Ion Collisions," Vol. 1 and 2 (1980). (R. Boch ed.) North-Holland, Amsterdam.
Heisenberg, J., and Dick, I. (1970). *Phys. Lett.* **32B**, 249.
Hendrie, D. L., Glendenning, N. K., Harvey, B. G., Jarvis, O. N., Duhm, H. H., Sandinos, J., and Mahoney, J. (1968). *Phys. Lett.* **26B**, 127.
Henley, E. M., and Yu, D. V. L. (1964). *Phys. Rev.* **133B**, 1445.
Hodgson, P. E. (1963). "The Optical Model of Elastic Scattering." Clarendon Press, Oxford.
Hodgson, P. E. (1971). "Nuclear Reactions and Nuclear Structure." Clarendon Press, Oxford.
Holt, J. R., and Young, C. T. (1950). *Proc. Phys. Soc.* (*London*) **78**, 833.
Horowitz, J., and Messiah, A. M. L. (1953). *J. Phys. Rad.* **14**, 695.
Horowitz, J., and Messiah, A. M. L. (1954). *J. Phys. Rad.* **15**, 142.
Iano, P. J., and Austern, N. (1966). *Phys. Rev.* **151**, 853.
Jackson, D. F. (1970). "Nuclear Reactions" Methuen, London.
Jarvis, O. N., Harvey, B. G., Hendrie, D. L., and Mahoney, J. (1967). *Nucl. Phys.* **A102**, 625.
Jean, M. (ed.) (1969). *In* "Proceedings of the International School of Physics, 'Enrico Fermi'," Course XL, 1967. Academic Press, New York.
Johnson, R. C., and Santos, F. D. (1967). *Phys. Rev. Lett.* **19**, 364.
Johnson, M. B., Owen, L. W., and Satchler, G. R. (1966). *Phys. Rev.* **142**, 748.
Kahana, S., and Baltz, A. J. (1977). *In* "Advances in Nuclear Physics" (M. Baranger and E. Vogt, eds.). Plenum, New York.
Kalckar, F., and Bohr, N. (1937). *K. Danske Vidensk. Selsk. Mat.-fys. Medd.* **14**, 10.
Kalio, A., and Koltveit, K. (1964). *Nucl. Phys.* **53**, 87.
Kapur, P. L., and Peierls, R. (1938). *Proc. R. Soc.* (*London*) **A166**, 277.
Kawai, M., Terasawa, T., and Izumo, K. (1964). *Nucl. Phys.* **59**, 289.
Kelley, J., *et al.* (1980). *Phys. Rev. Lett.* **45**, 2012.
Kerman, A. K., McManus, H., and Thaler, R. M. (1959). *Ann. Phys.* (*New York*) **8**, 551.
Kisslinger, L. S., and Sorensen, R. A. (1960). *K. Danske Vidensk, Selsk. Mat.-fys. Medd.* **32**, (9) 1.
Koeling, T., and Malfliet, R. A. (1975). *Phys. Rep.* **22C**, 181.
Korner, H. J. Morrison, G. C., Greenwood, L. R. and Siemssen, R. H. (1973). *Phys. Rev.* **C7**, 107.
Kovar, D. G., Becchetti, F. D., Harvey, B. G., Puhlhofer, F., Mahoney, J., Miller, D. W., and Zisman, M. S. (1972). *Phys. Rev. Lett.* **29**, 1023.
Kovar, D. G., *et al.* (1973). *Phys. Rev. Lett.* **30**, 1075.
Kozlowsky, B., and de-Shalit, A. (1966). *Nucl. Phys.* **77**, 215.
Kunz, P. D., and Rost, E. (1973). *Phys. Lett.* **47B**, 130.
Kuo, T. T. S., and Brown, G. E. (1966). *Nucl. Phys.* **85**, 40.
Kuo, T. T. S., and Brown, G. E. (1968). *Nucl. Phys.* **A114**, 241.
Lane, A. M. (1964). "Nuclear Theory." Benjamin, New York.
Lassela, K. E., Hull, M. H., Ruppel, M., McDonald, F. A., and Breit, G. (1962). *Phys. Rev.* **126**, 881.
Lee, S. Y., Takigawa, N., and Marty, C. (1978). *Nucl. Phys.* **A308**, 161.
Levier, R. E., and Saxon, D. S. (1952). *Phys. Rev.* **87**, 40.
Levin, F. S. (1966). *Phys. Rev.* **147**, 715.
Levinson, C. A., and Banerjee, M. K. (1957a). *Ann. Phys.* (*New York*) **2**, 471.
Levinson, C. A., and Banerjee, M. K. (1957b). *Ann. Phys.* (*New York*) **2**, 499.
Levinson, C. A., and Banerjee, M. K. (1958). *Ann. Phys.* (*New York*) **3**, 67.
Lichtenstadt, J., Heisenberg, J., Papanicolas, C. N., Sargent, C. P., Courtemanche, A. N., and McCarthy, J. J. (1978). *Phys. Rev. Lett.* **40**, 1127.
Lichtenstadt, J., Heisenberg, J., Papanicolas, C. N., Sargent, C. P., Courtemanche, A. N., and McCarthy, J. J. (1979). *Phys. Rev.* **C20**, 497.
Lin, C. L., and Yoshida, S. (1964). *Prog. Theoret. Phys.* (*Japan*) **32**, 885.

Lindgren, R. A., Gevoce, W. J., Bacher, A. D., Love, W. G., and Petrovich, F., 1979. *Phys. Rev. Lett.* **42**, 1524.
Lippmann, B. A. (1956). *Phys. Rev.* **102**, 264.
Lippmann, B. A., and Schwinger, J. (1950). *Phys. Rev.* **79**, 469.
Love, W. G., and Franey, M. A. (1981). *Phys. Rev.* **C24**, 1073.
Love, W. G., and Parish, L. J. (1970). *Nucl. Phys.* **A157**, 625.
Love, W. G., and Satchler, G. R. (1967). *Nucl. Phys.* **A101**, 424.
Love, W. G., and Satchler, G. R. (1970). *Nucl. Phys.* **A159**, 1.
Love, W. G., Terasawa, T., and Satchler, G. R. (1977). *Phys. Rev. Lett.* **39**, 6.
Love, W. G., Franey, M. A., and Petrovich F. (1983) *In* "Spin Excitations in Nuclei," ed. by F. Petrovich, G. E. Brown, G. Garvey, C. D. Goodman, W. G. Love, and R. A. Lindgren. Plenum, New York.
MacFarlane, M. H. (1969). Proceedings of the International School of Physics, "Enrico Fermi" (M. Jean, ed.) Course XL, 1967 (Academic Press, New York).
MacFarlane, M. H., and French, J. B. (1960). *Rev. Mod. Phys.* **32**, 567.
MacGregor, M. H., Arndt, R. A., and Wright, R. M. (1969). *Phys. Rev.* **182**, 1714.
Mackintosh, R. S., and Glendenning, N. K. (1971). *Nucl. Phys.* **A168**, 575.
McMahon, C. Alex, and Tobocman, W. (1973). *Nucl. Phys.* **A202**, 561.
Madsen, V. A. (1966). *Nucl. Phys.* **80**, 177.
Madsen, V. A., and Tobocman, W. (1965). *Phys. Rev.* **139B**, 864.
Madsen, V. A., *et al.* (1972). *Phys. Rev. Lett.* **28**, 629.
Madsen, V. A., *et al.* (1976). *Phys. Rev.* **C13**, 548.
Mahaux C. (1979). *In* "Microscopic Optical Potentials" (H. V. von Geramb, ed.), Springer-Verlag Lecture Notes in Physics Series. Springer-Verlag, Berlin.
Mayer, M. G. (1949). *Phys. Rev.* **78**, 16.
Melkanoff, M. A., Moszkowski, S. A., Nodvik, J., and Saxon, D. S. (1956). *Phys. Rev.* **101**, 507.
Messiah, A. (1962). "Quantum Mechanics," Vols. I and II. North-Holland, Amsterdam.
Migdal, A. B. (1959). *Nucl. Phys.* **13**, 655.
Miller, W. H. (1970). *J. Chem. Phys.* **53**, 1949.
Miller, W. H. (1974). *Adv. Chem. Phys.* **25**, 69.
Mills, R. L., Sessler, A. M., Mozkowski, S. A., and Shankland, D. G. (1959). *Phys. Rev. Lett.* **3**, 381.
Moshinsky, M. (1959). *Nucl. Phys.* **13**, 104.
Moszkowski, S. A., and Scott, B. S. (1960). *Ann. Phys. (New York)* **11**, 65.
Moszkowski, S. A., and Scott, B. S. (1961). *Ann. Phys. (New York)* **14**, 107.
Mott, N. F., and Massey, H. S. W. (1933). "The Theory of Atomic Collisions," 1st ed. Oxford University Press, Oxford. In the second edition (1949), see p. 144.
Nagarajan, M. A. (1972). *Nucl. Phys.* **A196**, 34.
Nagarajan, M. A. (1973). *Nucl. Phys.* **A209**, 485.
Nagarajan, M. A. (1982). *Phys. Scripta* **25**, 463.
Nagav, M., and Tonizuka, Y. (1971). *Phys. Lett.* **37B**, 383.
Newns, H. C. (1960). *Proc. Phys. Soc. (London)* **A76**, 489.
Nörenberg, W., and Weidenmüller, H. A. (1976). "Introduction to the Theory of Heavy-Ion Collisions," Springer-Verlag Lecture Notes in Physics Vol. 51. Springer-Verlag Berlin.
Oothoudt, M., Hintz, N. M., and Vedelsby, P. (1970). *Phys. Lett.* **32B**, 270.
Osterfeld, F., Wambach, J., and Madsen, V. A. (1981). *Phys. Rev.* **C23**, 179.
Penny, S. K., and Satchler, G. R. (1964). *Nucl. Phys.* **53**, 145.
Perey, F. G. (1974). *In* "Nuclear Spectroscopy and Reactions," (J. Cerny, ed.), Part B, p. 137. Academic Press, New York.
Perey, F. G., and Buck, B. (1962). *Nucl. Phys.* **32**, 353.
Perey, F. G., and Saxon, D. S. (1964). *Phys. Lett.* **10**, 107.

Petrovich, F. (1970). Thesis, Michigan State University.
Petrovich, F., and Love, W. G. (1981). *Nucl. Phys.* **A354**, 499c.
Petrovich, F., McManus, H., Madsen, V. A., and Atkinson, J. (1969). *Phys. Rev. Lett.* **22**, 895.
Petrovich, F., McManus, H., Borysowicz, J., and Hammerstein, G. R. (1977). *Phys. Rev.* **C16**, 839.
Petrovich, F. *et. al.* (1980a) *Phys Lett.* **91B**, 27.
Petrovich, F., Love, W. G., Picklesimer, A., Walker, G., and Siciliano, E. (1980b). *Phys. Lett.* **95B**, 166.
Petrovich, F., Brown, G. E., Garvey, G., Goodman, C. D., Love, W. G., and Lindgren, R. A. eds., (1983a). "Spin Excitations in Nuclei," Plenum, New York.
Petrovich, F., Carr, J. A., Halderson, D., Love, W. G., and Kelly J. J. (1983). To be published.
Philpott, R. J., Pinkston, W. T., and Satchler, G. R. (1968). *Nucl. Phys.* **A119**, 241.
Preston, M. A. (1962). "Physics of the Nucleus." Addison-Wesley, Reading, Massachusetts.
Preston, M. A., and Bhaduri, R. K. (1975). "Structure of the Nucleus." Addison-Wesley, Reading, Massachusetts.
"Proceedings of the International Conference on Reactions between Complex Nuclei, Nashville, Tennessee" (1974). (R. L. Robinson *et al.*, eds.). North-Holland, Amsterdam.
"Proceedings, Macroscopic Features of Heavy-Ion Collisions" (1976). Argonne National Laboratory, Argonne, Illinois.
"Proceedings of the Symposium on Classical and Quantum-Mechanical Aspects of Heavy-Ion Collisions," Heidelberg, (1974). Springer-Verlag, Berlin.
Pruess, K. (1977). *Nucl. Phys.* **A278**, 124.
Pruess, K. (1979). *Nucl. Phys.* **A330**, 209.
Racah, G. (1942). *Phys. Rev.* **62**, 438.
Racah, G. (1943). *Phys. Rev.* **63**, 367.
Rao, C. L., Reeves, M., and Satchler, G. R. (1973). *Nucl. Phys.* **A207**, 182.
Rawitscher, G. H. (1967). *Phys. Rev.* **163**, 1223.
Raynal, J. (1967). *Nucl. Phys.* **A97**, 572.
Reynolds, R. M., Maxwell, J. R. and Hintz, N. M. (1967). *Phys. Rev.* **153**, 1283.
Ried, R. (1968). *Ann. Phys.* (*New York*) **50**, 411.
Roberts, A. (1972). *Nucl. Phys.* **A196**, 465.
Robson, D., and Koshel, R. D. (1972). *Phys. Rev.* **C6**, 1125.
Rodberg, L. S., and Thaler, R. M. (1967). "Introduction to the Quantum Theory of Scattering." Academic Press, New York.
Rost, E. (1962). *Phys. Rev.* **128**, 2708.
Rowley, N., and Marty, C. (1976). *Nucl. Phys.* **A266**, 494.
Satchler, G. R. (1964). *Nucl. Phys.* **55**, 1.
Satchler, G. R. (1966). *Nucl. Phys.* **77**, 481.
Satchler, G. R. (1967a). *Nucl. Phys.* **A92**, 273.
Satchler, G. R. (1967b). *Nucl. Phys.* **A95**, 1.
Satchler, G. R. (1974). *In* "Reactions between Complex Nuclei," (R. L. Robinson, F. K. McGowan, J. B. Ball, and J. H. Hamilton, eds.). North-Holland, Amsterdam.
Sawaguri, T., and Tobocman, W. (1967). *J. Math. Phys.* **8**, 2223.
Schmittroth, F., Tobocman, W., and Golestaneh, A. A. (1970). *Phys. Rev.* **C1**, 377.
Scott, D. K., *et al.* (1975). *Phys. Rev. Lett.* **34**, 895.
Slanina, D. and McNanus, H. (1968). *Nucl. Phys.* **A116**, 271.
Smith, S. M., Moazed, C., Bernstein, A. M., and Roos, P. G. (1968). *Phys. Rev.* **169**, 951.
Smith, S. M., Roos, P. G., Bernstein, M. A., and Moazed, C. (1970). *Nucl. Phys.* **A158**, 497.
Stamp, A. P. (1966). *Nucl. Phys.* **83**, 232.
Stapp, H. P., Ypsilantis, T. J., and Metropolis, N. (1957). *Phys. Rev.* **105**, 302.
Stovall, T., and Hintz, N. (1964). *Phys. Rev.* **135**, B330.

Takimasa, T., Tamura, T., and Udagawa, T. (1979) *Nucl. Phys.* **A321**, 269.

Werby, M. F., Strayer, M. R., and Nagarajan, M. A. (1980). *Phys. Rev.* **C21**, 2235.

Talmi, I. (1952). *Helv. Phys. Acta* **25**, 185.

Tamura, T. (1965). *Rev. Mod, Phys.* **37**, 679.

Tobocman, W. (1954). *Phys. Rev.* **94**, 1655.

Tobocman, W. (1959). *Phys. Rev.* **115**, 98.

Tobocman, W., and Gibbs, W. R. (1962). *Phys. Rev.* **126**, 1076.

Tobocman, W., and Kalos, M. (1955). *Phys. Rev.* **97**, 132.

Toth, K. S., *et al.* (1976). *Phys. Rev.* **C14**, 1471.

Towner, I. S., and Hardy, J. C. (1969). *Adv. Phys.* **18**, 401.

True, W. W. (1963). *Phys. Rev.* **130**, 1530.

True, W. W., and Ford, K. W. (1958). *Phys. Rev.* **109**, 1675.

Valatin, J. G. (1958). *Nuovo Cimento* **7**, 843.

Vinh Mau, N. (1970). *In* "Theory of Nuclear Structure," p. 931. IAEA, Vienna.

Vinh Mau, N., and Bouyssy, A. (1974). *Nucl. Phys.* **A229**, 1.

Vinh Mau, R. (1979). *In* "Mesons in Nuclei" (M. Rho and D. Wilkinson, eds.), Vol. I, p. 151. North-Holland, Amsterdam.

Vinh Mau N., and Bouyssy, A. (1976). *Nucl. Phys.* **A257**, 189.

von Geramb, H. V., Brieva, F. A., and Rook, J. R. (1979). "Microscopic Optical Potentials" (H. V. von Geramb, ed.), Lecture Notes in Physics Series Springer-Verlag, Berlin.

von Geramb, H. V., ed. (1979). "Microscopic Optical Potentials" (1979). (H. V. von Geramb, ed.), Springer-Verlag Lecture Notes in Physics Series. Springer-Verlag, Berlin. See especially the work reported by and references to the Liege Group (J.-P. Jeukenne, A. Lejeune, and C. Mahaux), the Orsay Group (N. Vinh Mau and A. Bouyssy), and the Oxford Group (F. A. Brieva, H. R. Kidwai, and R. Rook).

Watson, K. M. (1953). *Phys. Rev.* **89**, 575.

Wolfenstein, L. (1956). *In* "Annual Review of Nuclear Science," Vol. 6. Annual Reviews, Inc., Palo Alto, California.

Wong, C., Brown, V. R., Madsen, V. A., and Grimes, S. M. (1979). *Phys. Rev.* **C20**, 59.

Woods, R. D., and Saxon, D. S. (1954). *Phys. Rev.* **95**, 577.

Yen, S. *et. al.*, 1981. *Phys. Lett.* **105B**, 421.

Yoshida, S. (1961). *Phys. Rev.* **123**, 2122.

Yoshida, S. (1962). *Nucl. Phys.* **33**, 685.

Index

F

G

H

I

J

K

L

M

N

O

P

Q

R

S

T

V

W

Z